Handbook of Exploration Geochemistry

VOLUME 3
Rock Geochemistry in Mineral Exploration

HANDBOOK OF EXPLORATION GEOCHEMISTRY

G.J.S. GOVETT (Editor)

Handbook of Exploration Geochemistry

VOLUME 3
Rock Geochemistry in Mineral Exploration

by

G.J.S. GOVETT
Professor of Geology
School of Applied Geology, University of New South Wales,
Kensington, N.S.W., Australia

ELSEVIER SCIENTIFIC PUBLISHING COMPANY
Amsterdam — Oxford — New York   1983

ELSEVIER SCIENCE PUBLISHERS B.V.
Molenwerf 1
P.O. Box 211, 1000 AE Amsterdam, The Netherlands

Distributors for the United States and Canada:

ELSEVIER SCIENCE PUBLISHING COMPANY INC.
52, Vanderbilt Avenue
New York, N.Y. 10017

First edition 1983
Second impression 1985

Library of Congress Cataloging in Publication Data

Govett, G. J. S.
  Rock geochemistry in mineral exploration.

  (Handbook of exploration geochemistry ; v. 3)
  Bibliography: p.
  Includes indexes.
  1. Geochemistry. 2. Geochemical prospecting.
I. Title. II. Series.
QE515.G747   551.9   81-15148
ISBN 0-444-42021-5   AACR2

ISBN 0-444-42021-5 (Vol. 3)
ISBN 0-444-41932-2 (Series)

Printed in The Netherlands

PREFACE

The growth of the use of geochemistry as an exploration method has resulted in two related practical problems — the wealth of accumulated data is becoming too great to be examined in reasonable depth within a single volume; and geologists and geochemists working in industry, particularly outside of the main cities of North America and Europe, find it difficult to keep abreast of new developments reported in the technical literature. These problems have been discussed with colleagues at various meetings around the world and subsequently with the editorial staff of Elsevier Scientific Publishing Company, leading to the concept of a series of separate volumes — *The Handbook of Exploration Geochemistry* — in which various aspects of exploration geochemistry could be examined in detail by specialists in particular fields. My role as editor of the series is largely confined to persuading my colleagues to give their time and knowledge to the task and maintaining, as far as possible, a common approach and format for the volumes. The first volume in the series was published in 1981 (*Analytical Methods in Geochemical Exploration* by W.K. Fletcher); this is the third volume, and five more volumes are in various stages of writing.

The objective of the Handbook series is rather different from that of a broad, general text covering the entire field of exploration geochemistry. The series of separate volumes is primarily designed to present the available data on a particular subject relevant to mineral exploration in a form that will be useful to the practising exploration geologist and geochemist; at the same time the volumes contain enough material to provide a reference for researchers and postgraduate students. Two volumes deal with the broad subjects of analytical methods and data processing; the rest of the volumes deal with specific types of exploration surveys.

This is the first volume on one of the operational techniques — rock geochemistry. Whereas the book is, of necessity, a reflection of my own views and approach to exploration rock geochemistry, it is based not only on my work but on the published work of others that is available in English or in English translation. The material selected for inclusion has been chosen to illustrate specific results and general conclusions. More than 200 line drawings have been used to present the data; most of those derived from published sources have been re-drawn for this book and in many instances represent a modified interpretation or presentation to complement data or conclusions from other sources.

I have avoided the use of the term "lithogeochemistry" because I believe that it is etymologically unsound; also it conflicts with Soviet usage (see Govett, 1978). I believe that simple, descriptive terms should be used wherever possible (e.g., soil geochemistry rather than pedogeochemistry). I have, hopefully, carried this belief in simplicity into the presentation of this book. Theoretical considerations have been kept to the minimum required for understanding and predicting results.

This approach has posed problems of presentation because an understanding of theoretical geochemical processes is vital to data interpretation and to the development of techniques of application. The principles of element distribution in the earth's crust are therefore briefly reviewed (in Chapter 2) and, because of their importance, are treated more comprehensively in a separate appendix (Appendix 1 by Dr. P.C. Rickwood) which also includes a review of the status of knowledge of the crustal abundance of elements.

A similar problem of presentation arose with respect to providing relevant geological data. Exploration rock geochemistry cannot be divorced from a thorough understanding of geological (and especially ore genesis) processes. To keep the book to a reasonable size the geological data are necessarily brief; however, the breadth of treatment bears no relation to the importance of the geological processes (the reader should refer to the literature cited for fuller details).

A major deficiency in exploration rock geochemistry studies — the lack of data on "barren" deposits (e.g., pyrite deposits devoid of minerals of economic interest) — is reflected in the material included in the book. Obviously hydrothermal or other mineralizing processes that are deficient in ore elements will nevertheless give rise to alteration and broad geochemical changes similar to those that do have associated ore elements. There is a great need for detailed geochemical studies of such barren systems. More generally, there has been a reluctance among exploration geochemists to publish negative results from investigations that are considered unsuccessful (in terms of finding an ore body); the publication of such negative data would greatly assist in the development of exploration rock geochemical techniques.

Those who know me well may be surprised that I have not discussed exploration costs; this is a deliberate omission because so many qualifications are needed for comparisons that the data become misleading or useless. Survey costs vary enormously from country to country and from locale to locale; apparent costs also vary from one company to another, even within the same country, depending on the type of personnel used and company accounting procedures. Furthermore, one activity is virtually impossible to cost — the interpretation of rock geochemical surveys — although I believe that much more of the exploration dollar should be spent on data interpretation.

During the course of writing this book (over a period of four years) I have become indebted to many people around the world who have agreed to my using their material and who have, in some cases, even supplied additional information. I must especially mention my many research students at the University of New Brunswick and the University of New South Wales over the past dozen years.

I am particularly indebted to my colleague at the University of New South Wales, Dr. P.C. Rickwood, for writing Appendix 1. I am also grateful for the comments of Dr. J.A. Coope, Dr. E.M. Cameron, Dr. R.G. Garrett, and Dr. W.K. Fletcher who have read a number of the chapters; they are not, of course, responsible for any of the views presented. The illustrations owe much to the draughting and photography work of Mr. J. Ross and Mr. R. McCulloch of the University of New Brunswick and, especially, the recent work of Mrs. M. Horvath and Mr. G. Small of the University of New South Wales. I would also like to thank my secretaries Mrs. J. Heydon (for typing much of the early draft) and Mrs. M. Valentine (for undertaking the tedious task of typing the tables and proofreading). I am immensely grateful to M.H. Govett for her indefatigable editorial assistance and for her work involved in presenting an acceptable manuscript to the publishers. Finally, I wish to thank Drs. F.W.B. van Eysinga (formerly of Elsevier Scientific Publishing Company) for being receptive to the initial concept of a handbook series, and Drs. H. Frank (of Elsevier Scientific Publishing Company) for his subsequent encouragement — and patience — during the preparation of this book.

G.J.S. GOVETT
Sydney, N.S.W.

April 1981

CONTENTS

XII

PART IV. SUMMARY AND CONCLUSIONS

INTRODUCTION

Exploration rock geochemistry is concerned with the detection of primary dispersion patterns around, or associated with, mineral deposits. The term "primary dispersion" was defined by James (1967) to describe the distribution of elements in unweathered rock — whether the mineral deposit is epigenetic or syngenetic. The spirit of this approach is followed here, and the term primary dispersion is used to describe the distribution of elements in rock that has arisen through ore-forming or rock-forming processes, regardless of whether the dispersion patterns are directly or indirectly associated with the ore-forming processes. This definition allows the inclusion of dispersion patterns such as geochemical halos in hanging wall rocks that clearly post-date the formation of volcanic-sedimentary massive sulphide rocks and geochemical halos in weathered rocks that are clearly related to both igneous and surficial weathering processes. In some cases of extreme weathering the distinction between primary and secondary (surficial) processes may become rather blurred, but this does not detract from the usefulness of the definition. All other geochemical terms used here have their conventional meanings.

Although an attempt has been made to present a comprehensive survey of rock geochemical data, the guiding principle throughout the selection of material for inclusion has been "is it useful in exploration?"; if data did not satisfy this basic criterion they were excluded — however interesting they may have been in other contexts. A second criterion for inclusion of material was "are there sufficient data on geology and mineralization, sampling procedures, and analytical techniques used to enable an independent assessment of the results to be made?"; it will be evident in the text that this criterion was less rigorously applied than the first criterion, particularly in the case of Soviet sources, in order to give a broad coverage. Only a few examples of the use of isotopes and fluid inclusions are included. A final restriction has been that the sources were available in English.

Much of the data was necessarily derived from published sources; these were supplemented, where appropriate, by the writer's own research and consulting activities and by the work of his research students at the University of New Brunswick (Canada) and the University of New South Wales

(Australia). To facilitate comparison and correlation most of the material from other workers has been re-plotted to a common format. In many cases where the data in the published source were adequate, alternate or additional interpretations have been made; it is hoped that the authors' original data and views have not been compromised by this approach.

Exploration rock geochemistry — as any other geochemical exploration method — is only one of many techniques that can be used in mineral exploration. Accordingly, in Chapter 1 the application of exploration geochemistry to the general mineral exploration sequence is considered.

In Chapter 2 there is a discussion of crustal abundance of elements and the factors that control their distribution in rocks and minerals since ore bodies represent a marked change from conditions of normal geochemical equilibrium, and an understanding of the normal conditions of element distribution in rocks is essential to the recognition of deviations due to mineralizing processes (these matters are discussed in more detail in Appendix 1). The problems of sampling and analysis — common to all earth materials — pose special problems in rocks because of variations in grain size, texture, and preferential distribution of elements between different minerals; these problems are also discussed in Chapter 2.

Obviously correct interpretation of rock geochemical data is fundamental to the successful application of the technique to exploration, and details of interpretative procedures are considered throughout this book (there is also a separate volume in the *Handbook of Exploration Geochemistry* series on this topic; see Howarth, 1982). Chapter 3, the final chapter in Part I, is devoted to the consideration of some of the broad, general problems of anomaly recognition in all types of geochemical surveys. The chapter also describes some of the specific interpretative techniques used in the U.S.S.R. for rock geochemical data since they differ from those employed by most English-speaking geochemists.

The organization of most of the rest of the book is designed to conform with exploration practice. The primary subdivision is:

Part II: Regional Scale Exploration — large-scale geochemical responses that are capable of discriminating between productive and barren terrain.

Part III: Local and Mine Scale Exploration — geochemical responses around individual deposits that can be detected up to 1—2 km away and the geochemical responses in the immediate wallrock of the deposits.

The subdivisions within Parts II and III are based largely on deposit type and associated geological environment. There are some drawbacks to this classification. For example, volcanic-sedimentary massive sulphides are discussed in both Part II and Part III; vein-type deposits are considered in various degrees of detail in three separate chapters. Moreover, some deposits cannot be classified easily, and the general form of the mineralization and the geochemical response was used to determine the classification. It could be argued that it would have been scientifically more appropriate to make the primary subdivisions on the basis of deposit type (e.g., to consider all scales of response around massive sulphides in one section). It was concluded, however, that the objective of making the book user-oriented was best served by adopting the exploration sequence as the basis of classification.

In Part II, Chapter 4 deals with deposits of plutonic association and is essentially concerned with geochemical discrimination between potentially productive and barren intrusions, regardless of whether the genetically associated mineralization is copper porphyry or vein and replacement deposits within and around the intrusions. The special problem of recognizing tin-bearing intrusions is discussed in Chapter 5. Chapter 6 deals with the identification of discrete regional scale spatial halos around vein and replacement deposits rather than with the recognition of geochemical signatures indicative of mineralization in particular rock units. The final chapter on regional scale exploration discusses the recognition of geochemical signatures indicative of mineralization in particular volcanic units and regional scale spatial halos.

Part III is concerned with geochemical halos around individual porphyry-type deposits (Chapter 8), vein and replacement deposits (Chapter 9), and volcanic-sedimentary massive sulphides (Chapters 10, 11, and 12). The relatively large amount of space devoted to massive sulphides reflects both the amount of published data available and the concentration of the writer's own work on massive sulphide deposits.

In Parts II and III details (as far as possible) are given on relevant geology, size and grade of mineralization, numbers of samples collected, sample interval or density, analytical techniques used, and the halo dimension or other responses of individual elements or element combinations. These data are compiled in the final chapter (Chapter 13, Part IV) to provide a summary of the most useful elements for the different scales of exploration for various types of deposits and to indicate the dimensions of halos to be expected.

The case history data and the interpretations presented in Parts II and III demonstrate that rock geochemistry can be used effectively at all stages of exploration for a wide range of deposits. The compilation in Chapter 13 should provide a general framework for those who use or intend to use rock geochemical techniques in the field and for those who are engaged in research to develop new or improved applications of rock geochemistry to mineral exploration. A word of warning, however, seems appropriate — the usefulness of the compilation in Chapter 13 depends upon a careful study of the individual case histories which comprise the bulk of the book, and as new information becomes available the compilations should be reviewed and, hopefully, expanded.

PART I. GENERAL PRINCIPLES

Chapter 1

GEOCHEMISTRY IN THE EXPLORATION SEQUENCE

INTRODUCTION

Mineral exploration is financially a high-risk enterprise — or gamble — where the odds of success are commonly quoted as 1000 to 1. On the basis of the present level of geological knowledge the risk factor in investing in mineral exploration cannot be eliminated, but efforts to reduce it have been made at both the corporate planning stage and in the application of the best technical ability available.

The broad pattern of decision-making is summarized in Fig. 1-1. The first, and key, decision is which commodity should be the exploration target. There are many subsequent places in the progressively more costly sequence of exploration where decisions to proceed or stop a particular programme have to be made on the basis of exploration data and financial considerations.

The three main factors governing a decision to undertake a particular exploration programme are economic, political, and geological. Some of the decisions are simple and straightforward, as in the case when a company needs to increase its reserves for corporate survival. Others are much more complex and depend on an assessment of whether in a potential exploration area long-term government policies can be expected to be — and remain — favourable to mineral development. Fascinating, and important, as the political and economic factors may be, this book deals with the geological factor in exploration — and specifically with the role of rock geochemistry in mineral exploration. As a framework for the discussion of rock geochemistry as a specific exploration technique this chapter briefly reviews the general principles of exploration geochemistry and discusses the main types of geochemical surveys and their role in the general sequence of mineral exploration.

PRINCIPLES UNDERLYING GEOCHEMICAL EXPLORATION

The use of exploration geochemistry in the search for mineral deposits is based on the fundamental premise (substantiated by a vast amount of

8

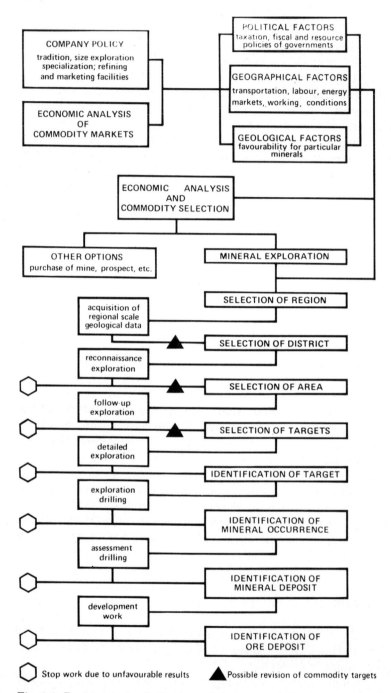

Fig. 1-1. Decision and activity stages from commodity selection to finding an ore deposit.

empirical data) that the chemical composition of materials of the earth's crust will be different in the vicinity of a mineral deposit from the chemical composition of similar materials where there is no mineral deposit. These differences are sought by systematic measurements of chemical parameters — the determination of one element or dozens of elements, as well as the determination of such factors as pH, Eh, and conductivity in samples of rocks, soils, stream waters and sediments, vegetation, or even the atmosphere.

It is intuitively obvious why there should be chemical differences near a mineral deposit — a stream draining an oxidizing copper sulphide deposit would be expected to contain abnormal amounts of Cu; residual soils would similarly be expected to contain abnormally high contents of Cu over such a body, as would vegetation growing in these soils. The scientific, and from an interpretative viewpoint more useful, basis for these expectations may be derived from the second law of thermodynamics stated simply in Le Chatelier's Principle: if a change occurs in one of the factors under which a system is in equilibrium, the system will tend to adjust itself as far as possible to annul the effect of the change.

The chemical composition of the earth's crust must be viewed as a reflection of equilibrium conditions at the time of formation of the materials, modified by an attempt to reach equilibrium with current conditions at any particular place. The concentration of millions of tonnes of sulphide minerals must inevitably be accompanied by a serious disturbance of geological equilibrium that will be reflected in physical and chemical abnormalities in the host rocks. Conceptually this disturbance in equilibrium should be detectable by differences from normal conditions in the composition of the enclosing rocks. Similarly, the composition of materials in the secondary environment (soils, stream waters and sediments) will also reflect the presence of the disturbing factor — a mineral deposit — by being different from normal. In exploration geochemistry the normal chemical composition is referred to as "background" and the abnormal chemical composition is referred to as "anomalous". On the basis of this simple concept a whole range of different geochemical techniques are used in the search for mineral deposits through the systematic collection and analyses of samples of various materials of the earth's crust.

MAIN TYPES OF GEOCHEMICAL SURVEYS

A broad review of the whole spectrum of geochemical surveys is given in the second edition of Hawkes and Webb (1962) by Rose et al. (1979). The main types of geochemical surveys are:

— rock surveys, which include sampling surface rock, vein material, and drill core and underground workings;

— drainage surveys, including sampling streams and lake sediments, stream waters, and groundwater;

— soil surveys, which include sampling surface and deep soil, residual soil, and transported soil;

— biogeochemical surveys, which include sampling of leaves and stem growth of vegetation;

— geobotanical surveys (interpretation of the distribution of types of vegetation and of stress effects on vegetation).

Some types of surveys do not readily conform to this classification. For example, sampling of weathered bedrock (common in areas of deep weathering such as are found in Australia) may be considered either as a soil or rock technique; gossan geochemistry (also very common in Australia) may be classified with rock or soil surveys; the use of volatiles, particularly when measured in situ, do not obviously fall into any of the defined categories (in fact, they are dealt with in a separate volume in this series).

SEQUENCE OF GEOCHEMICAL EXPLORATION

The *Handbook of Exploration Geochemistry*, of which this volume is a part, was organized to treat each type of survey separately (in addition to providing volumes on analytical and data processing techniques that are common to all types of surveys). The following description of the sequence of geochemical exploration is included here to provide a framework for all types of surveys — not just the rock geochemical surveys dealt with in this volume.

The aim of all mineral exploration is to progressively diminish the size of a search area in which an ore body may be found until a target that can be tested by drilling is defined. As a generalization, this demands increasingly more detailed — and more expensive — techniques as an exploration programme proceeds. An operational objective throughout the programme is to achieve the maximum probability of discovery at the lowest possible cost. Given the enormous variation in geological conditions it is not possible to define an invariable exploration programme that will be applicable to all situations. It is possible, however, to define a generalized sequence of exploration (this was shown schematically in Fig. 1-1 above).

The progression to the pre-drilling stage can be divided into fairly well-defined phases, although the units tend to be loosely and interchangeably used in practice. The following areal terms should be regarded only as orders of magnitude:

region: greater than 5000 km^2 ;

district: 500—5000 km^2 ;

area: 5—50 km^2 ;

target: less than 5 km^2 .

Obviously all exploration programmes do not consist of the complete cycle outlined in Fig. 1-1, and it is possible to enter the cycle part way through using existing knowledge. For example, in a known mining area a particular lithological structural unit may be known to be potentially ore-bearing and, therefore, the reconnaissance phase may be omitted. Similarly, extensions to an individual deposit may be sought by starting with detailed exploration. It is important to recognize, however, that at whatever phase in the cycle exploration is begun, the logical sequence of increasingly detailed work should be followed through to the end, with a decision as to whether to proceed to the next stage made at the end of each stage on the basis of the geological knowledge gained and economic decisions based on that knowledge.

The sequence of pre-drilling exploration shown in Fig. 1-1 and described below is intended to be illustrative and not comprehensive. Obviously, other techniques may — and should — be used at various phases. No matter what techniques are used, it is regarded as mandatory that geological observations be made during all phases of exploration at a scale consistent with the particular geochemical technique being used. Moreover, although detailed soil or rock sampling have as their aim the definition of a drilling target, in most cases it is prudent to attempt to confirm the target by other means such as ground geophysics, pitting, and trenching. The following sections discuss briefly the use of exploration geochemical techniques in the various exploration stages under the two broad headings — regional scale exploration, and local and mine scale exploration — adopted for the organization of this book.

Regional scale exploration

The first step in any exploration programme is the selection of the region where the work is to be done; this is normally an economic-political decision, although it is obviously influenced by knowledge of favourable geological conditions. In practice geochemical surveys are seldom used to select a region. If they are used, low-density stream sediment surveys of one sample per 200 km² would be suitable (see Garrett and Nichol, 1967 and Armour-Brown and Nichol, 1970).

The distinction between a region and a district is necessarily somewhat arbitrary. In general a district should be regarded as a more discrete and definite entity than a region. The presence of known mineralization or favourable geological conditions for a particular type of mineralization will usually guide the selection of a district, and the choice may be dictated by the type of target sought (for example, if the desired target is massive lead-zinc sulphide deposits, the Bathurst district of northern New Brunswick would be a more logical choice than southern Alberta in a Canadian exploration programme). Stream sediment surveys with a density of one sample per 5—20 km² are suitable for the definition of a district (see Bradshaw et al., 1972); in areas of large numbers of lakes a density of one lake sediment sample per 26 km² offers promise (Allan et al., 1973). Rock geochemical surveys, particularly those designed to discriminate between ore-bearing and barren plutons would be appropriate in selecting districts for further work (see below and Part II of this book).

Reconnaissance exploration is the first active field-oriented exploration within a chosen district to select an area for subsequent work. It is probably the most frequently misconceived phase of exploration. It is *not* the objective — nor the expectation — of this phase of exploration to find an actual ore body. Reconnaissance work is a preliminary survey to evaluate the potential of geographic areas and to locate the general disposition of areas of interest. Thus, whatever technique is used, reconnaissance surveys should be designed to delimit anomalous areas within a district or a region which warrant further work. The important consideration is the scale (and therefore the cost) of an operation capable of identifying target areas.

The primary geochemical technique in use today for reconnaissance surveys is the drainage survey with a sample density of one sample per 3–6 km² (see Bradshaw et al., 1972); provided multi-element determinations are made, this scale of sampling should enable geological interpretations of considerable detail, even to the extent of locating the general position of geological contacts. Lake sediments sampled at a density of one sample per 2.6 km² (as described by Davenport et al., 1975) offer a comparable technique in districts with an adequate number of lakes to provide a sampling programme. Till sampling in some glaciated areas can also effectively be used as a reconnaissance technique (see Szabo et al., 1975). In areas of reasonable outcrop rock surveys at a sample density of one sample per 5 km² appear to be suitable for volcanic-sedimentary sulphide deposits (Govett and Pwa, 1981); data given in Chapter 4 indicate that reconnaissance-scale rock sampling may be useful over large plutons to delimit favourable areas.

The number of elements determined in a reconnaissance survey may be fewer than needed for the selection of a region or district where a geochemical survey is used; there is usually some idea of the geological and geochemical environment of a chosen district, and the type of mineralization sought is more certainly known. The choice of elements in a reconnaissance survey will be dictated by both the geological and geochemical environment of the district and by the interpretative requirements.

An example of the use of rock geochemistry in the selection of districts and areas is the definition of geochemical characterization — as distinct from the definition of discrete spatial halos — of large intrusions to determine those that are potentially productive in terms of copper porphyry deposits or that are genetically associated with vein or replacement deposits of tin, base metals, and precious metals. Regional trends within a large intrusion may also indicate broad geographic locations of deposits on one side or another of an intrusion. Particular stratigraphic zones (especially in volcanic sequences) can similarly be identified by an anomalous geochemical character when compared with otherwise similar but barren stratigraphic horizons.

In regional scale surveys the recognition of both potentially mineralized intrusions and volcanic cycles depends upon the identification of a characteristic geochemical signature (for example, low K:Rb and high Rb:Sr ratios in tin-bearing granites); spatial trends are relatively unimportant. On the other hand, regional scale spatial patterns of increasing or decreasing element concentrations are important in the recognition of large mineralized districts or areas which characteristically (but not necessarily) contain many deposits.

Local and mine scale exploration

In the selection and identification of targets geochemical responses of up to a few kilometres in extent with well-defined spatial components are sought. Since the techniques that will be used after this phase are detailed and expensive, it is essential that target selection is designed to locate the targets as accurately as possible and to restrict them to the smallest possible areal extent (although the objective of this phase of exploration is not to locate an actual deposit, a deposit may be found under favourable circumstances).

The objective of local scale exploration is to define mineralization as the focus of discrete anomalies of individual elements or combinations of elements. The scale of response — and hence the scale of sampling — generally varies as a function of the type of mineralization. For example, anomalous aureoles of hundreds of metres extent are common around massive sulphide deposits, whereas wallrock anomalies around narrow veins normally extend for only a few tens of metres.

Detailed stream sediment sampling, with a sample interval of 30—300 m along a stream (with samples from each bank at each sample station where the bank material is suitable) is normally capable of closely defining the location of anomalous targets; at this sample level detailed geological interpretations of the geochemical data may become difficult due to the swamping effects of highly anomalous conditions. Rock, and possibly soil, geochemistry can be used if the drainage system is inadequate, the nature of the target and the environment give a fairly extensive dispersion pattern, and the size of the anomalous area is not too large. For comparable areal coverage, however, the number of samples required for rock and soil sampling is generally greater than for a drainage survey, and the costs of the programme are correspondingly higher.

Once a target has been selected the next phase in the exploration programme — identification of a target — involves two stages. The first stage is to find the approximate location of a mineralized zone; the second stage (which may be regarded as a refinement of the first stage) is to attempt to accurately define the extent of the zone. Soil, and increasingly, rock surveys are the most widely used techniques for detailed exploration. Widely-spaced traverses (up to about 300 m) sampled at intervals of 20—80 m are used.

Once a mineralized zone has been located, progressively more detailed sampling — perhaps with as little as 10 m between samples — is used to fill in and define the limit of mineralization. In areas of thick overburden vertical profile sampling may also be undertaken. Biogeochemical and geobotanical techniques are also used for target identification.

The actual identification of a mineral occurrence and the progressive definition of it to an ore deposit is achieved largely through drilling (possibly preceded by deep trenching). To date exploration rock geochemistry has not been used extensively in exploration drilling, partly because of inadequate analytical facilities (rapid reporting of results is necessary to assist in drilling) and also because there is a general lack of appreciation of the potential value of the technique at the drilling stage. Distinctive geochemical aureoles have now been documented for many types of deposits, and different elements show characteristic patterns of enhancement or depletion with proximity to mineralization. This zonation of elements would obviously enhance the interpretation of drill core and allow better predictions to be made of the location of the mineralization sought.

CONCLUSIONS

At each phase of the exploration sequence the use of specific geochemical exploration techniques will obviously be dictated by the nature of the target, the type of terrain, the type of sample material available, and the relative costs of alternative methods. Traditionally low-density stream sediment surveys have been used in regional scale exploration programmes for the selection of regions and districts; rock geochemical surveys, however, could also be used for district selection. Higher-density stream sediment surveys and rock surveys are used to select areas once a district has been delineated. On the local scale — in the search for and identification of targets — detailed stream, soil, and rock surveys are all used. Rock geochemical techniques hold out considerable promise for use in the drilling stage.

Mineralogical alteration zones in host rocks around mineral deposits have long been recognized and used as ore indicators. With the recent improvement of the understanding of ore genesis, there has been a much wider appreciation of the usefulness of rock geochemical techniques. The work reviewed in this book indicates that all of the main types of metallic mineral deposits have an associated characteristic geochemical signature on a greater or lesser scale for one or more of a small group of about two dozen elements; specific deposit types can be detected by an anomalous response of one or more elements from an even more restricted suite of less than ten elements. Further improvement in rock geochemical techniques that will come from an increasingly better understanding of the fundamental processes of element dispersion and from continued development of interpretative procedures should make rock geochemistry an even more potent exploration tool in all phases of the exploration sequence.

Chapter 2

CRUSTAL ABUNDANCE, GEOCHEMICAL BEHAVIOUR OF THE ELEMENTS, AND PROBLEMS OF SAMPLING

ABUNDANCE OF THE ELEMENTS

Geochemistry, as one of the disciplines of the earth sciences, is the study of the distribution and amounts of elements in the earth and the formulation of the principles that govern this distribution. The *relative* abundance of elements in the solar system is quite well known and accords broadly with the atomic structure of the elements; similarly, the *relative* abundance of the major elements, at least in the crust of the earth, is not a matter of serious dispute, and differences from the abundance of the elements in the solar system (e.g., relative enrichment of light elements) are plausibly explained by differentiation of the lithosphere. There is, however, considerable dispute concerning the *absolute* abundance of elements — particularly the trace elements.

The earliest attempts at estimating the abundance of elements (Clarke, 1889) were based on simple averaging of available analytical data; the compilations of Clarke and Washington (1924), based on 5159 analyses, are the most famous example of this approach. These calculations, and others like them, were immediately criticized — and continue to be criticized — on two grounds: (1) that the relative abundances of rock types must be considered (basic rocks were over-represented in the Clarke and Washington data); and (2) that the results obtained from averages of analyses of rocks that happened to be available were biased because unusual rock types tend to be over-represented.

Various attempts were made to overcome these difficulties. Thus, Knopf (1916) made calculations of element abundances on the basis of the areal extent of various igneous rock types of the Appalachian and Cordilleran regions of North America. Vogt (1931) attempted to derive the average composition of igneous rocks by using published average values for different rock types and weighting them in proportion to their relative abundance. Based on the reasoning that all sedimentary and metamorphic rocks are derived ultimately from igneous rocks, attempts were also made to derive abundances from a judicious mixture of igneous rocks; Vinogradov's (1962)

TABLE 2-I

Estimates of major element composition (wt. %) of Precambrian shield areas

Element	Finnish Shield (Sederholm, 1925)	Canadian Shield (Shaw et al., 1967)	Australian Shield (Lambert and Heier, 1968)	Average crystalline shield (Poldervaart, 1955)
Si	31.53	30.36	31.32	31.04
Al	7.74	7.74	7.68	8.20
Fe	3.32	3.09	3.35	3.44
Mg	1.02	1.35	1.51	1.21
Ca	2.42	2.94	2.86	2.72
Na	2.27	2.57	1.85	2.60
K	2.95	2.57	2.49	2.74
Mn	0.03	0.05	0.15	0.08
Ti	0.25	0.31	0.30	0.36

TABLE 2-II

Estimates of major element composition (wt. %) of the lithosphere (more extensive data are given in Table A-I, Appendix 1)

Element	Average 5159 analyses (Clarke and Washington, 1924)	2:1 felsic/ mafic mixture (Vinogradov, 1962)	1:1 granite/ basalt mixture (S.R. Taylor, 1964)	Computer balance, igneous rock (Horn and Adams, 1966)	Continental lithosphere, 1:2 granitic/ basaltic rocks (Beus, 1976)
Si	27.64	29.5	28.15	28.50	27.7
Al	8.12	8.05	8.23	7.95	8.1
Fe	5.10	4.65	5.63	4.22	5.7
Mg	2.10	1.87	2.33	1.76	2.4
Ca	3.63	2.96	4.15	3.62	4.3
Na	2.85	2.50	2.36	2.81	2.3
K	2.60	2.50	2.09	2.57	1.8
Mn	0.09	0.11	0.10	0.09	0.09
Ti	0.63	0.45	0.57	0.48	0.6

calculation was based on a mixture of two parts granite and one part basalt, and S.R. Taylor's (1964) estimate was based on a mixture of 1:1 granite and basalt. Many of the calculations by other workers (see Table 2-I) were based on shield areas which, because of their low erosional levels, tend to be more acidic than other areas. Some comparative data for the major elements derived by a number of different mehods are given in Table 2-II; the data clearly show the surprisingly wide variation in the estimates.

Percentage differences in estimates of trace and minor elements are even greater than those for the major elements (see Table 2-III). This is scarcely surprising, given the problems of adequately sampling and measuring elements with low concentrations (see below), but it does render the use of

TABLE 2-III

Estimates of abundance (ppm) of selected trace and minor elements in the lithosphere (more extensive data are given in Table A-I, Appendix 1)

Atomic No.	Element	2:1 felsic/ mafic (Vinogradov, 1962)	1:1 granite/ basalt (S.R. Taylor, 1964)	Continental crust (Lee Tan and Yao Chi-lung, 1970)	Igneous rocks (Horn and Adams, 1966)
3	Li	32	20	22	32
9	F	660	625	470	715
16	S	470	260	330	410
17	Cl	170	130	100	305
27	Co	18	25	18	23
28	Ni	58	75	61	94
29	Cu	47	55	50	97
30	Zn	83	70	81	80
33	As	1.7	1.8	1.7	1.8
34	Se	0.05	0.05	0.06	0.05
37	Rb	150	90	90	166
38	Sr	340	375	470	368
42	Mo	1.1	1.5	1.1	1.25
47	Ag	0.07	0.07	0.07	0.15
48	Cd	0.13	0.2	0.15	0.19
50	Sn	2.5	2.0	1.6	2.49
51	Sb	0.51	0.2	0.45	0.51
52	Te	0.001	—	0.0004	—
56	Ba	650	425	400	595
74	W	1.3	1.5	1.2	1.4
79	Au	0.004	0.004	0.004	0.004
80	Hg	0.08	0.08	0.08	0.33
82	Pb	16	12.5	13	15.6
83	Bi	0.009	0.17	0.003	—

the term "clarke" — defined by A.E. Fersman in 1923 (in Beus, 1976) as the average percentage of an element in a geochemical system, e.g., the earth — rather pointless as a means to assess enrichment or depletion in rocks for either theoretical or exploration purposes. The "clarke" will depend on whose data is being used. Appendix 1 discusses the various estimates in more detail.

In a number of chapters in this book the "average" composition of a particular rock type is referred to simply as a reference point to assess variations in the chemical composition of barren and productive rock units. As will be clear from the examples cited, although there are distinct common trends in element content in a particular rock type as a function of proximity to mineralization, the *absolute* element content varies markedly from one geographic area to another. It therefore appears to be imprudent to adopt any particular estimate of the absolute abundance of an element, even in a restricted rock type, as being indicative of normal or background conditions. Clearly, geochemical knowledge of the composition of even the upper lithosphere is imperfect.

Very large-scale variation in the chemical composition of the earth's crust has given rise to the concept of geochemical provinces — defined as regions that exhibit a characteristic and distinctive geochemical composition that differs significantly from the average. Such differences are normally best displayed in igneous rocks, but may also be recognizable in sedimentary and metamorphic rocks.

A special case of geochemical provinces occurs where the geochemical abnormality is reflected by relatively large concentrations of elements in economic proportions, leading to an abnormal abundance of ore deposits of a particular type or a particular element. The concept and the term for this relative concentration — metallogenic provinces — was discussed in the early part of this century by Finlayson (1910a), De Launay (1913), Gregory (1922), and Spurr (1923). More recently Turneaure (1955), Smirnov (1959), Bilibin (1960), Radkevich (1961), Petraschek (1965), Dunham (1973), and Guild (1974) have developed various concepts of epochs or periods when ore formation seemed to be at a maximum. A number of subdivisions or provinces (e.g., belts, zones, ore districts, fields) have been proposed, but there is no concensus on the subject. Guild (1974) has related the concept of metallogenic provinces to the main tectonic features of the earth's crust, dividing the provinces into those formed at or near plate margins and those formed within plates. Examples of the first subdivision are the Kuroko ores, the copper and copper-molybdenum deposits of North America, and the polymetallic sulphides of western North America and the Urals; examples of the second type include the Zambian-Katanga copper deposits and the Mississippi Valley lead and zinc deposits.

In spite of the considerable work done on developing the concept of metallogenic provinces — both as a theoretical subject and as an exploration tool — there is no precise definition of the term. For example, R.G. Taylor (1979) has pointed out that whereas a number of writers have defined a Southeast Asian "tin province", there could just as well be a series of adjacent tin provinces in Southeast Asia, for example in Malaysia where the eastern and western parts of the country are geologically different.

The usefulness of the concept of geochemical and metallogenic provinces in rock geochemical exploration programmes is severely limited by the lack of adequate geochemical data to relate the two. For example, it is not known, on a regional scale, whether the rocks as a whole may be enriched or depleted in an ore element, or how they differ in other geochemical characteristics. There is some evidence that volcanic rocks host to copper massive sulphides are depleted in Cu (Chapter 7); on the other hand, the rocks of the Bear Province of the northwestern Canadian Shield — which is a uranium metallogenic province — are enriched in U (Rose et al., 1979). In the present state of knowledge the concept of metallogenic provinces which are defined on empirical evidence of known ore deposits is obviously of greater practical value for exploration.

CLASSIFICATION AND GEOCHEMICAL BEHAVIOUR OF THE ELEMENTS

Goldschmidt (1937) classified the elements into three main categories largely on empirical grounds and by analogy with phases in meteorites and ore smelting:

(1) *siderophile* — elements that have a weak affinity for sulphur and oxygen and are preferentially enriched in metallic phases; dominantly metallic bonding

(2) *chalcophile* — elements that have a strong affinity for sulphur; dominantly covalent bonding

(3) *lithophile* — elements that have a strong affinity for oxygen and are concentrated in silicate minerals; dominantly ionic bonding.

This classification of the elements is illustrated in Table A-V (Appendix 1). The classification is for a three-phase system, whereas the crust of the earth is essentially a two-phase system (the metallic phase is missing). The classification is, therefore, an indication of tendencies; except for some elements with very strong siderophile tendencies (e.g., Au), the siderophile elements must, of necessity, enter a sulphide phase if such a phase is available.

The formation of crystalline minerals involves the building of space lattices of atoms or ions in a regular arrangement; only those particles that are of a size appropriate to a particular lattice can enter that lattice. Therefore crystals act as a sorting mechanism, allowing certain particles to enter and excluding others of unsuitable size. The chemical character (especially the charge) and bonding characteristics are also important; the effective size of an ion depends upon the nature of the bonds between it and its neighbours, upon its electron configuration, and upon its coordination (i.e., the number and arrangement of neighbouring particles).

Goldschmidt (1937) formulated some general rules governing the incorporation of ions into particular crystal structures that are remarkably effective, especially if used with modern calculations of ionic radii (e.g., Whittaker and Muntus, 1970). These rules are as follows:

— Ions of similar size and charge have an equal chance of incorporation in a particular structure if bond types are compatible; e.g., W^{6+} (0.50 Å) and Mo^{6+} (0.50 Å). The substitution of a trace element for a major element is referred to as "camouflage".

— For two ions of different radii but with the same charge, the smaller ion is preferentially incorporated at higher temperatures; thus the ion Rb^+ (1.81 Å) and Cs^+ (1.96 Å) substitute for K^+ (1.68 Å) in late-crystallized potassium feldspar.

— For two ions with similar radii but with different charges, the ion with the higher charge takes precedence at higher temperatures; for example, Li^+ (0.82 Å) is reluctantly admitted to Mg^{2+} (0.80 Å) minerals and is concentrated in late-crystallized magnesium minerals. A trace element that replaces a major element is "captured" if it has a higher charge, and it is "admitted" if it has a lower charge.

These simple rules are equally applicable to major element pairs. For example, early-crystallized plagioclase is calcium-rich (Ca^{2+}, 1.08 Å), and late-crystallized plagioclase is sodium-rich (Na^+, 1.10 Å); early-crystallized olivine is enriched in Mg^{2+} (0.80 Å) relative to Fe^{2+} (0.86 Å). This kind of substitution of one element for another is called "diadochy" and always refers to a specific mineral structure. Two elements may well be diadochic in one structure but not in another. There is also a limitation on how widely the sizes of the two ions can differ and still be diadochic; Ahrens (1964) suggested that the limit on size differences is 15% to 20%.

If the dominant bond character of two cations with a particular anion is significantly different, the size-charge rules tend to fail. Broadly, the greater the covalent character of the cation-anion bond, the less likely is the cation to substitute for another cation with a high proportion of ionic bonding. It should be noted that bonding between cations and anions in any particular structure is never wholly ionic nor wholly covalent; there is a "resonance" between the two types, and the proportion of each determines the dominant bond types. Various attempts to quantify the bond character have been made on the basis of electronegativities (the power of an atom in a molecule to attract an electron); ionization potentials (a measure of the energy required to remove an electron from an atom or an ion); and electron affinities (the energy required by an atom or an ion to attract an electron). Pauling (1960), for example, suggested that if the difference in electronegativities between a cation and an anion is greater than 2.1 the bond type will be dominantly ionic; if the electronegativity difference is less than 2.1 the bond type will be dominantly covalent.

The preferred type of bonding explains why Cu^+ does not substitute for Na^+, despite the similarity of size and charge; according to Pauling's rules the Na^+—O bond is 83% ionic, whereas the Cu^+—O bond is only 71% ionic. Similarly, although the size of Ag^+ does not differ from K^+ as much as Rb^+ and Cs^+, Ag will not substitute for K because of its stronger tendency to covalent bonding. Therefore, both Cu and Ag are chalcophile in character. It will be noted in later chapters that anomalous halos of Cu have been found to be in the sulphide form in some cases; Zn, on the other hand, displays both lithophile and chalcophile properties, and it apparently substitutes quite readily in ferromagnesian minerals (in many cases Zn halos are clearly related to the distribution of ferromagnesian minerals).

This brief and simplified consideration of element distribution is intended only to provide a background for some of the geochemical relations discussed in the chapters that follow. A concise account is given in Krauskopf (1967); more detailed discussions may be found in the references quoted above and in Shaw (1953); Goldschmidt (1954); Ringwood (1955); Gordy and Thomas (1956); Fyfe (1964); Burns and Fyfe (1967); and in Appendix 1.

SAMPLING

In taking rock samples from, for example, a granitic intrusion, the implied expectation is that the sample will be mineralogically and chemically similar to the whole intrusion. Because even the most optimistic geologist does not expect an intrusion to be homogeneous, it is normal to take a number of samples in the expectation that the range of composition encountered in the samples will reflect the range of composition in the intrusion. The samples are then crushed and split, ground and split again in the laboratory, and finally a gram or less is analyzed.

On the face of it, this is an audacious procedure. To take an example from Chapter 7, an average of 1 kg sample of rhyolite was collected from each 5 km² of an area in northern New Brunswick, and a 1-g subsample was analyzed. The objective was to determine whether mineralization existed within 250 m of the surface within 5 km². In effect, it was hoped that an analysis of a 1-g sample would provide information on a population of about 1.6×10^{16} g. Clearly, much of the reliability of analytical data depends upon the nature of the original sampling.

At the field sampling level there is obviously wide scope for bias. A particular phase of rock may preferentially outcrop because of its weathering characteristics and, therefore, may be preferentially sampled; at the outcrop level of sampling further bias may arise because of the mechanical ease of breaking the rock along bedding or fracture planes. Assuming that appropriate precautions are taken to avoid these kinds of bias, there remains the practical problem of collecting a large enough sample to ensure that the constituent sought is present in representative amounts.

As a generalization, the larger the grain size of the rock, the larger is the volume that should be collected; similarly, the more sporadic the distribution of a constituent of interest and the lower its concentration, the larger is the volume of sample needed. It is recommended in subsequent chapters that 1 kg of sample is adequate for most exploration purposes. The basis of this recommendation is examined below.

The standard deviation, S, of taking a sample of n grains from a large population containing $P\%$ of a mineral of interest is given by:

$$S = \sqrt{nP(100 - P)} \tag{1}$$

The relative standard deviation, R, expressed as a percentage of P is:

$$R = \frac{S}{nP} \times 100 \tag{2}$$

Therefore:

$$R = \frac{\sqrt{nP(100 - P)}}{nP} \times 100 \tag{3}$$

An expression to determine the number of grains, n, required for particular values of R and P is obtained by rearrangement of equation (2):

$$n = \frac{(100 - P) \times 10^4}{PR^2} \tag{4}$$

The volume, $v \, cm^3$, of sample required for equidimensional grains of diameter d (cm) is:

$$v = \frac{\pi d^3}{6} \left[\frac{(100 - P) \times 10^4}{PR^2} \right] \tag{5}$$

If a density of $2.75 \, g/cm^3$ is assumed, the weight of sample in grams, g, is:

$$g = \frac{2.75\pi d^3}{6} \left[\frac{(100 - P) \times 10^4}{PR^2} \right] \tag{6}$$

The total variance, V_t, is given by the sum of the variance of the field sampling, V_s, the variance of taking an aliquot for analysis, V_{ss}, and the analytical variance, V_a; i.e.:

$$V_t = V_s + V_{ss} + V_a \tag{7}$$

The analytical variance is fixed for a particular analytical technique for a particular mineral; Rickwood (1979) has made the point that it is wasted effort to use an analytical technique that has a variance far lower than the sampling variance. He suggests that a useful guide to sampling is:

$$V_s = V_{ss} \leqslant \tfrac{1}{2} V_a \tag{8}$$

Since $V_s = S_s^2$ and $V_a = S_a^2$, from equation (2):

$$S_s^2 = \frac{R_s^2 n^2 P^2}{100^2} \tag{9}$$

and:

$$S_a^2 = \frac{R_a^2 n^2 P^2}{100^2} \tag{10}$$

Therefore, from equations (9) and (10):

$$V_s = \frac{R_s^2 n^2 P^2}{100^2} \tag{11}$$

$$V_a = \frac{R_a^2 n^2 P^2}{100^2} \tag{12}$$

From equations (8), (11), and (12):

$$\frac{R_s^2 n^2 P^2}{100^2} = \frac{1}{2} \frac{R_a^2 n^2 P^2}{100^2} \tag{13}$$

$$R_s^2 = \tfrac{1}{2} R_a^2 \tag{14}$$

In this treatment the numerical value of the variance is therefore equal to the numerical value of the relative standard deviation. Therefore, the relative standard deviation of the analytical technique, R_a, can be inserted into equation (6):

$$g = \frac{2.75 \pi d^3}{6} \left[\frac{(100 - P) \times 10^4}{P R_a^2} \right] \tag{15}$$

The variation in required sample weight as a function of R_a for $P = 1\%$ for various grain sizes, as derived from equation (15) is shown in Fig. 2-1; the variation in sample weight as a function of P for various grain sizes and for $R_a = 10\%$ is shown in Fig. 2-2. In both these diagrams the dimensions of a cube of rock in terms of the length of the cube edge, L are also shown (L is derived from the cube root of equation (5)).

The main conclusions from these graphs are that over the normal range of exploration geochemistry analytical precision, a sample of 1 kg is adequate for grain sizes of 1—2 mm, provided that the constituent of interest is present in amounts in excess of about 0.2%. For large grain sizes (greater than 5 mm) sample weights of 1 kg are also adequate, provided that the constituent of interest is present in amounts greater than about 3%. It will be noted that the required sample weight increases sharply as R_a decreases below 5% (Fig. 2-1); similarly, the required sample weight for any particular value of R_a decreases rapidly for proportions of P greater than 50%. The common — and recommended — practice of taking chip samples to make an aggregate sample weight of 1 kg will tend to decrease the sample variance.

The second possible major source of error is in the choice of sample aliquot and grain size taken for actual analysis (assuming that sample reduction and splitting is carefully carried out so that the variance of this step is very small compared to analytical variance). This problem is discussed in a companion volume to this series (Fletcher, 1981) and is therefore treated only briefly here.

As may be expected, for trace elements in rare mineral grains the chance

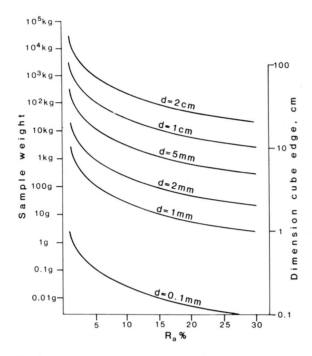

Fig. 2-1. Variation of minimum sample weight (density $= 2.75\,g/cm^3$) and in size of sample cubes (L = length of cube edge) as a function of relative standard deviation of analysis (R_a) for selected grain sizes (diameter = d) for 1% ($P = 1\%$) of a constituent mineral.

of such a grain occurring in any one analyzed aliquot becomes increasingly remote as the sample weight decreases. Ingamells and Switzer (1973) state that the relative standard deviation varies inversely with the square root of the sample weight. As a guide to an appropriate grain size to ensure a reasonable chance of analyzing a representative subsample, Kleeman (1967) suggests that the number of particles at any stage of subsampling (including taking an aliquot for analysis) should be 10^6 to 10^7. If a specific gravity of $2.75\,g/cm^3$ is assumed, this means that if a 1-g aliquot is taken the sample must be ground to 170 mesh (A.S.T.M.). During crushing and grinding the total amount of sample should not be less than 160 g at 35 mesh and not less than 8 g at 80 mesh. More information on subsampling can be derived from the above equations.

Given the influence of the original sample size, the mesh size of the analyzed sample, and the weight of the aliquot on the recorded analytical result — plus the probable wide variation in sampling and subsampling practice that probably occurred in the samples used to derive the crustal abundances discussed above — it is surprising that the agreement in the estimates (particularly for trace elements) is as good as it is. In exploration

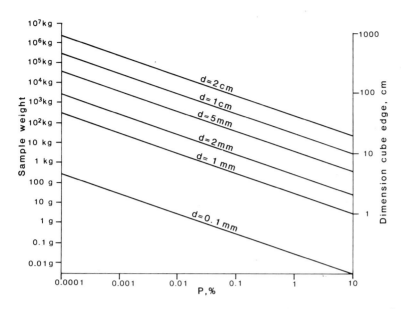

Fig. 2-2. Variation in minimum sample weight (density $= 2.75 \, \mathrm{g/cm^3}$) and in size of sample cubes (L = length of cube edge) as a function of amount of constituent mineral (P) for selected grain sizes (diameter $= d$) for relative standard deviation of analysis of 10% ($R_a = 10\%$).

geochemistry an attempt is being made to make economic judgements on the basis of analyzed samples, and it is obviously imperative that all stages of sampling be carefully considered. Routine analytical precision in AAS techniques is now probably greater than many samples deserve (the coefficient of variation is generally less than 10%). More reliable and more useful data can be derived from rock samples if the size of the samples is adequate to meet the requirements of particular determinations.

Chapter 3

RECOGNITION OF GEOCHEMICAL ANOMALIES

INTRODUCTION

A companion volume in this series (Howarth, 1982) is devoted entirely to data processing and interpretation; therefore a detailed discussion of procedures would be superfluous here. Nevertheless, some general concepts of geochemical anomalies — not necessarily restricted to rock geochemistry — are reviewed in this chapter to provide a background to the philosophy of data interpretation in the chapters that follow.

One of the greatest problems in the interpretation of geochemical surveys is to discriminate between "anomalies" that are due to mineral deposits and those that are due to other causes; the ultimate objective is to discriminate between anomalies due to non-economic and economic mineral deposits. In Chapter 1 an anomaly was defined conceptually as a chemical abnormality reflecting a disturbance of the normal (background) chemical equilibrium caused by a mineralizing event. In its simplest expression, the objective of the interpretation of geochemical data is the recognition of such an abnormal concentration of an element in a sample associated with mineralization compared to the concentration of an element in similar sample material from a geologically comparable but non-mineralized situation.

Unfortunately, the presence of a mineral deposit in an area under investigation is not the only factor that can disturb chemical equilibrium. A simple — and common — case is the effect which a zone of slightly impeded drainage in an area of otherwise well-drained soils may have on the content of elements in soils. Clearly, in this case, there are two background situations — well-drained soils, and poorly-drained soils. A competent exploration geochemist will recognize this fact and interpret the data accordingly. Other factors that can disturb chemical equilibrium are not so easily recognized, either due to lack of visible evidence (such as minor variations in the degree of fractionation of a volcanic sequence) or simply because of inadequate knowledge of fundamental geochemical processes.

Regardless of interpretative difficulties and deficiencies, the practical application of exploration geochemistry is based on the search for geochemical anomalies in samples of the earth's crust to assess the mineral potential of

30

an area and, more specifically, as an indication of the presence of individual mineral deposits. Much of the early practice of exploration geochemistry in soils and stream sediments relied on large differences in element concentration between samples; this approach is still adequate to detect near-surface and large mineralized areas that give rise to intensive and extensive halos of elements associated with mineralization. The identification of anomalous samples has become progressively more difficult as the scope of exploration geochemistry has expanded to include the search for deeply-buried deposits, the identification of broad targets with relatively few samples, and, especially, the application of rock geochemistry to exploration.

In defining, and therefore recognizing, an anomaly it is necessary to consider the value of the measured parameter, the physico-chemical nature of the sample material, the physico-chemical environment from which the sample is taken, and the spatial distribution pattern of the measured variable. An appropriate definition of an anomaly is: *an abnormally high or low content of an element or element combination, or an abnormal spatial distribution of an element or element combination in a particular sample type in a particular environment as measured by a particular analytical technique.*

BACKGROUND AND THRESHOLD

In conventional interpretation an anomaly cannot be recognized unless background can be defined. The content of any element in any geological material cannot be represented as a single number. Apart from variations due to error in sample preparation and analysis that are inevitably present in the reported value for a particular sample (see Fletcher in the first volume in this series), there is an inherent variability in the content of an element even in a small and petrologically homogeneous granite or in B-horizon soil samples overlying an homogeneous rock. In statistical terms the recognition of an anomalous sample depends upon establishing the statistically probable range that occurs in the background samples — background population — and calculating (for some defined confidence level) an acceptable upper limit — threshold — of background fluctuation. Any sample that exceeds this threshold (which is the lowest anomalous value) is then regarded as possibly anomalous and belonging to a separate population. Similarly, for negative anomalies threshold defines the *lower* limit of background fluctuation.

Background is a term that tends to be loosely used in exploration geochemistry; most commonly it implies the abundance of an element in a particular material. Conventionally abundance is represented by the *average* content of an element (see discussion in Chapter 2). This approach is reasonable for a normally distributed population; however, the frequency distribution of elements is commonly positively skewed, and the arithmetic average is clearly biased by scattered high numbers. Therefore, although the arithmetic mean is a good estimate of the abundance of an element, it is not necessarily a good estimate of the most commonly occurring concentration of that element.

The limitations of using the arithmetic mean for interpretation of geochemical data are illustrated schematically in Fig. 3-1. If a 1-cm-thick slab of granite measuring 10 cm^2 is divided into 10 cubes of 1 cm^3 and 25 systematic samples on a grid pattern are taken as indicated in Fig. 3-1, the arithmetic mean (45.6) is a close estimate of the Cu in the whole slab (the abundance, 43.6, is given by the mean of all 100 cubes). As the frequency distribution shows, however, the mean is a poor indicator of the concentration that

40	30	30	40	60	10	30	30	20	50
40	90	50	30	40	50	10	40	30	40
70	40	60	60	50	30	40	10	40	20
40	70	90	60	50	50	40	30	70	60
40	20	50	10	70	60	30	90	30	40
20	70	40	20	20	30	50	80	60	40
80	30	50	100	30	80	20	30	60	40
50	60	40	20	30	80	30	30	20	60
50	50	90	10	40	30	30	50	30	30
30	50	30	30	40	30	80	20	40	50

Samples for n=25

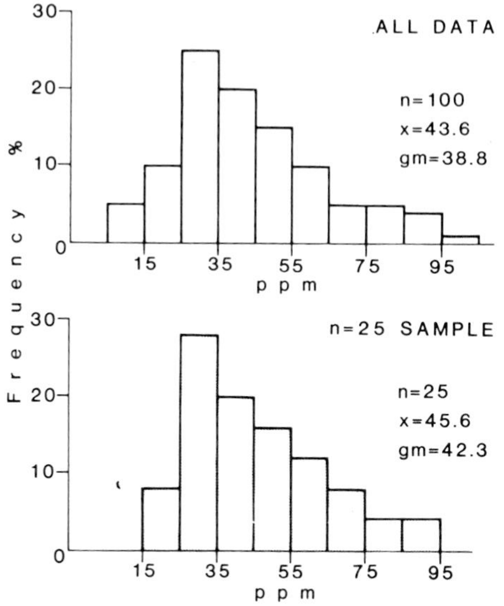

Fig. 3-1. Plan of simulated random distribution of Cu in equidimensional cubes of slabs of rock, and frequency distributions of the whole population ($n = 100$) and of a systematic sample ($n = 25$) of the whole population.

most commonly occurs — the mode. In Fig. 3-1 the mode is 30 ppm for both the whole population and for the 25-sample data set. In determining background for positively skewed data — the most common situation — a better approximation to the most commonly occurring value is the geometric mean (the antilog of the arithmetic average of the $\log_{10}$ of the values). The geometric mean reduces the importance of a few high values in a sample group and therefore is numerically less than the arithmetic mean, making it a useful indicator of background for most geochemical data.

SPATIAL DISTRIBUTION OF DATA

If the same sample data used in the above example has a systematic spatial distribution (as shown in Fig. 3-2 where the same numbers are arranged in a different pattern) the frequency distribution of the whole population of 100 samples if obviously exactly the same as it is for the random spatial distribution shown in Fig. 3-1. The shape of the frequency distribution of the 25 samples taken from the population in Fig. 3-2 differs slightly from

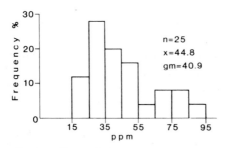

20	50	30	40	40	60	80	80	90	100
40	60	20	30	50	50	70	80	90	90
50	30	40	10	60	40	70	90	80	80
10	20	50	30	30	40	60	60	70	70
40	60	30	40	50	30	40	20	70	60
30	20	40	20	40	60	20	50	30	40
30	50	40	30	30	30	50	10	30	50
30	40	10	40	50	30	60	40	30	30
20	50	30	60	30	40	50	30	20	40
10	40	30	30	50	30	40	30	50	20

Samples for n=25

n=25
x=44.8
gm=40.9

Fig. 3-2. Plan of systematic spatial distribution of a random population of Cu contents (same population as used in Fig. 3-1), and frequency distribution of a systematic sample of the population.

the frequency distribution of 25 samples taken from the random spatial distribution of Fig. 3-1 (there is a break in the distribution between 55 and 65), but the essential features — mode, arithmetic mean, and geometric mean — are similar. Thus, the spatial difference between the data in Figs. 3-1 and 3-2 would not be obvious from the statistical data, whereas simple visual inspection of the distribution of the 25 samples in Fig. 3-2 would immediately reveal a *spatial pattern*.

Instead of regarding the 100 numbers as representing cubes of a 10-cm^2 granite slab, let them represent the Cu content in samples of grid soil, rock, or biogeochemical surveys in two different geographic areas; in one area there is an essentially random distribution of values, and in the other area there is a systematic spatial concentration of high values in the northeast corner. Obviously, since the actual numbers are exactly the same in both cases — only their spatial distribution differs — the frequency distributions must be exactly the same. Only when a spatial parameter is introduced will a difference between the two areas become obvious.

The central limit theorem states that, regardless of the form of the frequency distribution of a set of numbers, the frequency distribution of the means of randomly selected subsets of the numbers will tend to normality. A logical extension of this theorem is that, regardless of the form of the frequency distribution of analyses of individual samples in a geochemical grid, the frequency distributions of the means of systematic blocks of samples will tend to normality *if* the spatial distribution of the individual samples is random (see Govett et al., 1975). If the individual samples have a systematic spatial distribution, the frequency distribution of the means of systematic blocks of samples will *not* tend towards normality; it will tend towards a polymodal distribution.

This effect is illustrated in Figs. 3-3 and 3-4 for the hypothetical geochemical surveys discussed above. The data for both surveys are averaged in the same way by calculating the means of 4-sample groups (as shown in Fig. 3-3). The difference between the hypothetical areas is now obvious. The frequency distribution of the 4-sample means for the spatially random distribution is close to normal (Fig. 3-4); the frequency distribution of the 4-sample means of the spatially systematic distribution is polymodal and shows two distinct populations. Thus, the positively skewed distribution of the individual samples represents heterogeneous populations in both cases; the only difference is that the lack of homogeneity in one case has no spatial significance (on the sampling scale used), whereas it does in the other case (a practical application of this approach to a soil survey is given in Chork and Govett, 1979).

In this example there is little need to calculate a background or threshold value; the spatial distribution of the data is adequate to indicate an anomalous situation (Fig. 3-3). If the same 100 values represented analytical values of samples from a regional stream sediment survey over a large area (so that

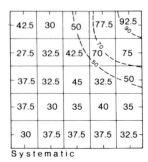

Systematic

Random

Fig. 3-3. Plan of 4-sample block averages for systematic spatial distribution (from Fig. 3-2), and random spatial distribution (from Fig. 3-1).

only one sample was taken from each stream) the problems of interpretation would be much more difficult, although the above principles would still apply. Under such circumstances — if no independent assessment of background and anomalous conditions is possible through an orientation survey — interpretation is facilitated by calculation of threshold.

A widely used assessment of threshold for data from a background population that is normally distributed (or can be mathematically transformed to essentially conform to a normal distribution) is the mean plus 2 standard deviations (i.e., only one sample in 40 is expected to exceed this level by chance alone). A more stringent threshold would be the mean plus 3 standard deviations (i.e., only one sample in 667 is expected to exceed this level by chance alone). These methods of calculating threshold are the classic approach suggested by Hawkes and Webb (1962) ". . . as a *disciplinary guide* . . . " (p. 29, my italics) to interpretation. The work of Hawkes and Webb has been misquoted and mis-used ever since the book was published by a wholesale use of the mean plus 2 or 3 standard deviations for any and all types of geochemical surveys.

As an illustration of the consequences that can arise through the routine application of the mean plus 2 standard deviations principle of threshold calculation, consider its application to the data used in Fig. 3-1 above. The

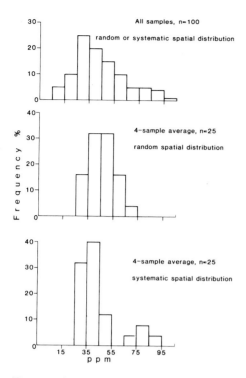

Fig. 3-4. Comparison of frequency distributions of 100 individual samples (random or systematic spatial distribution), and averages of 4-sample blocks of random and systematic spatial distributions (see Fig. 3-3).

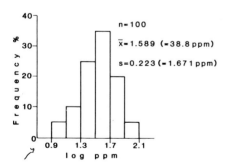

Fig. 3-5. Frequency distribution of $\log_{10}$ transformed data from Fig. 3-1.

frequency distribution is positively skewed; therefore to apply parametric statistics, the data must be transformed to approximate a normal distribution. A common practice is to convert the raw data to $\log_{10}$; this is done in Fig. 3-5. The frequency distribution that results has a slightly negative skew, but it is much closer to a normal distribution than the arithmetic data.

The arithmetic mean is 1.589 (38.8 ppm, which is also the geometric mean); the standard deviation is 0.223 (1.671 ppm). The threshold, as defined by the mean plus 2 standard deviations is 38.8×1.671^2 (or the antilog of $1.589 + 2 \times 0.233$); this is 108 ppm. This value exceeds the highest value of the data set. Therefore, if this method of calculating a threshold value is used, no sample would be considered as anomalous; however, as shown above, if there is a spatial pattern to the values (Fig. 3-2) there is very clearly an anomalous group of samples.

THE "LAW OF LOGNORMALITY"

The commonest form of frequency distribution, as noted above, is positively skewed; this observation led Ahrens (1954) to propose his "law of lognormality". Despite his qualifiers to its universal applicability and the opposition to the "law" (e.g., Chayes, 1954; Miller and Goldberg, 1955; Aubrey, 1956; Durovic, 1959; Vistelius, 1960), exploration geochemists commonly log-transform data prior to applying statistical tests — such as in the calculation of threshold illustrated above. Although this practice is *statistically* correct, in many cases it can give rise to misleading geochemical conclusions (as in the above example).

Log-transformation has the effect of diminishing the importance of high values (i.e., anomalous values) relative to low values (i.e., background values). It is therefore a self-defeating exercise in some types of data interpretation. Where adequate control data are available the samples in the tail of a distribution can be shown to belong to a separate (anomalous) population that overlaps the main (background) population (Govett and Pantazis, 1971; Govett, 1972; Govett et al., 1975; Chapman, 1976). Even on Ahrens' (1954) own data it can be shown that the distribution of Pb in Canadian granites can be "normalized" by taking the cube root of the values (Chayes, 1954); the distribution of Zr in Canadian granites could as well obey a square root law as a log law and, even more plausibly the distribution could belong to two dominantly normal distributions that overlap (Govett, 1974).

POPULATION PARTITION

If an a priori assumption is made that positively skewed distributions represent a mixture of two or more overlapping populations unless the contrary can be proved, an alternative approach to log-transformation is to attempt to separate the populations — to estimate the threshold of the lower (background) population. A graphical solution is the easiest approach, utilizing the fact that the cumulative percentage frequency of a normally distributed population plots as a straight line on arithmetic probability paper (similarly, a log-normal population plots as a straight line on logarithmic probability paper). Deviations from a straight line indicate a mixture of two or more populations. There are various techniques for estimating geochemical threshold from probability paper (e.g., Tennant and White, 1959; Lepeltier, 1969; Bölviken, 1971; Parslow, 1974; and A.J. Sinclair, 1974); a comprehensive description on the use of probability paper and an elegant exposition on the partitioning of populations is given by A.J. Sinclair (1976).

The 100-sample data set used above is plotted on arithmetic probability paper in Fig. 3-6 and is partitioned into its component populations according to Sinclair's (1976) procedure. An estimate of the proportions of the two populations is given by the inflection in the curve. For the whole data set this occurs at the 90th cumulative percentile (shown on curve T in Fig. 3-6), indicating that the sample group comprises 10% of a higher population (A) and 90% of a lower population (B). The 90-ppm level represents 99% of the total set; it also represents $(1/10 \times 100) = 10$ cumulative percent of population A since there is no significant contribution from population B. Thus, at 90 ppm population A is represented by $(100 - 10) = 90$ cumulative percent (point 1 in Fig. 3-6). The 80-ppm level represents 95% of the total data set, and hence population A is $(5/10 \times 100) = 50$ cumulative percent (point 2 in Fig. 3-6). Similarly, by selecting points on the lower part of the total data curve the distribution of population B can be calculated. For example, the 15-ppm level represents 10% of the total data set, which

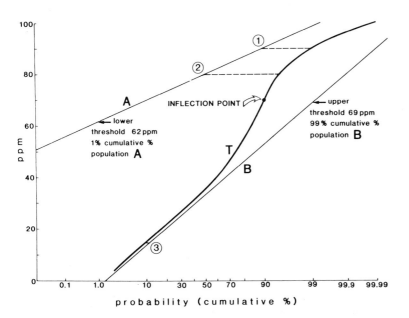

Fig. 3-6. Cumulative frequency distribution of the 100-sample data set (from Fig. 3-1), and its partition into two component populations A and B.

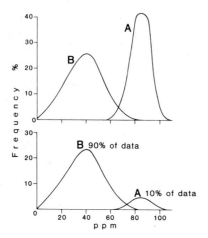

Fig. 3-7. Forms of frequency distributions of theoretical populations A and B from Fig. 3-6 (top), and frequency distributions of mixed populations A and B from Fig. 3-6 (bottom).

is $(10/90 \times 100) = 11.1$ cumulative percent of population B (point *3* in Fig. 3-6).

The two populations thus calculated are illustrated as conventional frequency distributions in Fig. 3-7 which clearly shows the overlap of the two populations between 60 ppm and 80 ppm (where samples may belong to either anomalous or background populations). Useful thresholds can be visually derived from Fig. 3-7: samples greater than 80 ppm are first priority targets, whereas samples between 60 ppm and 80 ppm are second priority targets; samples with less than 60 ppm are regarded as background.

A more rigorous approach recommended by A.J. Sinclair (1976) is to arbitrarily choose a threshold near the upper limit of population B (98 or 99 cumulative percentiles) and near the lower limit of population A (1 or 2 cumulative percentiles). For the data shown in Fig. 3-6 the 99th and 1st cumulative percentiles for population B and A, respectively, are selected. This gives an upper threshold of 69 ppm (read from the 99th cumulative percentile for population B) and a lower threshold of 62 ppm (read from the 1st cumulative percentile of population A). This divides the whole data set into three groups: group I, greater than 69 ppm; group II, less than 69 ppm and greater than 62 ppm; and group III, less than 62 ppm (Table 3-I). In group I, 10% of the total data exceed 69 ppm (this is read from curve *T* in Fig. 3-6); 92% of population A exceeds 69 ppm (read from curve *A* in Fig. 3-6); and 1% of population B exceeds 69 ppm (defined). Since there are 100 samples in the total data set, the corresponding numbers of samples are 10 for the total data (10% of 100); about 9 for population A (total number of population A samples is 10% of 100, and 92% of these exceed 69 ppm); and about one sample for population B (total number of population B

TABLE 3-I

Summary of graphical partitioning of 100 samples of hypothetical data set shown in Fig. 3-1

	Total data		Population A		Population B	
	%	N	%	N	%	N
Group 1 >69 ppm	10	10	92	9.2	1	0.9
Group 2 <69 >62 ppm	4	4	7	0.7	2	1.8
Group 3 <62 ppm	86	86	1	0.1	97	87.3

samples is 90% of 100, and 1% of these exceeds 69 ppm). Similar calculations can be made for group III (less than 62 ppm), and group II parameters can be obtained by the difference between group I and group II. The results are summarized in Table 3-I.

In this particular example there is not a large overlap between the two populations; a threshold of 69 ppm would include 9 out of the 10 anomalous samples and would include only one background sample. Because of this a threshold of about 65 ppm could be chosen by inspection of the histogram of the 25 samples from the data set shown in Fig. 3-2 and from the 4-sample average of the systematic distribution in Fig. 3-4. Visual selection of threshold is much more difficult in many real life examples; overlap may be as much as 30% and there may be two background populations. Fuller discussion of these problems, with examples, is given in a companion volume in this series (Howarth, 1982).

ANOMALOUS PATTERNS

In situations where geochemical responses are weak — as is commonly the case for deeply-buried deposits — a more useful approach is to seek anomalous *patterns* rather than anomalous values. In this type of approach the important consideration is not the absolute content of an element — or even its content relative to some defined background in a particular sample — but the *relative* content of elements (or any measured parameter) in adjacent samples.

A schematic illustration of this concept is shown in Fig. 3-8. All the values for individual samples fall within the range of plus or minus two standard deviations from the mean of an approximately normally distributed sample group. On the basis of the classic definition of an anomaly discussed above, none of the samples would be classified as anomalous. There is, however, an

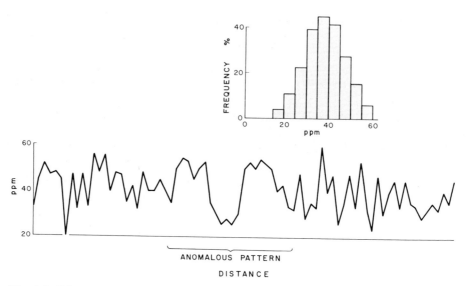

Fig. 3-8. Schematic illustration of frequency and spatial distributions of measured parameters in soil showing an anomalous spatial pattern (Govett, 1977).

unmistakable anomalous *pattern* within the *spatial* distribution of the data. This approach to anomaly recognition through patterns in univariate data has been used successfully by Govett in the interpretation of geochemical data in soils overlying deeply-buried deposits (Govett, 1976a; Govett and Chork, 1977) and by Howarth (1982) for multivariate data.

SOVIET INTERPRETATIVE PROCEDURES

Soviet geochemists use rock geochemistry extensively, but some of their interpretative procedures are not immediately obvious in the translated literature, in part because of the use of many unfamiliar terms. Therefore, a brief explanatory review of selected procedures used with rock geochemistry is given here (based largely on Ovchinnikov and Grigoryan, 1971; Ovchinnikov and Baranov, 1972; Grigoryan, 1974; and Beus and Grigoryan, 1977).

A great deal of Soviet rock geochemical work is characterized by the use of simple arithmetic manipulations of multi-element data based on a general zoning sequence of trace elements. Grigoryan (1974) has stated that despite large variations in the composition of hydrothermal orebodies and differences in geological conditions, there is a remarkably uniform zoning pattern from top (*supra-ore elements*) to bottom (*sub-ore elements*) as follows: Ba—(Sb,As,Hg)—Cd—Ag—Pb—Zn—Au—Cu—Bi—Ni—Co—Mo—U—Sn—Be—W. "Additive halos" (addition or subtraction of different element values standardized to respective background for each element), "multiplicative halos" (multiplication or division of element values), and ratios of supra-ore to sub-ore elements are generally used rather than computer-based statistical techniques (such as discriminant analysis or factor analysis). The use of multi-element ratios has the advantage of minimizing analytical variations (most of the Soviet analyses are semi-quantitative spectrographic) and reducing the effect of local reversals in zoning sequences. A comparison of the results obtained by discriminant analysis and the multiplicative approach for rock geochemical data was given by Govett (1972).

The dimension of halos — which is inherent in all geochemical data but not used by western geochemists except in a descriptive sense — is specifically incorporated in Soviet quantitative interpretations as "linear productivity"; this is the product of the width of the anomaly in metres and the average content of an element in percent. Linear productivity therefore has the dimensions of m%.

The terms and concepts described above are illustrated schematically in Fig. 3-9. The elements As and Ag give supra-ore halos, and the elements Cu and Co give sub-ore halos. The average element content, the width of the corresponding anomaly, and calculated linear productivity is shown in Table 3-II. The contrasting linear productivities for Ag (a supra-ore element) and for Co (a sub-ore element) are shown in Fig. 3-9; also illustrated is the linear productivity of the multiplicative ratios. The enhancement of the anomaly is impressive. The linear productivities of the multiplicative ratios show more than 4 orders of magnitude difference between the supra-ore (level I) and sub-ore (level IV) halos; this contrasts with a maximum of 2 orders of magnitude difference between level I and level IV for linear productivities of individual elements and a maximum difference of only 2.5 times in simple ratios between the supra-ore and sub-ore elements. This type of approach is obviously of immense value in determining from surface or drill data whether an anomaly represents a buried deposit or merely a root zone of a deposit that has been eroded.

A variety of other terms are encountered in the Soviet literature on rock geochemistry. Some of these are defined below.

Coefficient of contrast. The coefficient of contrast is the ratio of an element (or other parameter) at the surface to the element (or other parameter) at depth. It is used to determine the zoning sequence of elements in cases where the linear productivity varies monotonically with depth. In Table 3-II the sequence of elements based on element content is, from top to bottom, Ag—As—Co—Cu; based on linear productivities the sequence is As—Ag—Cu—Co. Obviously Ag and As, and Co and Cu, show similar behaviour, and maximum contrast is obtained by (Ag x As)/(Co x Cu) as shown in Fig. 3—9.

Zonality index. This index is used in cases where the linear productivities do not vary monotonically with depth. The zonality index is rather more complicated to calculate than the coefficient of contrast. The linear pro-

42

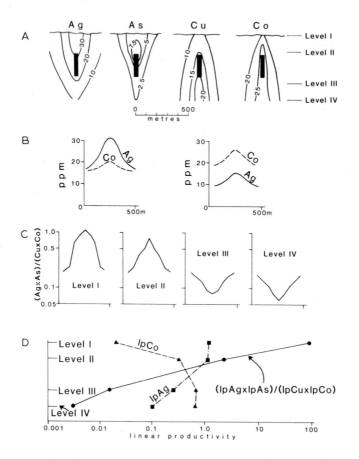

Fig. 3-9. Schematic illustration of anomaly enhancement through the use of multiplicative halos and linear productivities. A. Distribution of supra-ore elements (Ag, As) and sub-ore elements (Cu, Co). B. Profile of Ag and Co along levels I and IV. C. Multiplicative halos (Ag × As)/(Cu × Co) at the different levels. D. Linear productivities for Ag and Co, and ratios of linear productivities for Ag, As, Cu, and Co. (Modified from Govett, 1977 and Govett and Nichol, 1979.)

ductivities are first "normalized" to correspond to the same order of magnitude as the maximum value. Using the data in Table 3-II as an example, the maximum linear productivity is 1.18 for Ag at level I. The maximum linear productivity of As is 0.213 at level II, and therefore all As linear productivities are multiplied by 10. By similar reasoning, the linear productivities of Cu and Co are also multiplied by 10. This is shown in Table 3-III. The zonality index is the ratio of the normalized linear productivity to the sum of the normalized linear productivities. The zonal sequence is given, from top to bottom, by the order of the level in which the maximum zonality index occurs. Thus, the sequence from level I to level IV is (Ag-As)—Co—Cu.

TABLE 3-II

Mean concentration for As, Ag, Cu, Co, width of anomaly, and linear productivity (lp = mean × width × 10^{-4}) at different depth levels for data in Fig. 3-10

Level	Ag			As		
	mean (ppm)	width (m)	lp (m%)	mean (ppm)	width (m)	lp (m%)
I	23.7	500	1.18	3.9	400	0.156
II	23	500	1.15	7.1	300	0.213
III	18.4	130	0.24	2.5	100	0.025
IV	10.0	100	0.10	2.5	50	0.0125
I/IV	2.37		11.8	1.56		12.4

Level	Cu			Co		
	mean (ppm)	width (m)	lp (m%)	mean (ppm)	width (m)	lp (m%)
I	10.0	100	0.1	20.0	40	0.08
II	11.7	200	0.334	23.7	140	0.33
III	16.0	360	0.576	23.1	300	0.69
IV	15.0	400	0.600	22.3	300	0.67
I/IV	0.67		0.17	0.9		0.12

Variability index. Since zonal order calculated by the various methods given above is not entirely consistent, an index is used to refine the zonality. The variability index is calculated for each element as the sum of the ratios of the largest zonality index to the other zonality indices at other levels. Using the example of Ag in Table 3-III, the largest zonality index (0.44) is at level I; this is divided by the zonality index for each of the other levels, and the ratios are added. Thus the variability index for Ag is:

$$(0.44/0.43) + (0.44/0.09) + (0.44/0.03) = 20.58$$

The corresponding variability indices for As, Cu, and Co are 27.37, 8.85, and 10.92, respectively. The zonal sequence, from top to bottom, is therefore As—Ag—Co—Cu.

Anomaly ratio. This ratio is equivalent to the English term "contrast"; it is the ratio of an anomalous value to background.

Gradient of concentration. This is the tangent of the acute angle between the trend line of increasing concentration of an element towards mineralization and the horizontal (distance) axis.

Mobility of an element. The term refers to the reciprocal of the gradient of concentration.

TABLE 3-III

Zonality index for data in Fig. 3-10. Linear productivities are from Table 3-II. These are "normalized" by multiplying the factors given in brackets for each element. Zonal sequence from level I to level IV is Ag (0.44*)—As (0.52*)—Co (0.39*)—Cu (0.37*)

Level	Linear productivities (lp)				Normalized linear productivities (lp)				Zonality index ($lp/\Sigma lp$)			
	Ag	As	Cu	Co	Ag (×1)	As (×10)	Cu (×10)	Co (×10)	Ag	As	Cu	Co
I	1.18	0.156	0.100	0.08	1.18	1.56	1.0	0.8	0.44*	0.38	0.06	0.05
II	1.15	0.213	0.334	0.33	1.15	2.13	3.34	3.3	0.43	0.52*	0.21	0.19
III	0.24	0.025	0.576	0.69	0.24	0.25	5.76	6.9	0.09	0.06	0.36	0.39*
IV	0.1	0.0125	0.600	0.67	0.10	0.125	6.00	6.7	0.03	0.03	0.37*	0.38
Sum					2.67	4.065	16.1	17.7				

CONCLUSIONS

Probably the greatest single hinderance to interpretation of exploration geochemical data is a misconception of the nature of anomalies, and especially the well-entrenched expectation that mineralization will be revealed by an abnormally *high* content of an element. Thus, Levinson (1974) defined the objective of exploration geochemistry to be to " . . . find some dispersion of elements as compounds sufficiently above normal to be called an anomaly" (p. 1). This statement ignores the common existence of negative anomalies. The example shown in Fig. 3-10 illustrates a negative zinc anomaly in soils overlying a mineralized vein in Greece; many other examples for rock geochemical surveys are given in later chapters.

A geochemical anomaly — whether positive or negative — is not an abso-

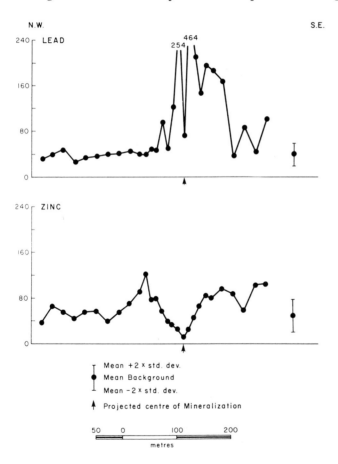

Fig. 3-10. Distribution of Pb and Zn in B-horizon soils over lead-zinc mineralization in Greece.

lute measure; in many cases it varies as a function of the analytical technique used (Fletcher, 1981). This type of variation is dramatically illustrated by some stream sediment data from the Philippines where a stream draining a disseminated copper deposit showed an anomaly of 4 times threshold for the total Cu content on the minus 80-mesh fraction of a sample, whereas cold-extractable Cu on unsieved samples showed an anomaly of 32 times the corresponding threshold for cold-extractable Cu (Govett and Hale, 1967). Similarly, it is demonstrated in Chapter 11 that rock geochemical anomalies can be enhanced through the use of partial instead of total digestion techniques.

The nature of an anomaly will also vary as a function of sample type. Obvious examples are the differences in geochemical response obtained in a plus 10-mesh and a minus 80-mesh soil sample, or the differences in metal content in A-, B-, and C-soil horizons. In rock geochemistry variations in texture of the same petrological type of rock can cause marked variations in the geochemical response (Chapter 7).

The concept of a geochemical anomaly — and consequently the approach to sampling patterns, analysis, and data interpretation — must take cognizance of the nature of the target and, especially, the scale of the survey. The design of a sampling pattern to locate discrete mineralization within a well-defined anomaly becomes a statistical exercise to determine the appropriate sampling interval for the expected area of measurable dispersion for some defined probability of success. The sampling pattern required to define a narrow copper vein deposit with a linear dispersion pattern for Cu 20 m wide is different from that required to define a copper porphyry deposit with an equidimensional dispersion pattern with a diameter of 200 m. Both can easily be calculated, and in an area of residual soil with good geochemical response single element analysis may be adequate; interpretation is likely to be relatively simple.

To discriminate between a mineralized and a barren pluton or to discriminate between a potentially mineralized and a barren volcanic sequence on a regional scale with rock samples is quite a different kind of problem. In such cases the objective is to be able to geochemically characterize a tract of country of thousands of square kilometers. The sampling *pattern* is almost irrelevant because the only determining factor is the *number* of samples per unit area required to recognize a statistically significant geochemical difference between two sample sets. A density of one sample per 5 or more km^2 may be adequate, but multi-element analysis will almost certainly be required, and interpretation will probably be complex. Simple enrichment or depletion of single elements are rarely adequate, and discrimination between barren and mineralized rocks depends on recognition of variations in the geochemical character of the rock (similar comments are also applicable in many cases to local detailed rock geochemical surveys).

A major problem in recognizing geochemically abnormal (i.e., possibly mineralized) intrusions or volcanic sequences is the difficulty of deciding what is geochemically normal. Although in many of the interpretations discussed in succeeding chapters on regional scale exploration, comparisons are made of necessity with the chemical composition of "average" rocks, the data presented in Chapter 2 showed that "average" element concentrations are singularly unreliable. It will be evident from the case histories discussed in this book that geochemical deviations from "normal" are relatively uninformative, and that only variation trends within a region, district, or area are of practical value.

PART II. REGIONAL SCALE EXPLORATION

REGIONAL SCALE EXPLORATION FOR DEPOSITS OF PLUTONIC ASSOCIATION — DISCRIMINATION BETWEEN PRODUCTIVE AND BARREN INTRUSIONS

ELEMENT CONCENTRATION AND GEOCHEMICAL SPECIALIZATION

For many years prior to the present upsurge of interest in rock geochemistry both geologists and geochemists sought chemical criteria that would serve to discriminate productive from non-productive host rocks; most of this work was directed towards the recognition of potentially productive felsic plutons as a reconnaissance technique. It can be argued that ore deposits genetically associated with intrusions leave their signature in the plutons in one of two ways: the pluton is enriched in the ore elements because the original magma was abnormally high in these elements, leading to the segregation of the mineral deposits; or, the pluton is depleted in the ore elements because the mineral deposits result from concentration of normal amounts of ore elements during cooling of the magma. There is a considerable volume of published work in the U.S.S.R. on the topic, especially on what the Russians term "metallogenetically specialized" intrusions; these are defined by Barsukov (1967) as containing concentrations of ore elements four to five times the clarke value in unaltered rocks that are nevertheless genetically related to ore occurrences (i.e., ore deposits are related to *primary* magmatic features).

The majority of early work was concerned with establishing that an abnormal content of one or more of the ore elements in intrusions was a criterion for recognition of productive plutons. The evidence for most elements is ambiguous, and no universal pattern of enrichment or depletion has been established; only in the case of Sn does there appear to be a fairly consistent enrichment in tin-bearing felsic rocks (see Chapter 5). Part of the practical problem lies in both small scale and large scale petrologic variations in felsic rocks — particularly the amount of mafic minerals present — and the partitioning of elements between individual minerals (see Chapter 2).

Tauson (1967) pointed out the difficulty of determining the characteristic trace element content of primary magmas. The concentration measured in samples collected from outcrop and drill core reflects not only the amount in the primary magma, but also the results of contamination of the magma due to assimilation of country rock, metasomatic processes, and post-magmatic processes — especially those associated with volatile emanations such as fluorine. In the case of surface samples, the effects of weathering can be an added complication. Tauson concluded that differentiation processes in granite magmas do not lead to a marked enrichment of elements in the felsic differentiates, and that the trace element content in unaltered rocks of a given genetic type is fairly constant, regardless of geologic setting and age; he believes that any enrichment of ore elements that does occur is due to very late-stage or post-magmatic processes. Thus, trace elements are either dispersed within the rock-forming minerals or are residually concentrated as mineralizing solutions. The proportion of the elements in each state depends on the composition of the original magma, the size and depth of the intrusion, and the tectonic conditions of crystallization.

This conclusion is illustrated by data from eastern Siberia in Table 4-I; ore deposits of Sn, Be, Mo, Pb, and Zn occur chiefly associated with Mesozoic granites, but Tauson and Kozlov (1973) stated that there is no significant difference — even for Sn — in the average ore element content between the Mesozoic granites and the dominantly barren Proterozoic and Palaeozoic granites. Barsukov (1967) similarly concluded that Pb, Zn, Mo, and W have a uniform distribution in granitoids, and that their concentration cannot be correlated with the presence or absence of mineralization. Many other authors (Blaxland, 1971; Flinter et al., 1972; Sheraton and Black, 1973) have also concluded that there is no correlation between the base metal contents in granites and the ore potential of the granites.

On the other hand, Beus (cited in Beus and Grigoryan, 1977) took the contrary view that a magma that has above-normal concentrations of ore elements must give rise to solid products (on crystallization) that also have high contents of ore elements. There are considerable published data to support this view (e.g., Allen et al., 1976; Edwards, 1976; Ahmad, 1977). Clearly there is a conflict of evidence, and individual cases will be discussed later in this chapter.

For exploration purposes (as pointed out by Beus and Grigoryan, 1977) if an abnormal distribution of elements in a rock can be shown to be spatially

TABLE 4-I

Average content (ppm) of some ore elements in palingenic granites in eastern Siberia; ore deposits are mainly in Mesozoic granites (from Tauson and Kozlov, 1973, based on 10,000 samples from 20 magmatic complexes)

Age	Sn	Be	Mo	Pb	Zn
Proterozoic	5.6	4.3	1.8	22	66
Palaeozoic	3.9	2.3	1.6	16	57
Mesozoic	5.4	4.7	2.0	26	44

related to mineralization, it is irrelevant whether it arose because of a primary feature of the magma or because of post-magmatic processes. The authors define a term, "geochemical specialization", that has a more general application than the rather restricted term "metallogenetic specialization". Geochemical specialization covers *all* geochemical characteristics of an ore-bearing rock (including metamorphic and sedimentary rocks) that distinguish it from barren but otherwise similar rocks; the term has no genetic implications regarding the primary character of the magma.

An initial broad aim of determining the ore potential of felsic intrusions is the recognition of a specific *type* of granite commonly associated with particular types of ore deposits. Tauson (1974) defined nine types of granite: four derived from differentiation of basalts; four derived from palingenic melting of crustal matter; and one derived from partial melting of highly metamorphosed rock. Compositional data for four of Tauson's nine granite types are given in Table 4-II. The plagiogranite and ultra-metamorphic granites have a poor ore potential; the palingenic granite has potential for Au, Cu, Mo, and lesser potential for Sn and W; the plumasitic leucogranite has potential for Pb, Zn, W, Nb, Ta, and, characteristically, Sn.

The data in Table 4-II show that the Sn content is not very different among the four granites, but there are very large differences in Li, Rb, Sr, and Ba. High Li contents, an increase in Rb relative to K (low K:Rb ratio),

TABLE 4-II

Content of some elements in granitoids (from Tauson and Kozlov, 1973)

Element	Plagiogranite (tholeiite series)	Ultrametamorphic leucocratic granite	Palingenic granite (calc-alkaline series)	Plumasitic leucocratic granite
K (%)	0.5	5.4	3.3	4.0
Na (%)	3.2	2.5	2.9	2.8
F (ppm)	150	140	600	3000
Li (ppm)	2	8	36	97
Rb (ppm)	4	140	140	400
Sr (ppm)	190	420	300	100
Ba (ppm)	180	1600	750	200
Be (ppm)	0.6	0.6	3.5	6.8
W (ppm)	0.7	0.7	2.0	4.1
Mo (ppm)	1.3	1.4	1.6	1.4
Pb (ppm)	4.4	14	25	30
Zn (ppm)	70	43	45	57
Sn (ppm)	2.7	2.6	5.3	6.3
K:Rb	1250	386	236	100
Ba:Rb	45	11.4	5.4	0.5
Rb:Sr × 100	2.1	33.3	46.7	400

and decreases in Sr relative to Rb (high Rb:Sr ratio) are indications of extreme fractionation (S.R. Taylor, 1965); therefore they are indicators of the possibility of volatile and rare metal-rich magma fractions. Also, since Rb shows a stronger relation to F than to K (F is generally accepted as an important constituent of mineralizing fluids) and since Rb migrates with F and Ba has a weak relation with F (but tends to accumulate in primary high temperature minerals), a magma differentiating with a high volatile content will have a low Ba:Rb ratio (Tauson and Kozlov, 1973).

CHARACTER OF FREQUENCY DISTRIBUTIONS

The difficulties of making a satisfactory interpretation on the basis of abundance of elements in plutons should be clear from the discussion above. The greater significance of variance and positive skewness of the distribution of the elements has been pointed out many times in the context of rock geochemistry (Beus, 1969; Bolotnikov and Kravchenko, 1970; Cameron and Baragar, 1971; Govett and Pantazis, 1971; Garrett, 1973; Tauson and Kozlov, 1973; Govett et al., 1975) — the greater the variance and skewness of the distribution the greater is the likelihood of mineralization.

The distribution frequency of an element depends on the geochemical character of the element, its mode of occurrence in minerals, the distribution of minerals in rocks, and the conditions of formation and subsequent history of the rock. It is argued by Tolstoi and Ostafiichuk (1963) that similarity of populations as an expression of a common origin is more readily deduced from a similarity of form of frequency distributions than from a similarity of abundance. This concept is illustrated in Fig. 4-1 for three genetically related massifs (Krykkuduk, Ashchikol'sk, and Akkuduk) and one massif (Tasadyr) that is not related to the former three massifs; only the Ashchikol'sk massif is mineralized.

The Tasadyr massif is different in age, composition, and chemical nature from the Krykkuduk and Ashchikol'sk massifs. Nevertheless, the mean Co content (23 ppm) in the Tasadyr porphyritic granite is the same as in the genetically different Krykkuduk granodiorite, whereas the mean Co contents of the genetically related Krykkuduk and Askchikol'sk granodiorites are quite different. The frequency distribution and modal values of Co in the latter two massifs are similar, however, and both are distinctly different from the Tasadyr massif. Thus the *form* of the distribution curves gives a far more reliable indication of the similarities and differences of several populations.

It is also possible that a mineralizing event can be recognized by the form of frequency distributions. The frequency distribution of Cu in the three genetically related granodiorite massifs is also shown in Fig. 4-1. The modal values for the three intrusions are essentially the same, but the distribution of Cu in the Ashchikol'sk intrusion has a strong positive skew; this intrusion,

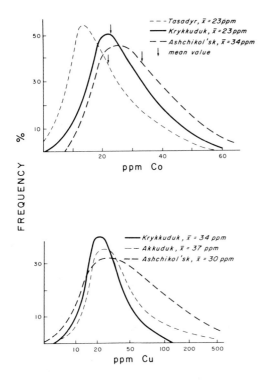

Fig. 4-1. Frequency distribution of Co in granite intrusions and Cu in granodiorite intrusions. (Redrawn with permission from Tolstoi and Ostafiichuck, 1963, *Geochemistry*, 1963, figs. 2 and 3, p. 988.)

TABLE 4-III

Distribution of Li and Sn in barren and potentially ore-bearing Mesozoic palingenic granites in central Transbaikalia (from Tauson and Kozlov, 1973)

	Li (ppm)		Sn (ppm)	
	$\bar{x}$	σ^2	$\bar{x}$	σ^2
Barren	47	100	4.3	0.7
Potentially ore-bearing	64	710	7.4	12.6

unlike the other two, has many areas of copper mineralization (as noted above).

The types of population differences can, of course, generally be recognized by a greater variance in the data. This is illustrated in Table 4-III which gives the means and variance of Li and Sn for barren and potentially ore-bearing granites in an area of tin mineralization in central Transbaikalia. The mean contents of the elements show only a small increase in the potentially ore-bearing granite, but the variance is significantly greater. Tauson and Kozlov (1973) regard this as typical; they attributed the greater variance in ore-bearing plutons to extensive migration of volatiles and ore-forming elements to the apical part of intrusions just prior to crystallization which causes considerable differences in trace element content within a single intrusion.

CASE HISTORIES BASED ON WHOLE ROCK ANALYSES OF FELSIC INTRUSIONS

Canadian Cordillera

The most comprehensive regional study so far published is the investigations by Garrett (1971a; b, 1973, 1974) of techniques to assess the ore potential of 74 Cretaceous granitoid plutons over an area of 31,000 km^2 in the Yukon and Northwest Territories in Canada. The vast majority of the plutons are granodiorite or quartz monzonite; alaskite, granite, syenite, and quartz diorite also occur. Base metal mineralization is widespread, although it is probably not genetically related to the granites. Tungsten occurs in the eastern part of the area as scheelite skarns with chalcopyrite; the most important occurrences are the producing Cantung deposit and the Amax Mt. Allen prospect (21 in Fig. 4-2). Tungsten also occurs with gold in the northwest as scheelite skarns and in quartz veins and as lesser wolframite in veinlets and pegmatites. Duplicate samples were taken at a minimum of 15 sites in each pluton; the samples were analyzed for 22 elements. The location of the plutons and the type of mineralization associated with them is given in Fig. 4-2.

Recognition of petrological variation is of critical importance to the use of metal variation in assessing the ore potential of plutons. Garrett (1973) used an elegant statistical approach to this problem in his interpretations of the distribution of Cu, Pb, and Zn. Using *R*-mode principal component analysis of the major element variation, he selected two significant factors that were identified as the degree of differentiation and the content of K (which he related to the degree of porphyricity of the rock). By regressing the principal component scores against the base metal content he was able to calculate the residual base metal unaccounted for by the principal component factors.

The general principles of Garrett's (1973) approach are illustrated here with reference to Zn. There is a general concentration of high mean Zn values in the northwest of the area; given the strong positive skew of frequency distributions, plutons with a mean Zn content of greater than 90 ppm must be considered as anomalous (Fig. 4-3). Reference to Figs 4-3 and 4-4 shows that there are 10 plutons with mean Zn greater than 90 ppm; of these, five are associated with mineral occurrences.

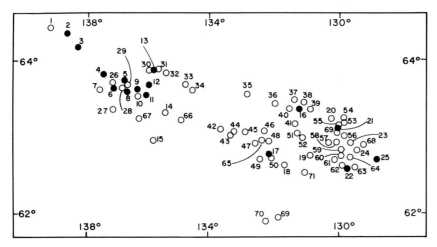

Fig. 4-2. Cretaceous granitoids in the Canadian Cordillera sampled by Garrett, 1973. Black circles are granitoids containing mineralization of the following type: 2 = Ag-Pb; 3 = Cu-Sb; 4 = Au; 5 = Au-Pb; 6 = Au; 8 = Pb-Zn; 9 = W, Au-Pb; 11 = no data; 12 = Zn-Ag-Pb; 13 = W; 16 = Cu-W; 17 = Cu-Zn; 21 = W; 22 = Cu-Zn-W; 25 = Sb (compiled from Garrett, 1971a, 1973, 1974).

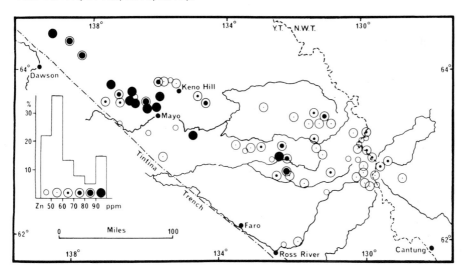

Fig. 4-3. Distribution of mean Zn content in granitoids in Northwest Territories (N.W.T.) and Yukon Territory (Y.T.), Canada. (Reproduced with permission from Garrett, 1973, *Geochemical Exploration 1972*. Institution of Mining and Metallurgy, 1973, fig. 5, p. 211.)

A low maximum Zn content associated with a high mean Zn content indicates a low variance. Conversely, a high maximum Zn content in conjunction with either a low or a high mean Zn content indicates a high variance or positive skewness. High variance and positive skewness is characteristic of

Mean residual score	Maximum residual score < 2 SD	Maximum residual score 2-3 SD	Maximum residual score > 3 SD
<0.25 SD	·	◆	◆
0.25–0.75 SD	△	▲	▲
0.75–1.25 SD	○	●	●
>1.25 SD	□	■	■

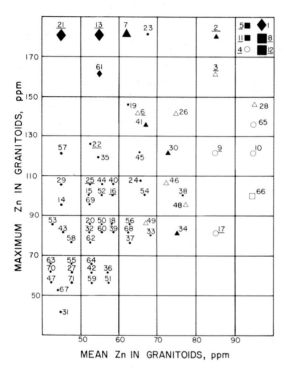

Fig. 4-4. Mean and maximum Zn, and mean and maximum residual Zn scores for granitoids in Cretaceous acidic rocks of the Canadian Cordillera. Numbers refer to locations in Fig. 4-2; underlined numbers are granitoids containing known mineralization (derived from Garrett, 1973).

mineralized areas. A consideration of the skewness of the distribution (as indicated by maximum Zn content) reduces the interest in some of the plutons with more than 90 ppm Zn (intrusions 10, 28, 65, 66) and increases the interest in some of those with less than 90 ppm mean Zn (intrusions 21, 13, 7, 23, 2 in Fig. 4-4).

Of the 11 plutons with maximum Zn greater than 170 ppm, eight are associated with mineralization. Examination of the mean residual scores (which are assumed to be indicative of the content of Zn unaccounted for by differentiation and porphyritic factors) shows that most of the plutons in the east have near to expected Zn contents; and the number of plutons defined as anomalous on the basis of a mean Zn content greater than 90 ppm in the northwest is almost halved on the basis of having mean residual scores of less than 0.75 standard deviation units (Figs. 4-3 and 4-4).

There are six plutons with maximum residual scores greater than 3 standard deviation units; four of these (intrusions 8, 12, 13, and 21) are associated with known mineralization. There are nine plutons with maxima between 2 and 3 standard deviation units, and five of them are associated with mineral occurrences. Plutons 3, 6, 16, 17, 22, and 25 — which are mineralized — do not have anomalous Zn but are anomalous in one of the other base metals.

By a combination of residual scores, actual metal content, and the degree of skewness of Cu, Pb, and Zn, Garrett was able to show that most plutons associated with mineralization could be recognized, and some additional plutons with no known mineralization were identified for investigation. On the basis of the Zn data used as an example here, it is obvious that pluton 1 is a prime target (see Fig. 4-4).

In the western part of the area alluvial gold production appears to be associated with plutons rich in W and poor in Mo — although exceptions to these relations occur. Garrett (1974) observed that the principal areas of alluvial gold production occur in association with the West Ridge, Scheelite Dome, and Potato Hills stocks (26, 9, and 13 in Figs. 4-2 and 4-4); they are characterized by regionally abnormal contents of Hg in the plutons (Table 4-IV). Garrett determined that the regional differences are significantly

TABLE 4-IV

Mean contents of Hg (geometric mean), W, and Mo in some granitoids in the Yukon, Canada. Numbers in brackets refer to locations in Fig. 4-2 (data from Garrett, 1971b, 1974)

Pluton	Hg (ppb)	W (ppm)	Mo (ppm)
Mount Christie (24)	3	1.3	0.7
Keele Peak (20)	4	1.7	3.9
Vancouver Creek (6)	3	1.0	0.3
Stewart Crossing (27)	5	1.2	0.7
Tombstone Mountain (2)	7	1.5	0.6
Rusty Peak (45)	9	1.3	0.3
Two Buttes (14)	9	29.5	3.4
Lansing Range (35)	18	1.0	1.6
Potato Hills (13)	19	9.7	0.7
Scheelite Dome (9)	13	5.0	0.3
West Ridge (26)	15	5.2	0.9
Mount Allan (21)	—	3.5	1.5

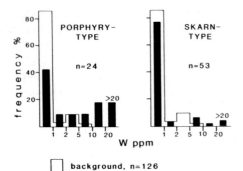

background, n=126

Fig. 4-5. Frequency distribution of W in mineralized and barren granitic intrusions, northern Canadian Cordillera. (Redrawn with permission from Govett and Nichol, 1979, *Geophysics and Geochemistry in the Search for Metallic Ores. Geological Survey of Canada, Economic Geology Report*, 31, 1979, fig. 15-2, p. 343; original data from Garrett, 1971a, b.)

greater than differences due to sampling or analytical error. The data given in Table 4-IV show that the Lansing Range pluton contains very high Hg; this is attributed by Garrett (1974) to contamination from the high Hg content (34 ppb) of Palaeozoic shales into which the pluton is intruded. A further point of note is that the region is a Hg-poor province compared with the global average (which is about 62 ppb for granitoids; Jonasson and Boyle, 1972).

The abundance of W in the various granitoids is not, in all cases, a good indication of tungsten mineralization. The form of the frequency distributions in mineralized and barren intrusions is, however, quite different. In barren plutons few samples exceed 5 ppm W, and none exceed 10 ppm W, whereas at least 10% of the samples from mineralized plutons exceed 10 ppm W, and the distributions are more positively skewed (Fig. 4-5). This is interpreted as indicating a tungsten mineralizing event superimposed upon the primary magmatic tungsten distribution. The plutons that are host to porphyry-type disseminated mineralization — and therefore presumably closely allied to primary magmatic processes — are more clearly distinguishable from barren plutons than those associated with more local skarn-type mineralization (Fig. 4-5).

Other examples

Data on other regions are not nearly as comprehensive as those of Garrett. The following are some examples of approaches used and results obtained by a number of workers in other areas.

Conclusions from whole rock analyses of the granites of southwestern England (which are considered in more detail below on the basis of element distribution in mica and feldspar) are contradictory. The Carboniferous

TABLE 4-V

Copper content of granites in Devon and Cornwall, England compared to global averages (from Ahmad, 1977)

Granite	Cu (ppm)
Land's End and Carnmenellis, $n = 30$	73 ± 18
St. Austell, Bodmin Moor, Dartmoor, $n = 48$	21 ± 9
Average granite (S.R. Taylor, 1968)	10
Average low-Ca granite (Turekian and Wedepohl, 1961)	10

TABLE 4-VI

Distribution of aqua-regia-extractable Cu and Zn in intrusive rocks in British Columbia, Canada (from Warren and Delavault, 1960)

Number of localities	Number of samples	Cu (ppm) avg.	range	Zn (ppm) avg.	range	Cu:Zn avg.
Barren, 13	32	3.0	0.5—13	49	14—136	0.06
Cu-positive, 4	28	29	3—120	50	3—124	0.58
Zn-positive, 4	27	12	1—24	117	35—200	0.10

batholith intrusion in Devon and Cornwall is exposed as isolated granite masses of Dartmoor, Bodmin Moor, St. Austell, Carnmenellis, and Lands End. Apart from its fame for tin mining, it was also an extremely important copper mining area during the 19th century. Production from hydrothermal lodes (mostly in the country rock marginal to the granites) amounted to approximately 660,000 tonnes from the Cornish mines and 65,290 tonnes from the mines of Devon (Ahmad, 1977). The content of Cu in the granitic rocks is given in Table 4-V; it is clearly greater than in average granites. Moreover, it is also much higher in the granites of the richest copper-mining districts of Lands End and Carnmenellis than in the remainder of the granites. Edwards (1976), however, suggested that the Cornish granites are enriched in Pb and Zn, but not in Cu. Since Edwards' data are largely based on the studies of Ahmad this contradiction cannot be resolved; moreover, the validity of some of Edwards' data computations have been questioned by Wilson and Jackson (1977).

An early study that deserves mention is that of Warren and Delavault (1960) on intrusive rocks in British Columbia, Canada. On the basis of aqua-regia-extractable Cu and Zn in samples from acidic rocks over about 130,000 km^2 they concluded that it was possible to discriminate between barren and mineralized areas. Their data are summarized in Table 4-VI; the differences between the barren and other plutons are indeed large. The significance of this work is that the anomalous plutons — whose economic potential was not

TABLE 4-VII

K:Rb and Mg:Li ratios for barren and mineralized granites host to Li, Be, Sn, W, Ta and Cs deposits (from Beus and Grigoryan, 1977)

Granitoid type	K:Rb	Mg:Li
Average granitoids	170	370
Average granite	170	90
Average for granitoids not related to Li, Be, Sn, W and Ta deposits	170	270
Average for granitoids related to Li, Be, Sn, W and Ta deposits	130	75
Average for biotite granites with pegamatitic deposits Li, Be, Ta and Cs	160	40
Average for biotite granites with pegmatitic and apogranite deposits of Ta	126	30

recognized when Warren and Delavault did their work — now have mines such as Bethlehem, Valley Copper, Highmont, Lornex, Craigmont, and Brenda.

Intrusions that have suffered post-magmatic hydrothermal or metasomatic alteration — and hence are possibly ore-bearing — may be expected to show different element contents when compared to similar intrusions that are not altered. Beus and Grigoryan (1977) suggested that Mg and Li can be used to identify granites that have suffered post-magmatic metasomatism due to the tendency of Mg to be depleted (because of its replacement by Fe in biotite) and Li to be concentrated. Data illustrating the trend of low Mg:Li ratios in granites that host Li, Be, Sn, W, Ta, and Cs deposits are given in Table 4-VII.

The Coed-y-Brenin low-grade chalcopyrite deposit is believed to be an example of a deposit in an intrusion that has suffered post-magmatic alteration. It lies within an altered dioritic intrusion in Cambrian rocks on the southeast margins of the Harlech Dome in North Wales and has the characteristics of a porphyry-type copper deposit. Allen et al. (1976) compared the unmineralized intrusions of the area with the low-K_2O intrusions of the Caribbean associated with copper porphyry mineralization as defined by Kesler et al. (1975a) as characteristic of an island arc environment. Allen et al. (1976) also cited the additional similarities of high Cu and Zn and low Pb contents. There are significant differences (at the 99% confidence level) between the barren and the mineralized intermediate intrusions on the south of the Harlech Dome: the mineralized intrusions have more K_2O, Cu, and Mo and less MnO, CaO, Na_2O, and Zn (these relations, except for Mo, are illustrated in Fig. 4-6). At the 95% confidence level there are significantly greater amounts of Ti, P, and Fe in the barren intrusions. The enhancement and depletion of elements are interpreted as a reflection of the effects of mineralization and associated hydrothermal alteration. The K_2O, Na_2O, and CaO variations are associated with phyllic alteration, whereas the Zn depletion possibly reflects Zn migration to the pyritic and propylitic zones.

The Coed-y-Brenin data should be treated with caution. Allen et al. (1976) drew attention to the fact that the data are not normally distributed, and, as

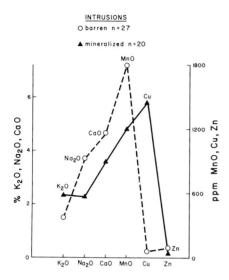

Fig. 4-6. Mean concentration of K_2O, Na_2O, CaO, Mn, Cu, and Zn in barren and miner-alized intermediate intrusions in Cambrian rocks of the Harlech Dome, Wales (compiled from Allen et al., 1976).

shown in Table 4-VIII, the differences in mean element content (except for Cu and Zn) between the mineralized and the unmineralized data sets are small; there is also considerable overlap in the ranges. Nevertheless, it seems that these intrusions in Cambrian strata on the southeast of the Harlech Dome could have been identified as promising for copper porphyry mineral-ization on the basis of analysis of a small number of samples for a variety of elements, especially K, Na, Ca, Mn, and Cu.

Simpson et al. (1977) used delayed neutron analysis and Lexan plastic fission track analysis to determine the content and distribution of U in Cale-donian granites in Scotland and Hercynian granites in southwest England. The analytical data are summarized in Table 4-IX. These data show that the Hercynian granites are considerably enriched in U compared to the Cale-donian granites (which contain about average U for granites). Simpson et al. (1977) reported that on the basis of the two-tailed Kolmogorov-Smirnov test the difference in U content between the two granite suites is significant at the 99% confidence level; the Helmsdale granite is similarly significantly different from the other Caledonian granites.

Uranium mineralization in southwest England occurs in vein systems closely associated with the granites, and Simpson et al. (1977) concluded that mineralization is comagmatic with the granites. The U occurs essentially in primary minerals such as apatite, monazite, sphene, and zircon. In Cale-donian granites uranium mineralization is essentially in post-magmatic alter-ation zones along major faults and later molasse facies sedimentary rocks;

TABLE 4-VIII

Variations in composition of mineralized and unmineralized intrusions[*] (from Allen et al., 1976)

		Mineralized	Unmineralized
K_2O	$\bar{x}$	2.31	1.47
	σ	0.80	1.12
	R	1.12—2.77	0.28—5.36
Cu	$\bar{x}$	1450	57
	σ	1785	73
	R	17—7400	0—350
CaO	$\bar{x}$	3.58	4.85
	σ	1.17	1.80
	R	2.0—3.62	0.50—9.28
Na_2O	$\bar{x}$	2.27	3.67
	σ	1.40	1.82
	R	0.50—4.52	0.22—6.48
MnO	$\bar{x}$	0.12	0.18
	σ	0.04	0.09
	R	0.07—0.21	0.06—0.42
Zn	$\bar{x}$	36	92
	σ	14	67
	R	17—70	30—270
n		20	27

[*] $\bar{x}$ = mean; σ = standard deviation; R = range.

the U occurs mostly in secondary alteration minerals such as iron oxides and chlorite.

The uranium study by Simpson et al. (1977) demonstrates the importance of mineralogical control in trace element analysis. Combined measurement of U content and determination of its mineralogical site in granite serves to identify primary uranium provinces such as southwest England; also, by identifying the enrichment processes in provinces with low U content, this approach can assist in defining economic targets.

More detailed geochemical data on the Scottish Caledonian granites have been presented by Plant et al. (1980). The authors pointed out that the granites fall into three main suites: the older granites emplaced before and during the main Caledonian metamorphic-deformation event; the newer granites emplaced at a low structural level (the "forceful" granites); and the late discordant granites. Mineralization (which varies from a copper-molybdenum to a tin-uranium association) is essentially restricted to the late discordant granites and some of the youngest of the "forceful" suite. There is a geochemical evolution from the early to the late granites that is typified by an

TABLE 4-IX

Geometric mean and standard deviation for uranium content of granites in Scotland and
southwest England (from Simpson et al., 1977)

Age	Area	Number of samples	Geometric mean (ppm)	Standard deviation
Caledonian	Helmsdale	30	7.0	1.56
	Aberdeenshire	29	2.73	1.48
	Shetland	59	3.44	1.42
All Caledonian		118	3.92	1.72
Hercynian	southwest England	66	10.8	1.65

increase in large ion lithophile elements (U, Th, K, Rb, and Cs) and other incompatible elements (Li, Be, and Mo). There is virtually no overlap between
the late discordant and earlier granites for the ratios U:Th, K:Rb, K:Ba,
Ba:Rb, Ba:Cs, K:Sr, Rb:Sr, U:Sr, U:Zr, Rb:Zr, Sr:Y, and Sr:Li except for
the "forceful" Helmsdale granite. The latter has some of the most significant
mineralization (uranium, lead, barium, fluorine) in the region, but it is
suggested that the Th:U and K:Rb ratios have been decreased by hydrothermal alteration. There is no direct relation between the degree of evolution
of the magma and the extent of mineralization.

The Cairngorm granite (late discordant suite), for example, is the most
evolved of the granites and contains the highest U content, but no uranium
mineralization has been found. Indeed, the Cairngorm granite would be
claimed as a tin granite (see Chapter 5) on the basis of anomalous ratios for
K:Rb (91), Rb:Sr (6.79), Ba:Rb (0.57), and Mg:Li (10.4). These relations
also place it in the category of potentially mineralized granites according to
the data of Tauson (1974) and Tauson and Kozlov (1973) as discussed earlier
in this chapter. The Cairngorm granite is, in fact, mineralized in Sn-Nb-F-Pb
Zn-F-Li-Ag to a minor extent. The Helmsdale granite, however, which contains significant mineralization is scarcely anomalous in any of the ratios
listed above.

Plant and her co-workers suggest that only high-level granites (such as the
late discordant granites) intruded into much lower temperature wet sediments are likely to be mineralized. Disseminated mineralization in the
granite would arise through leaching from the sediments, and later vein-type
mineralization would occur by a hydrothermal concentration through a
meteoric flow of water through the granite. The reason advanced for the
lack of major mineralization in the later discordant Scottish Caledonian
granites is limited supply of water (especially a lack of brine) and the lack of
major faults to facilitate movement of water. The "forceful" Helmsdale
granite and also the Grudie granite (with significant Cu-Mo-Pb-F-Ba mineralization) are both less evolved than the late discordant granites: the authors

suggest that these granites are mineralized because they were intruded at a relatively high structural level and cut by major faults that permitted the circulation of meteoric water.

Notwithstanding some of the problems mentioned above in recognizing a geochemical signature for mineralized granites, those intrusions that are mineralized all have regional geochemical anomalies. For example, the Helmsdale granite (with U-Pb-Ba-F mineralization) is enriched in U, Pb, Ba, Mo, Be, Co, Mn, and B; the Grudie granite (with Cu-Mo-Pb-Fe-Ba mineralization) is enriched in Cu, Mo, Pb, Ba, and U; the Carn Chuinneag granite (with Sn-Fe and Mo-Th-F-Cu mineralization in faults) is enriched in Mo, Sn, and U; and the Cairngorm granite (with Sn-Nb-F-Pb-Zn-F-Li-Ag mineralization) is enriched in K, Li, Be, B, U, Rb, Y and elements of basic association, and is depleted in Ba, Sr, and Zr.

CASE HISTORIES BASED ON WHOLE ROCK ANALYSES OF MAFIC INTRUSIONS

There are far fewer published data on geochemical criteria for recognition of mineralized mafic and ultramafic intrusions than for the acidic rocks. Two studies relating to determination of potential for nickel mineralization are worth citing. Polferov and Suslova (1966) showed that intrusions containing nickel sulphides are richer in femic minerals and Mg, K, Ni, Co, S, and H_2O than similar barren intrusions. Data for Ni, Co, S, and H_2O are given in Table 4-X. The enrichments in the ore elements and H_2O identify particular intrusions within a region as ore-bearing and are also useful for local deposit-scale exploration.

Häkli (1970) used factor analysis on the results of analyses of Ni, Co, Cu, Zn, and S on 7751 rock samples from across Finland; the samples ranged in

TABLE 4-X

Ratio of Ni, Co, S, and H_2O to corresponding clarke values in barren and Cu-Ni mineralized mafic and ultramafic rocks (from Polferov and Suslova, 1966)

Region	Rock	Ratio to clarke for non-mineralized (1) and mineralized (2) rocks							
		Ni		Co		S		H_2O	
		1	2	1	2	1	2	1	2
Kola Peninsula	mafic	0.8	2.1	0.6	1.3	0.05	1.1	0.09	2.13
	ultramafic	1.0	2.3	0.6	1.0	0.06	0.8	2.38	4.03
North Baikal	mafic	1.3	5.9	0.4	2.5	0.07	0.1	1.16	3.55
	ultramafic	0.6	1.4	0.5	0.7	0.01	0.06	0.72	2.33
Noril'sk	mafic	1.2	1.9	1.2	1.1	0.3	1.0	—	—

composition from quartz diorite to dunite. The two most important factors that were identified (from an exploration point of view) are S-Cu and Ni. The former is presumed to be related to the formation of an iron-rich sulphide phase with variable Cu; the latter is presumed to indicate the affinity of Ni for sulphides and early magmatic mafic silicates. Contoured normalized factor scores for S-Cu and Ni in the ultramafic and mafic rocks are shown in Figs. 4-7 and 4-8. These were computed on the basis of the number of 1-km^2 cells within 20-km^2 areas that had factor scores greater than 1.0, divided by the total number of 1-km^2 cells in the entire country that had factor scores greater than 1.0; the relative frequencies thus obtained were contoured.

A comparison of Fig. 4-9 (which shows the distribution of the main zones

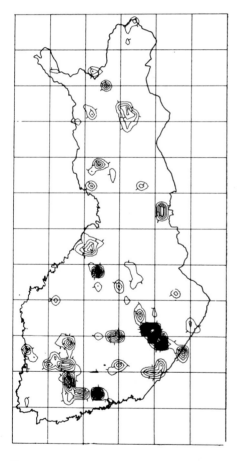

Fig. 4-7. Normalized distribution of the S-Cu factor scores in mafic-ultramafic rocks in Finland. (Reproduced with permission from Häkli, 1970, *Bulletin of the Geological Society of Finland*, vol. 42, 1970, fig. 3, p. 115.) The lowest contour and the contour interval is 2‰; grid size is 100 km^2.

68

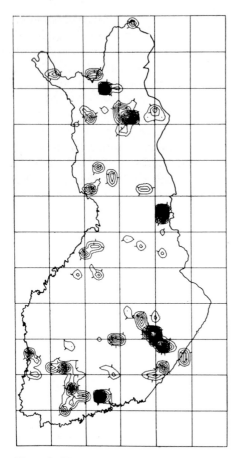

Fig. 4-8. Normalized distribution of the Ni factor score in mafic-ultramafic rocks in Finland. (Reproduced with permission from Häkli, 1970, *Bulletin of the Geological Society of Finland*, vol. 42, 1970, fig. 4, p. 116.) The lowest contour and the contour interval is $2^0/_{00}$; grid size is $100\ km^2$.

of base metal and nickel sulphide mineralization) with Figs. 4-7 and 4-8 reveals a close correlation between high factor scores and mineralized zones. A prominent NW-SE anomalous zone from Nivala to Parikkala falls within the prominant Ni-Cu-Zn zone in Fig. 4-9 which has a number of economic deposits. Another anomalous zone striking E-W from Pori to Lapeenranta includes numerous nickel deposits, and an anomalous zone extending SW and SE from Forssa in southern Finland also corresponds to a known mineralized belt. In Lapland and around the Ranua-Taivalkoski-Suomussalmi region in the north there are two more anomalous zones of factor scores that do not correspond closely to the known mineralized zones. Also, the Lapland mafic rocks are poorer in sulphur than elsewhere in Finland (Fig. 4-7) but have quite high Ni factor scores (Fig. 4-8). Häkli (1970) stated that this was due to high Ni in the sulphide phase (because sulphur is scarce), but it is

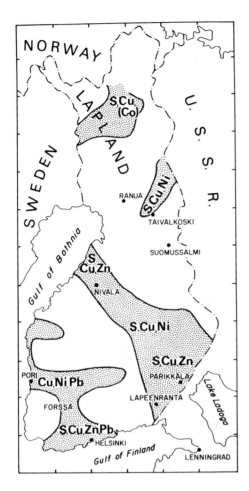

Fig. 4-9. Ore zones and metal provinces of Finland. Grid is 100 km^2 (derived from Mik-kola and Niini, 1968).

also possibly due to higher silicate Ni. The close correlation between high factor scores — particularly of Ni — and known mineralized belts containing significant economic nickel deposits clearly supports the validity of Häkli's approach to regional scale exploration for this type of mineralization.

THE USE OF MINERAL SEPARATES, HALOGENS, AND MINERAL-SELECTIVE LEACHES

Rationale for the approach

A number of studies have sought to circumvent some of the interpretative problems inherent in whole rock analyses by determining the distribution of elements in individual minerals, using mineral-selective leaches and measurement of "volatile" elements such as chlorine and fluorine. Most of the work on mineral phases has been done on biotite, although element distributions in feldspar, hornblende, magnetite, and other minerals have also been used. The rationale for the use of biotite is that its structure can readily accommodate minor elements in different states within the lattice; hence, biotite from a mineralized intrusion may contain characteristic minor element contents as a result of hydrothermal alteration associated with mineralization. Measurement of minor elements (such as Pb) in potash feldspar is done to characterize the primary magma; i.e., high Pb in the magma would lead to residual magmatic solutions being high in Pb, causing lead minerals to form and would also favour high Pb in potash feldspars (Pb^{2+} substituting for K^+).

The contrary view (Ingerson, 1954) is also possible; if conditions are such that Pb cannot enter a silicate structure, the Pb will instead be concentrated to form lead minerals. In this situation *low* contents of Pb in silicate minerals would indicate lead mineralization, whereas high contents of Pb in silicate minerals would indicate the absence of lead mineralization.

The basis of the measurement of halogens relates to their well-documented role (especially Cl) in the transportation of ore metals (Helgeson, 1964; Krauskopf, 1964; Barnes and Czamanske, 1967). The expectation, therefore, is that intrusions host to mineral deposits would be enriched in halogens. Roedder (1967, 1977) has pointed out that fluid inclusions in magmatic minerals in igneous rocks and in ore minerals contain saline solutions in which the amount of Cl present could account for most of the total Cl measured in whole rock analysis (although the amount of F present is low and is generally less than 10% of the total F present in rock). Since simple crushing and grinding of rocks leads to fracturing of inclusions and release of most of the fluids, measurements of halogens in water leaches of crushed rock have also been used in an attempt to differentiate between barren and mineralized intrusions.

Measurement of halogens and water in whole rock

Kesler et al. (1973) determined water-soluble Cl and F by ion-selective electrodes on samples from nine plutonic complexes in the Greater Antilles and Central America. The analytical results are shown in Fig. 4-10 and Fig. 4-11 for Cl and F, respectively. The Cerro Colorado (Panama) and Cala Abajo-Piedra Hueca (Puerto Rico) plutons are associated with copper porphyry mineralization; the Terre-Neuve (Haiti) and Minas de Oro (Honduras) intrusions have contact copper mineralization. Kesler et al. (1973) classed the remainder as unmineralized; however, the Virgin Islands intrusion (which was noted as having minor molybdenite veins by Kesler et al., 1973) was later classed as mineralized by Kesler (in Kesler et al., 1975a); it was described as having porphyry copper-type mineralization. Clearly neither the

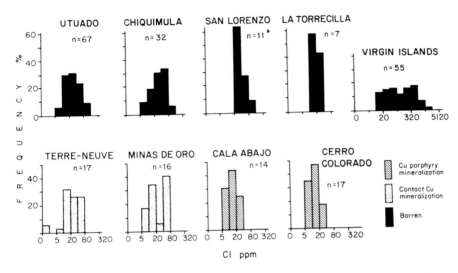

Fig. 4-10. Frequency distributions of H$_2$O-soluble chloride in intrusive rocks (derived from Kesler et al., 1973).

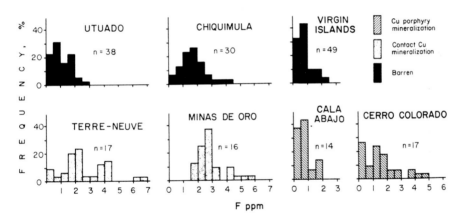

Fig. 4-11. Frequency distributions of F in intrusive rocks (derived from Kesler et al., 1973).

average content nor the distribution pattern of Cl discriminates between mineralized and barren plutons; the only discernible grouping is those plutons with porphyritic texture and apparent shallow level of emplacement (Cerro Colorado, Cala Abajo—Piedra Hueca, and La Torrecilla) which have the lowest Cl abundances. The F contents and ranges in mineralized intrusions tend to be greater than for barren intrusions, but one of the four mineralized plutons is clearly in the background range, and one of the three barren intrusions is clearly in the range of mineralized intrusions. Water-

TABLE 4-XI

Average ($\bar{x}$) Cl (ppm), F (ppm) and H$_2$O (%) and number of analyses (n) for mineralized and barren intrusions in Central America and the Caribbean (from Kessler et al., 1975b)

Intrusion and Location	Composition*	Mineralization**	Cl $\bar{x}$	n	F $\bar{x}$	n	H$_2$O $\bar{x}$	n
Mineralized								
1. Above Rocks, Jamaica	GD-E	Cu porphyry (C)	233	18	676	16	1.55	14
2. Cerro Colorado, Panama	GD-EP	Cu porphyry (A)	369	5	995	5	1.32	4
3. Terre Neuve, Haiti	GD-EP	Contact Cu (A)	205	18	679	18	1.27	16
4. Petaquilla, Panama	QD-EP	Cu porphyry (B)	90	5	253	5	1.75	5
5. Rio Pito, Panama	QD-EP	Cu porphyry (C)	162	8	443	8	1.22	6
6. Minas de Oro, Honduras	QD-E	Cu porphyry (B)	340	10	506	9	0.65	6
7. San Francisco, Honduras	QD-P	Cu porphyry (B)	88	8	370	8	1.64	9
8. El Yunque, Puerto Rico	QD-E	Cu porphyry (C)	145	11	311	11	1.19	8
9. Cuyon, Puerto Rico	QD-P	Cu porphyry (C)	160	8	267	8	1.60	10
10. Rio Vivi, Puerto Rico	QD-P	Cu porphyry (A)	118	17	262	18	0.88	6
11. Virgin Islands, Virgin Islands	QD-EP	Cu porphyry (C)	210	9	313	23	1.56	13
12. St. Martin, St. Martin	QD-EP	Cu porphyry (C)	273	3	384	4	1.03	15
13. Loma de Cabrera, Dominican Republic	QD-EP	Cu porphyry (C)	133	12	316	12	0.86	4
Average			176		142		1.30	12
Barren								
14. Bocas del Tovo, Panama	GD-EP	(D)	178	3	264	3	0.51	3
15. Chiquimula, Guatemala	GD-E		181	6	540	6	—	—
16. Utuado, Puerto Rico	GD-EP		295	11	674	13	1.26	9
17. Foliated Tonalite, Dominican Republic	QD-E		228	17	374	17	1.14	9
18. Rio Guayabo, Panama	SYD-E	(D)	160	8	320	8	—	—
19. Azuero, Panama	QD-E		108	6	318	6	0.50	6
20. El Bao, Dominican Republic	QD-EP	(D)	162	10	355	7	1.02	10
21. El Rio, Dominican Republic	QD-EP	(D)	128	12	418	12	0.74	12
22. Median, Dominican Republic	QD-E	(D)	161	25	481	25	1.62	20
23. Ciales-Morovis, Puerto Rico	GD-E	minor mineralization	134	8	595	8	0.67	8
24. San Lorenzo, Puerto Rico	QD-E	minor mineralization	151	19	406	21	1.15	14
Average			176		130		1.06	

*GD = granodiorite; QD = quartz diorite; SYD = syendiorite; E = equigranular; P = porphyritic; EP = equigranular and porphyritic.
**(A) = economic mineralization; (B) = significant mineralization not fully evaluated; (C) = varying size and grade, some drilling; (D) = minor exploration only.

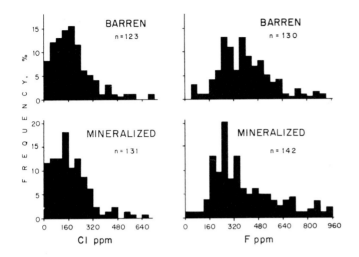

Fig. 4-12. Frequency distributions of Cl and F in barren and mineralized intrusions in the Caribbean and Central America (from Kesler et al., 1975b).

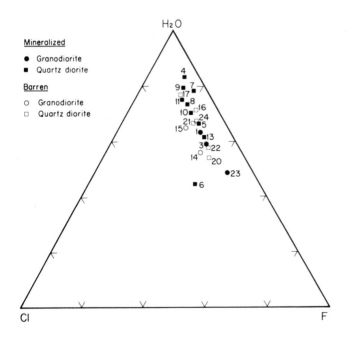

Fig. 4-13. Average Cl-F-H_2O composition of mineralized and barren intrusions in the Caribbean and Central America (from Kesler et al., 1975b). Numbers refer to intrusions listed in Table 4-XI.

soluble Cl is of no value in distinguishing mineralized plutons, and the best that can be said of water-soluble F is that the results are inconclusive.

The number of intrusions discussed above was expanded by Kesler and his co-workers, and total Cl and F (measured by selective-ion electrodes after fusion with NaOH) and H_2O were determined on samples of whole rock (Kesler et al., 1975b). The analytical results are given in Table 4-XI. There are no significant differences in Cl, F, or H_2O contents between mineralized and barren intrusions (frequency distributions for Cl and F are shown in Fig. 4-12). Each intrusion appears to have its own characteristic Cl-F-H_2O composition which bears no relation to the presence of mineralization (see Fig. 4-13). Moreover, the data in Table 4-XI show that F and possibly Cl are enriched in granodiorites compared to quartz diorites regardless of the presence or absence of mineralization; Kesler et al. (1975b) pointed out that there is good correlation between K and F in all intrusive rocks. Thus, the contents of Cl and F are a function of rock composition. Another feature noted by Kesler et al. (1975b) is the depletion of halogens and enrichment of water in high level, porphyritic intrusions. The conclusion of this important study is that neither absolute abundances of Cl, F, and H_2O nor any combination of them can serve to discriminate between mineralized and barren intrusions.

Halogens in biotite

Stollery et al. (1971) pointed out that since the lead-zinc-silver deposits associated with the granodiorite stock at Providencia (Zacatecas, Mexico) were derived from aqueous chloride ore-forming solutions, it was reasonable to expect that the stock itself might carry abnormal quantities of Cl. Analyses of 20 samples of biotite from the stock showed that the Cl content is 2 to 3 times the amount in biotite in German granites that were selected for comparison. There is complete overlap, however, in the upper range of Cl contents between the Providencia stock and the German granites (Fig. 4-14). Stollery and his co-workers strongly suggested that the Cl was fixed in the biotite during the magmatic stage and was not influenced by the hydrothermal mineralization associated with the lead-zinc-silver deposits. Determination of Cl in two whole rock samples gave 500 ppm and 900 ppm Cl, which is similar to the mean Cl content of 507 ppm in the mineralized granites of southwest England (Fuge and Power, 1969) and significantly greater than the mean of 202 ppm Cl in average granitic rocks (Johns and Huang, 1967).

An extensive follow-up of Parry's (1972) preliminary investigation on the distribution of Cl in biotite samples from plutons in the Basin and Range province of the U.S. was carried out by Parry and Jacobs (1975) and Jacobs and Parry (1976). The average Cl and F contents of some of the plutons sampled are given in Table 4-XII; frequency distributions of Cl and F in

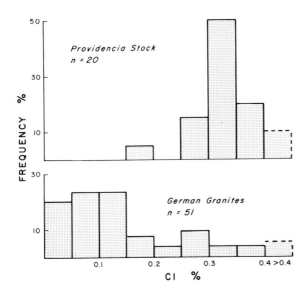

Fig. 4-14. Frequency distributions of Cl in biotite from the Providencia stock, Mexico, and German granites. (Redrawn with permission from Stollery et al., 1971, *Economic Geology*, vol. 66, 1971, fig. 6, p. 366.)

biotite from barren plutons and from plutons with Cu, Pb, Zn, Ag, and Au contact-metasomatic, replacement, and vein deposits are shown in Fig. 4-15.

The distribution of Cl in the barren stocks is quite similar to that in the German granites shown in Fig. 4-14 above; although the modal concentrations in the mineralized stocks are higher than in the barren stock, the distributions fall entirely in the barren range. Similarly, the F content is not useful in discriminating between barren and mineralized plutons — the highest F occurs, in fact, in non-mineralized plutons. The authors remarked that the highest Cl content occurs in biotite that coexists with hornblende, and that at any particular locality hydrothermal biotite contains more F than magmatic biotite (Parry and Jacobs, 1975); they also showed that Cl and F in apatite did not serve to identify mineralized plutons. In the 1976 study it was shown that variations of TiO_2, BaO, and mole fraction of phlogopite (defined as Mg/(Mg + Fe)) in magmatic, hydrothermal and replacement biotite are not diagnostic of mineralized plutons — although most of the measured parameters are distinctly different in different intrusions. The variation in element content in biotites of different origins in the same intrusion obviously leads to interpretative complications. On the other hand, in the 1976 study Jacobs and Parry point to this variation as being the only possible useful discriminator: the F and TiO_2 contents (as a function of the mole fraction of phlogopite) is quite different for magmatic, hydrothermal, and replacement biotite from the Santa Rita (New Mexico) pluton that is host to copper porphyry deposits, whereas there is no distinction between

TABLE 4-XII

Mean concentration of Cu, Pb, Zn, Cl, and F in biotite, and Pb in K-feldspar in various acidic intrusions of the Basin and Range Province, U.S.A.*

Type of mineralization and district	Biotite					Feldspar
	Cu (ppm)[1]	Pb (ppm)[1]	Zn (ppm)[1]	Cl (%)[2]	F (%)[2]	Pb (ppm)[3]
Porphyry Cu						
Bingham	1100 (11)	10 (11)	120 (11)	0.17	0.97	44 (12)
Ely-Liberty	780 (10)	8 (10)	210 (10)	0.15	2.73	12 (11)
Ely-Veteran				0.11		12 (4)
Cu, Pb, Zn, Ag, Au contact, replacement, fissure veins						
Tintic-Silver City	20 (7)	19 (7)	260 (7)	0.20	0.75	29 (10)
East Tintic district				0.14	1.59	
Clayton Peak	290 (9)	12 (9)	250 (9)	0.21	0.88	44 (14)
San Francisco	91 (9)	10 (9)	220 (9)	0.10	0.42	
Gold Hill	52 (8)	35 (8)	300 (8)	0.17	1.02	
Little or no known base or precious metals						
Last Chance	74 (9)	36 (9)	370 (9)	0.26	1.06	80 (10)
Little Cottonwood	10 (6)	13 (6)	450 (6)	0.05	0.73	55 (6)
Mineral Range	10 (4)	37 (4)	680 (4)	0.03	2.65	
Ibapah	8 (4)	24 (4)	450 (4)	0.01	1.13	
Whitehorse Pass	9 (5)	12 (5)	230 (5)	0.11	1.47	41 (6)
Iron Springs	7 (4)	8 (4)	180 (4)	0.09	1.09	15 (3)

* Numbers of samples shown in parentheses. Cu, Pb, and Zn were determined spectrographically; Cl and F were measured by electron microprobe. Compiled from: [1]Parry and Nackowski, 1963 (note: these are geometric means); [2]Parry and Jacobs, 1975; [3]Slawson and Nackowski, 1959.

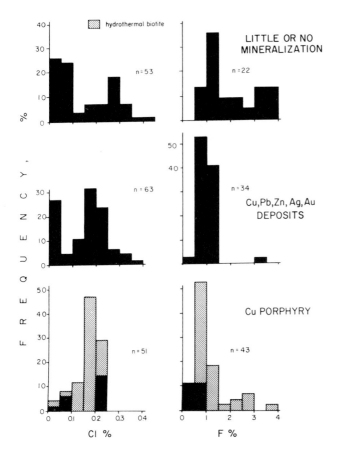

Fig. 4-15. Frequency distributions of Cl and F in biotite in intrusions in the Basin and Range structural province, U.S (compiled from Parry and Jacobs, 1975).

magmatic and replacement biotite in barren intrusions. A similar relation was found for the content of TiO_2 in biotite.

Kesler et al. (1975c) also examined the distribution of Cl and F (as well as H_2O, Cu, Zn, Fe, and Mg) in 92 samples of biotite from 35 intrusions in western North America and the Caribbean region; 54 samples are from strongly mineralized intrusions (such as Butte, Bingham, and Guichon Creek), and 38 samples are from intrusions considered to be barren. The distribution of H_2O does not differ between barren and mineralized intrusions; although the modal value of F in mineralized intrusions is markedly greater than in barren intrusions, there is a complete overlap in the two populations (Fig. 4-16). The frequency distribution of F is replotted in Fig. 4-17 at larger class intervals to correspond to the scale used by Parry and Jacobs (1975) shown in Fig. 4-15 above.

Comparison of the two data sets shows that the range and abundance of F

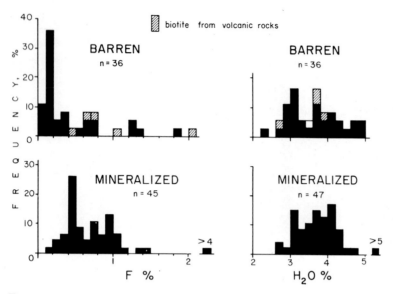

Fig. 4-16. Frequency distributions of F and H_2O in biotite from intrusions in western North America and the Caribbean (compiled from Kesler et al., 1975c).

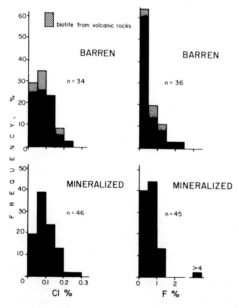

Fig. 4-17. Frequency distributions of Cl and F in biotite from intrusions in western North America and the Caribbean (compiled from Kesler et al., 1975c).

obtained by Kesler et al. (1975c) is considerably less in both mineralized and barren intrusions than that obtained by Parry and Jacobs (1975). The distribution of Cl in biotite in the study by Kesler and his co-workers is also

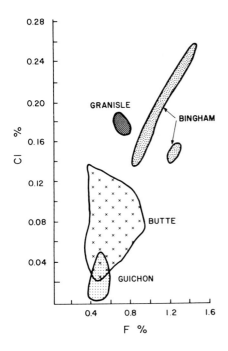

Fig. 4-18. Distribution of Cl and F in biotite from plutonic intrusions in western North America (modified after Kesler et al., 1975c).

shown in Fig. 4-17; there is no discernible difference between barren and mineralized intrusions (although Kesler and his co-workers stated that the abundance in the latter is slightly greater than in the former). A comparison of Fig. 4-15 with Fig. 4-17 also shows that the abundance of Cl obtained by Kesler and his co-workers is considerably lower in both barren and mineralized intrusions than that measured by Parry and Jacobs (1975). The higher Cl and F values obtained by Parry and Jacobs probably do not indicate analytical bias; their Cl and F results for Bingham are 0.17% and 0.97%, respectively, compared to 0.15% and 1.5%, respectively, obtained in the Kesler study (Bingham is the only locality common to both studies).

The marked differences between the data emphasize one of the major difficulties in using halogen abundances to recognize mineralized intrusions; each intrusion (as remarked in both studies) has its own characteristic Cl and F abundance; this is illustrated in Fig. 4-18 for Cl and F, and in Fig. 4-13 above for H_2O, Cl, and F. Moreover, the Cl and, especially, the F contents increase with increasing K; this can be seen in the triangular plot of H_2O-Cl-F in Fig. 4-13 above where both mineralized and barren granodioritic intrusions are generally enriched in Cl and F compared to mineralized and barren quartz diorites.

TABLE 4-XIII

Distributions of Pb and Zn in minerals of Caledonian granitoids of the Susamyr Batholith (from Tauson and Kravchenko, 1956)

Element	Mineral or rock	Porphyritic granodiorite*		Coarse-grained biotite granite*		Leucocratic medium-grained granite*	
		1	2	1	2	1	2
Pb (ppm)	quartz	4	1.4	2	0.6	3	1
	feldspar	40	24.1	28	18.7	35	22.8
	biotite	20	0.8	14	0.6	20	0.7
	hornblende	—	—	8	0.1	—	—
	magnetite	17	0.1	6	—	20	0.1
	total		26.4		19.8		24.6
	whole rock	26		20		30	
Zn (ppm)	quartz	7	2.5	11	3.3	7	2.2
	feldspar	10	6.0	12	7.8	24	15.6
	biotite	870	34.9	740	31.2	620	21.1
	hornblende	—	—	710	5.3	—	—
	magnetite	100	0.6	190	0.5	230	0.7
	total		44		48.1		39.6
	whole rock	40		52		40	

* 1 = in mineral; 2 = calculated mineral contribution to rock.

Distribution of metals in biotite and feldspar

Some experimental work of Tauson and Kravchenko (1956) on granitoids in the U.S.S.R. showed that most of the Zn occurs in biotite at concentrations 10 times greater than in whole rock (Table 4-XIII). Using a sulphide-selective leach, they removed an average of 80% of the total Zn from whole rock and 70—80% from biotite; they concluded that most of the Zn is in a non-silicate form and probably occurs as sphalerite. Most of the Pb in granitoids, however, occurs in feldspars (Table 4-XIII). The authors found that a sulphide-selective leach removed 30—50% of Pb from whole rock, but only 12—18% of Pb from K-feldspar; most of the Pb in K-feldspar is therefore present as isomorphous substitution in the silicate lattice. Since the content of Pb in other feldspars is about the same as in K-feldspar, and Pb^{2+} is not likely to substitute for Ca^{2+} or Na^+, it is assumed that the Pb in other feldspars and minerals must occur largely as submicroscopic inclusions of Pb minerals — probably galena.

Putman and Burnham (1963) showed that the Zn content of the ferromagnesian phase (chiefly biotite) of samples from seven major plutonic systems in Arizona is sufficient to account for all the zinc in the rock. On

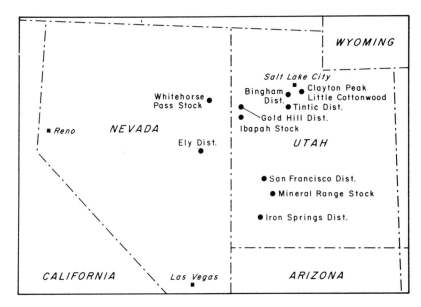

Fig. 4-19. Location of Basin and Range plutons listed in Table 4-XII. (Redrawn with permission from Parry and Nackowski, 1963, *Economic Geology*, vol. 58, 1963, fig.1, p. 1136.)

the other hand, this is not true of Cu, and a considerable amount occurs as chalcopyrite; wherever co-existing chalcopyrite is present, the Cu content of biotite is abnormally high.

Numerous studies have been made to try to use ore and other element variations in biotite and feldspar to identify mineralized intrusions. Parry and Nackowski (1963) attempted to correlate the contents of Cu, Pb, and Zn in biotite with productive intrusions in some Basin and Range intrusive stocks (Fig. 4-19). Later work on biotite from Basin and Range intrusions identified three genetic types of biotite (magmatic, hydrothermal, and replacement) and showed that Cl and F (discussed in the last section) and also Ti, Mg, and Ba contents varied according to the type of biotite (Parry and Jacobs, 1975; Jacobs and Parry, 1976). Notwithstanding the implications of this later work on the validity of the conclusions on Cu, Pb, and Zn, empirically there does seem to be a significant variation in Cu and Zn related to mineralization. Analytical data are given in Table 4-XII, and the distribution of Cu and Zn in biotite is plotted in Fig. 4-20. The productive districts have greater than 15 ppm Cu and less than 350 ppm Zn and are clearly separated from barren districts. There is a close negative correlation between Zn and the logarithm of Cu in productive stocks, and the non-productive intrusions lie a considerable distance from the regression zone passing through the productive intrusions. This could be an important discriminatory relation.

Although the Cu-Zn relation in biotite offers a convincing indication that

82

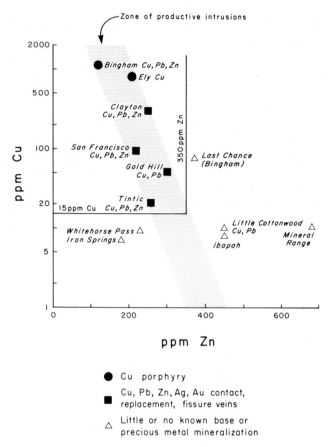

Fig. 4-20. Mean content of Cu and Zn in biotite from Basin and Range plutons (compiled from Parry and Nackowski, 1963).

significant productive intrusions can be discriminated from barren intrusions, the base metals are less reliable as indicators of specific metals produced; the exception is copper. Parry and Nackowski (1963) have pointed out that, in general, the greater the production of copper, the greater the amount of Cu in biotite and also the greater the standard deviation of the Cu in the samples. The Tintic district is an exception to this, but the authors considered that there may have been a sampling problem.

There is little relation between Pb in biotite and lead production; indeed, some of the highest and the lowest Pb contents are from districts that have produced no lead. The Zn content in biotite from districts that have produced zinc is less than 300 ppm, but there are also non-productive districts that have similarly low Zn in biotite. Parry and Nackowski (1963) also remarked that each intrusion or district has its own distinctive trace base metal population in biotite. Putman and Burnham (1963) also concluded that the trace element composition of ferromagnesium minerals was unique in each of the seven plutons they studied in Arizona.

Slawson and Nackowski (1959) showed that the Pb distribution in K-feldspars is generally distinctly different in each district. For the districts that they investigated that are common to the biotite work of Parry and Nackowski (1963), there is little obvious trend for Pb concentrations related to lead production. The highest Pb (80 ppm) is for the Last Chance Stock — which is barren — and one of the lowest Pb values (29 ppm) is for the Tintic district (Table 4-XII). Nevertheless, they claim that on the basis of all of their data, *generally* high Pb is correlated with significant lead production, although they admit notable reversals of the trend.

Work by Cuturic et al. (1968) on feldspars in Tertiary igneous rocks of the Balkan Peninsula showed that in the lead-zinc mineralized Dinarides the mean Pb content in granitoids and quartz monzonites was 40 ppm (range 26—59 ppm) compared to 37 ppm (range 32—44 ppm) in similar rocks of the copper mineralized Balkanides. The Pb content of K-feldspar in related quartz latites and latites was 43.5 ppm (range 18—57 ppm) in the Dinarides and 20 ppm (range 11—29 ppm) in the Balkanides. Whereas there is considerable overlap in values between the two areas, the data support Slawson and Nackowski's (1959) contention that lead mineralization is accompanied by enhanced Pb values in feldspars.

Bradshaw (1967) measured Cu, Pb, Zn, Rb, Sr, Zr, Mn, and Ca by X-ray fluorescence, Na and K by flame photometry, and Sn by emission spectroscopy in feldspar, biotite, and muscovite from five mineralized intrusions from Devon and Cornwall in southwest England (55 samples) and from eight non-mineralized intrusions elsewhere in England and Scotland (94 samples). These are listed in Table 4-XIV. The dominant metals produced from southwest England were tin, copper, zinc, and lead derived from vein deposits spatially and genetically related to the Hercynian granites that have intruded Devonian and Carboniferous slaty shales and mudstones. The mean contents of ore elements in feldspar, biotite, and muscovite in mineralized and non-mineralized intrusions are given in Table 4-XV. The mean contents of Sn and Pb are statistically significantly higher in the mineralized group than in the non-mineralized group; in addition, Zn is higher in feldspar and muscovite for the mineralized group. There is no significant difference between the two groups in the mean Cu content of any of the minerals.

In biotite higher Rb and Sn and lower Sr contents are characteristic of mineralized intrusions compared to barren intrusions. In feldspars mean contents of Zn, Rb (and to a lesser extent K, Mn, Pb, and Sn) are higher and mean contents of Ca and Sr are lower in mineralized than in non-mineralized intrusions. Although there are clear differences in mean element contents, there is also considerable overlap between the two groups of granites in terms of element content of individual samples, as illustrated for Zn in feldspar in Fig. 4-21. A further problem with these data is that the mineralized and the

TABLE 4-XIV

Name, rock type, and numbers of samples of intrusions examined by Bradshaw (1967)

Intrusion	Rock type	Number of samples
Mineralized		
Dartmoor	granite	8
Bodmin Moor	granite	3
St. Austell	granite	6
Carnmenellis	granite	18
Lands End	granite	20
Non-mineralized		
Weardale	granite	8
Shap	adamellite	4
Criffell-Dalbeattie	granodiorite	14
Etive, Cruachan	adamellite	13
Etive, Ballachulish	tonalite-adamellite	4
Etive, Allt na Larige	granodiorite	3
Strontian	granite	2
Strontian	granodiorite	7
Strontian	tonalite	5
Foyers	granodiorite	4
Foyers	tonalite	7
Cluanie	granodiorite	7
Aberdeen	adamellite	16

TABLE 4-XV

Mean contents of Sn, Pb, Zn, and Cu in feldspar, biotite and muscovite from mineralized and non-mineralized intrusions in Great Britain (from Bradshaw, 1967)

Mineral and element	Mean value (ppm)		Probability difference at 95% level
	mineralized	non-mineralized	
Feldspar			
Sn	29	15	yes
Pb	154	101	yes
Zn	14	6	yes
Cu	25	30	no
Biotite			
Sn	44	11	yes
Pb	31	20	yes
Zn	350	300	no
Cu	23	22	no
Muscovite			
Sn	52	28	yes
Pb	21	7	yes
Zn	130	240	yes
Cu	30	20	no

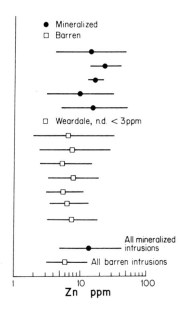

Fig. 4-21. Geometric means and standard deviations of Zn in feldspar from mineralized granites in southwest England and non-mineralized granites elsewhere in England and Scotland. (Redrawn with permission from Bradshaw, 1967, *Institution of Mining and Metallurgy, Transactions, Section B*, vol. 76, 1967, fig. 14, p. 145.)

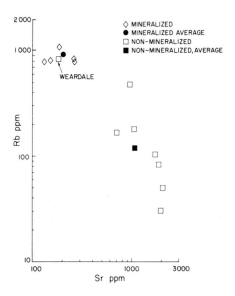

Fig. 4-22. Distribution of Rb and Sr in feldspar from mineralized granites in southwest England and non-mineralized granites elsewhere in England and Scotland (compiled from Bradshaw, 1967).

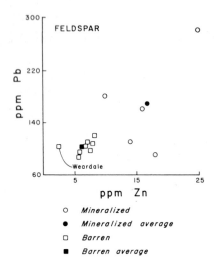

Fig. 4-23. Distribution of Pb and Zn in feldspar from mineralized granites in southwest England and barren granites elsewhere in England and Scotland (compiled from Bradshaw, 1967).

non-mineralized granites are not strictly comparable. The mineralized granites are more highly fractionated than the barren granites; only the barren Weardale intrusion is petrochemically similar to the mineralized intrusions. Whereas there are marked differences between the mineralized and barren groups in Pb and Sr contents in feldspars, as shown in Fig. 4-22, there is no distinction between the mineralized group and the Weardale granite. Nevertheless, there are distinctive differences in Pb and, especially, in Zn contents of feldspar between the mineralized intrusions and the Weardale granite (Fig. 4-23) that indicate a definite enrichment in these elements in feldspars of the mineralized granites; comparable data are not available for biotite because there was no adequate sample available from the Weardale granite. There are no differences in Cu content in biotite or feldspar between barren and mineralized intrusions in contrast to the results of Parry and Nackowski (1963) in the Basin and Range region (discussed above).

Although Bradshaw's study was based on relatively few samples, his data indicate that the mineralized granites of southwest England are distinctly different from barren intrusions due to enrichment in Pb, Zn, and Sn. This study also points out again the importance of adequate geological control since variations in the content of elements are related to varying degrees of fractionation in the granites.

A limited study by Lovering et al. (1970) of the distribution of Cu in biotite from intrusions in the Santa Rita and Sierrita Mountains in southern Arizona showed that the highest values of Cu occur in intrusions genetically and spatially related to mineralization. The Cu content of whole rock samples is also larger in the mineralized intrusion, but the contrast is not as great as that in bitotite.

TABLE 4-XVI

Copper content of biotite and whole rock, Santa Rita Mountains, Arizona (from Lovering et al. 1970)

Rock type	Number of samples	Cu (ppm) in biotite		Cu (ppm) in selected whole rock samples
		range	mean	
Barren				
Precambrian granodiorite and quartz monzonite	3	30—70	50	10
Cretaceous granodiorite	2	100—200	150	20
Paleocene quartz monzonite	4	70—150	90	10
Mineralized				
Palaeocene quartz latite porphyry (Cu, Pb, Zn, Ag)	4	700—7000	3400	50

The results of the study by Lovering and his co-workers are summarized in Table 4-XVI. The biotite samples are primary magmatic, and therefore the reservations due to the work of Parry and Jacobs (1975) on different genetic types of biotite should not apply. Subsequent work by Banks (1974) on biotite and its chlorite alteration products from this area indicates that anomalous Cu contents are associated with chlorite and not biotite. Banks (1974) suggested that an apparently high Cu content in biotite is due to contamination from chlorite. Banks admitted that it is possible that increasing copper content of chlorite may indicate proximity to mineralization, but he suggested that this is not a practical possibility because of the difficulty of obtaining clean chlorite separates and the extreme extent of the variation in Cu content both within and between chlorite grains.

Sulphide-selective leaches

Cameron et al. (1971) used a sulphide-selective leach (cold ascorbic acid and hydrogen peroxide) to determine the metal content of sulphide minerals to discriminate significantly mineralized from unmineralized ultramafic bodies of the Canadian Shield. Cameron and his co-workers started with the premise that for an ultramafic magma to give rise to significant copper-nickel sulphide mineralization the magma must be enriched in sulphur at an early stage of its crystallization; if this occurs, the solubility product of metal sulphides will be exceeded, and a sulphide liquid or crystal will separate. Accordingly, they analyzed 1079 samples from 16 ore-bearing ultramafic bodies (deposits with more than 4500 tonnes Ni-Cu), 5 mineralized bodies (less than 4500 tonnes Ni-Cu), and 40 barren bodies for total S and for Cu, Ni, and Co present in sulphide form. The results, summarized in Table 4-XVII, show that ore-bearing intrusions are enriched in S, and in sulphide-held Ni, Cu, and (to a lesser extent) Co, compared to both mineralized and

TABLE 4-XVII

Content of Cu, Ni, and Co in sulphide form (as determined by ascorbic acid-hydrogen peroxide cold leach) and total sulphur in ore-bearing, mineralized, and barren ultramafic rocks of the Canadian Shield (from Cameron et al., 1971)

		Ore-bearing	Mineralized	Barren
Number of intrusions		16	5	40
Number of samples		372	91	616
Cu (ppm)	$\bar{x}$	439	52.2	25.9
	s	1111	166	74.5
Ni (ppm)	$\bar{x}$	1875	842	579
	s	3577	687	483
Co (ppm)	$\bar{x}$	83.7	43.5	43.9
	s	84.5	45.4	33.3
S (%)	$\bar{x}$	0.582	0.177	0.059
	s	1.37	0.559	0.107

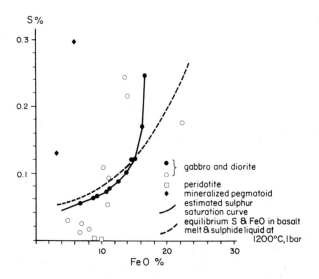

Fig. 4-24. Contents of S and FeO in Bushveld rocks compared to experimentally measured S and FeO contents in basaltic melt in equilibrium with sulphide liquid. (Redrawn with permission from Buchanan, 1976, *Institution of Mining and Metallurgy, Transactions, Section B*, vol. 85, 1976, fig. 1, p. 289.)

barren intrusions. Discriminant analysis applied to the data showed that Cu and S have about the same value in discriminating between barren and ore-bearing rocks, whereas Ni has less importance, and Co has little value. In terms of exploration, however, there are some situations where there are analytical (and therefore economic) advantages in avoiding S determinations.

The discriminant analysis based on Cu, Ni, and Co as variables showed that ore-bearing rocks — particularly those containing large deposits — could be distinguished fairly reliably.

The empirical results of the work by Cameron et al. (1971) are extremely promising (and it is surprising that they do not appear to have been followed up). Generally more successful interpretation would result from integration of petrological data — as in all rock geochemical studies. For example, the capacity of a mafic melt to dissolve S depends upon a number of factors, especially its FeO and SiO_2 content. Buchanan (1976) has pointed out that if the degree of S saturation at liquidus temperatures is established, it is possible to assess whether the rock is oversaturated with S and hence to assess the probability of it containing significant sulphide segregations. This principle is illustrated in Fig. 4-24 for gabbros and diorites from the Bushveld. Samples from mineralized pegmatoids associated with the diorites are clearly oversaturated with S, whereas ultramafic peridotites associated with the sequence are considerably undersaturated with S and are therefore not likely to have given rise to an immiscable sulphide melt.

CONCLUSIONS

Regional scale exploration for mineralization associated with plutonic rocks has been through various phases; particular approaches have been fashionable from time to time and have then been discarded. Specifically, halogens in whole rock and mineral separates (especially biotite) and base metals in biotite and feldspar have received considerable attention. These approaches have been shown to yield, at best, equivocal results. The content of total and water-soluble halogens in whole rock appear to have characteristic values for each major intrusion and show little consistent relation to the presence or absence of mineralization. The contents of halogens in biotite apparently vary, depending upon whether the biotite has a primary magmatic, hydrothermal, or replacement origin. This latter conclusion at least holds out the hope of recognizing hydrothermal biotite as a guide to ore. Ore metals in mineral separates — copper, lead, and zinc — appear to be marginally more successful, but recent work (at least in one case) has cast doubt upon whether anomalous Cu recorded in biotite really resides in biotite or in its alteration mineral chlorite which is present as a contaminant. The only really promising mineralogical approach is that of sulphide-selective leaching.

The largely negative results from the geochemistry of mineral separates should be regarded — from a practical point of view — with a profound sense of relief. It is extremely difficult, and therefore expensive, to obtain clean, monomineralic samples from rocks. Whole rock geochemistry, in contrast, is a simple and rapid procedure, and it is encouraging that the relatively few case studies of whole rock geochemistry available yield positive results.

Investigators may have turned to a study of element distribution in individual minerals when faced with the extreme variability of metal content within individual plutons and the lack of an invariable enrichment in ore elements in productive plutons. It is, of course, now recognized that a high variance and strong positive skewness in the distribution of elements is characteristic of mineralized plutons, and that the *form* of the frequency distribution of an element is generally more important than the absolute abundance of an element in discriminating between barren and productive plutons. The more closely the mineralization is related to magmatic processes the easier it is to recognize abnormal element distributions. For example, in the Canadian Cordillera tungsten distributions in plutons that are host to tungsten porphyry mineralization are more strongly skewed than in plutons associated with skarn mineralization. Similarly, in Britain granites that are host to comagmatic uranium mineralization are enriched in U relative to granites where uranium mineralization is post-magmatic.

An important consideration in whole rock geochemistry is to recognize and compensate for differences in element content that result from petrological variations; this generally means that a wide range of elements should be determined. The degree of fractionation is obviously significant insofar as it has been demonstrated that generally it is the highly evolved magmas that can give rise to genetically related mineralization. Extreme fractionation can be recognized in plutons by a high content of Li, a high Rb:Sr ratio, and low K:Rb and Ba:Rb ratios; such plutons are usually enriched in U, Th, Cs, Be, and Mo. Measurement of these elements will identify plutons with the greatest *potential* for being mineralized. Further refinement of the data to identify anomalous plutons among prospective plutons may be achieved in a manner proposed by Garrett (1973), i.e., by regressing pathfinder elements against elements indicative of fractionation and deriving the residual values.

The few studies published on mineralization in ultramafic intrusions indicate that the geochemical situation is much less complicated than in felsic plutons. Ultramafic intrusions that are host to nickel sulphides appear to the invariably enriched in Ni and S.

Practical guides for whole rock geochemical surveys to discriminate between barren and productive plutons on a regional scale are scarce in the literature. The practical objective of such surveys is to be able to classify plutons as potentially ore-bearing or barren. To achieve this in a quantitative manner demands that enough samples are collected to allow the data to be analyzed statistically and such variables as sampling error to be assessed. Obviously the greater the number of samples, the greater is the statistical reliability of the conclusion; on the other hand, a very large number of samples defeats the objective of reconnaissance exploration. The problem has been approached statistically by Beus and Grigoryan (1977) and Garrett (1979, 1982). To calculate the number of samples that must be collected to recognise a potentially mineralized pluton the characteristics of element distributions must be determined (by an orientation survey) in barren and mineralized plutons. The number of samples is given by:

$$N = (p, P_r > r)$$

where N is the number of samples required to yield at least r samples with element contents (or element ratios or other parameters) greater, or less, than some predetermined value with a probability of occurrence of p. Garrett (1982) has published tables for various probabilities (0.90, 0.95, 0.98, 0.99) associated with P_r; a shortened verison of the tables for the 0.95 probability level is given in Appendix 3. As an example, suppose it has been found that 10% of samples from plutons with tungsten mineralization exceed 10 ppm W, but that this value is never exceeded in barren plutons. At the 0.95 level, to ensure that at least one sample from a mineralized pluton exceeds the threshold of 10 ppm, 29 samples would have to be collected; to ensure that 3 samples exceed the threshold value, 61 samples must be collected. If 20% of samples from mineralized plutons exceed the defined threshold, then 14 samples would have to be collected to ensure that one sample exceeds the threshold, and 30 samples would have to be collected to ensure that 3 samples exceed the threshold.

The sampling problem, and the matters of sampling and analytical variability are discussed by Garrett (1982) and other authors in the companion volume edited by Howarth (1982) which should be consulted for details; analytical variability is also discussed by Fletcher (1981) in another volume in this series. Broadly, if at least 10% of the samples from a mineralized pluton may be expected to be anomalous, and 1—3 anomalous samples are considered adequate, at the 0.95 level 30—60 samples should be collected from each unzoned and homogeneous pluton; if a pluton has a number of distinct phases, then a similar number of sites should be sampled from each phase. A minimum sample weight of 1 kg, preferably as a number of smaller chip samples, should be taken. If the grain size of the rock is particularly large, then larger samples must be collected (see Chapter 2).

It is the author's opinion that whole rock geochemistry offers the best possibility of identifying potentially ore-bearing plutons on a reconnaissance scale. Mineral-selective leaching could possibly be very useful, but it requires much more study.

Chapter 5

REGIONAL SCALE EXPLORATION FOR DEPOSITS OF PLUTONIC
ASSOCIATION — IDENTIFICATION OF TIN GRANITES

INTRODUCTION

Tin seems to be the only element that generally — but not universally —
shows enrichment in intrusions associated with tin mineralization in many
parts of the world. Tin deposits and their host granites have received extensive study, resulting in a large literature on their occurrence and prospecting
characteristics (see R.G. Taylor, 1979).

There is a general concensus that tin deposits characteristically occur in
the more felsic granitoids and do not occur in granodiorites (Flinter, 1971;
Tauson, 1974; Hesp and Rigby, 1974, 1975). Various petrological discrimi-
nators have been proposed. Flinter et al. (1972) stated that lode and vein
tin deposits tend to occur in high-SiO_2 (greater than 72%) leucogranites and
disseminated tin deposits tend to occur in low-SiO_2 (less than 72%) meso-
granites; they also correlated tin-bearing granites with granites that have a
differentiation index greater than 85 and a petrological index of less than 4.
There is also agreement on the general content of tin that discriminates
between barren and ore-bearing granites. Barsukov (1957, 1967, 1969)
reported 16—30 ppm Sn in tin-bearing granites compared to 3—5 ppm in
barren intrusions; Ivanova (1963) reported 16—32 ppm Sn in mineralized
granites compared to less than 5 ppm in barren granites; Flinter (1971)
suggested 10—15 ppm as a threshold.

Unfortunately for the success of exploration for tin all tin-rich granites
are not host to economic tin deposits, and the problem is far from solved —
as witnessed by the continued flow of papers on the characterization of tin-
bearing granites. Re-interpretation of published data given below, however,
suggests that the problem is amenable to a geochemical approach.

CASE HISTORIES

Discrimination based on tin in rock and minerals

The broad nature of the problem of distinguishing tin-bearing granites on

94

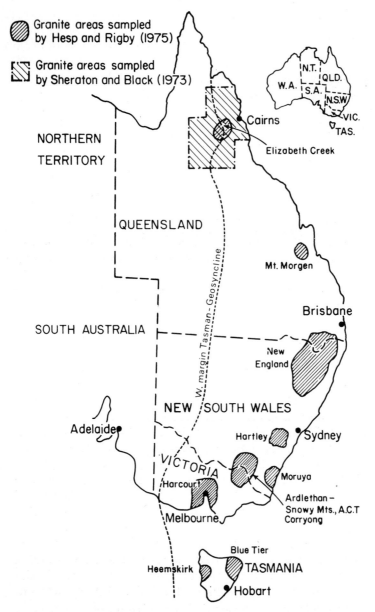

Fig. 5-1. Location of granitoids in eastern Australia (compiled from Sheraton and Black, 1973, and Hesp and Rigby, 1975).

the basis of their tin content is illustrated by the work of Hesp and Rigby (1975) on the occurrence of tin deposits in granite rocks of the Tasman Geosyncline of eastern Australia (Fig. 5-1). Typical ore deposits in these granites are of tin, tungsten, gold, and copper. The frequency distributions

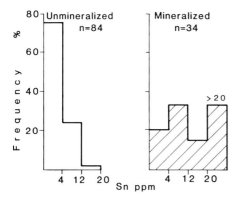

Fig. 5-2. Frequency distribution of Sn in mineralized and barren intrusions, Tasman Geosyncline, eastern Australia. (Redrawn with permission from Govett and Nichol, 1979, *Geophysics and Geochemistry in the Search for Metallic Ores. Geological Survey of Canada, Economic Geology Report*, 31, 1979, fig. 15-2, p. 342; based on data from Hesp and Rigby, 1975.)

of Sn in granites with cassiterite mineralizaton and in unmineralized granites are shown in Fig. 5-2; the former has a mean Sn content of 26 ppm compared to a mean content of 3.4 ppm in the latter. These results are based on samples collected from outcrops some distance from known cassiterite mineralization; however, the extent to which the higher Sn values are characteristic of the mineralized granites is not known since the sampling was biased to some extent towards the areas of tin mineralization rather than being representative of the granitoid rocks as a whole. The percentage of granites (defined as calc-alkali biotite granites) that are associated with cassiterite is plotted against the Sn content in Fig. 5-3. The higher the Sn content of the granite, the greater is the probability that it contains cassiterite mineralization; all granites with more than 20 ppm Sn have associated tin deposits, but 7 out of 70 granites with a Sn content of 0—4 ppm are also associated with tin deposits. Thus, although a high Sn content (more than 20 ppm) is a fairly certain indication of associated tin deposits, granites with low Sn contents cannot be excluded from consideration.

The lack of uniformity in the distribution of Sn in a granite intrusion seems to be a more reliable criterion for the recognition of potentially tin-bearing granites than the actual content of Sn. Bolotnikov and Kravchenko (1970) illustrated this conclusion in a study of the distribution of Sn in 83 samples collected from the Verkhneurmiyskiy granite in the Amur region of the U.S.S.R. that is spatially associated with skarn, griesen, and cassiterite mineralization. The frequency distribution of Sn (Fig. 5-4) clearly indicates multiple populations. The distribution can be partitioned (see Chapter 3) into two populations: $\bar{x} = 7.8$ ppm, $s = 3.7$ ppm; and $\bar{x} = 33$ ppm, $s = 8.5$ ppm. In fact, there is a marked difference in spatial distribution. The mean of 43

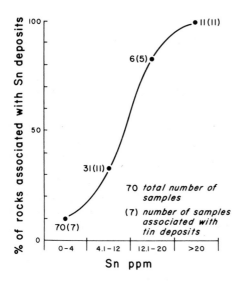

Fig. 5-3. Relation between Sn content of granites and tin mineralization, eastern Australia (from Hesp and Rigby, 1975).

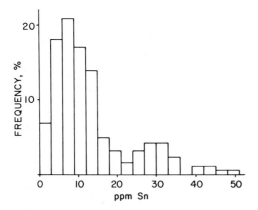

Fig. 5-4. Distribution of Sn in granites of the Verkhneurmiyskiy intrusion, U.S.S.R. (Redrawn with permission from Bolotnikov and Kravchenko, 1970, *Doklady Akademii Nauk SSSR*, vol. 191, 1970, American Geological Institute translation, fig. 1, p. 187.)

samples collected near the top of the pluton is 24.8 ppm, whereas the mean of the remaining 40 samples collected in a valley 850 m below is 5.8 ppm. This type of non-conformity of spatial distribution may generally be detected by high variance; this is illustrated in Table 5-I for the statistical differences in the Sn distributions in mineralized and in barren intrusions of the Tasman Geosyncline.

Barsukov (1957) suggested that biotite is the best geochemical indicator

TABLE 5-I

Mean, standard deviations, and coefficient of variation of Sn in mineralized and barren intrusions of the Tasman Geosyncline (Hesp and Rigby, 1975)

Granites	Mean ($\bar{x}$)	Standard deviation (s)	Coefficient of variation (cv) (%)
Barren granite	3.7	2.3	62
Mineralized granite	18.1	18.9	104
Barren adamellite	2.9	1.5	52
Mineralized	7.8	5.6	72

of a granite containing Sn. He stated that not less than 80% of the total Sn in an unaltered granite occurs in biotite. The content of Sn in biotite was 30—50 ppm in granites that were not associated with tin deposits, whereas in mineralized granites the Sn content of biotite was very much higher, reaching hundreds of ppm. At the Anchor Mines in Tasmania (see below) Groves (1972) found that the biotite in the mica granite directly associated with tin mineralization had an average Sn content of 556 ppm, whereas the biotite of the barren granite-adamellite contains an average of only 64 ppm Sn. In the barren rocks studied most of the Sn was in the biotite, but in the mineralized rocks less than 50% of the Sn was in biotite; in samples with

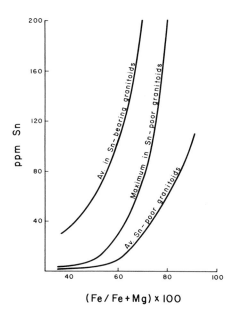

Fig. 5-5. Dependence of Sn content in biotite in granitoids on Fe:(Fe + Mg) ratio (compiled from Levashev et al., 1973).

TABLE 5-II

Content of selected elements, and element ratios in granites associated with tin mineralization at the Anchor Mine, Tasmania, Australia (from Groves, 1972)

Sample No.	K (%)	Mg (%)	F (%)	Li (ppm)	Rb (ppm)	Sr (ppm)	Sn (ppm)	K:Rb	Rb:Sr	Mg:Li
Barren porphyritic biotite granite-adamellite										
1	4.1	0.24	0.11	46	360	95	6	114	3.9	52
2	4.0	0.23	0.15	46	355	85	5	113	4.2	50
3	4.1	0.24	0.14	46	355	105	6	115	3.4	52
4	4.0	0.19	0.10	23	385	30	5	119	11.2	83
5	3.4	0.21	0.21	70	425	65	16	80	6.5	30
6	3.9	0.22	0.40	46	365	75	9	107	4.9	48
Average	3.9	0.22	0.14	46	365	75	9	107	4.9	48
Mineralized biotite-muscovite granite										
7	3.9	0.04	1.4	235	1105	4	1820*	35	276	1.7
8	4.2	0.03	1.4	232	1115	4	740*	38	279	1.3
9	3.9	0.03	0.11	46	840	6	66	46	140	6.5
10	4.2	0.03	1.3	209	1225	4	61	34	306	1.4
11	3.1	0.03	0.78	164	935	7	52	33	134	1.8
12	3.6	0.04	1.1	164	980	6	28	37	163	2.4
Average	3.8	0.04	1.02	186	1035	5	49	37	207	2.2

* = excluded from mean.

visible cassiterite less than 2% of the Sn was in biotite. Bradshaw (1967) found significant differences in the Sn contents in biotite between mineralized granites (average 44 ppm Sn) in Cornwall and in barren (average 11 ppm Sn) granites elsewhere in the British Isles, although the absolute levels were rather low; he also found a significant difference in the Sn content of feldspars between mineralized granites (average 24 ppm Sn) and barren granites (average 15 ppm). (Bradshaw's work is discussed in more detail in Chapter 4.)

Data presented by Hesp (1971) indicate that there is an apparent positive correlation of Sn in biotite with Fe^{3+} and Li^+, and a negative correlation with Fe^{2+}, Mg^{2+}, Mn^{2+}, and Ti^{4+}. He deduced a relation (called the "Tin Holding Capacity" of biotite) between the amount of Sn in biotite and its chemical composition:

$$(Fe^{3+} + Li^+)/(Fe^{2+} + Mg^{2+}) - (Ti^{4+} + Mn^{2+})/10.$$

Levashev et al. (1973) stated that Sn in biotite could be used reliably as a geochemical indicator for tin mineralization only in extremely silicic rocks; they derived a simpler relationship than Hesp that relates the Sn content of biotite to the ratio $Fe:(Fe + Mg)$ which is illustrated in Fig. 5-5.

Discrimination based on non-ore elements

A detailed study by Groves (1972) on the tin-bearing granites of the Anchor Mine in the Blue Tier Batholith in Tasmania is instructive. The batholith appears to be a high-level intrusion with a fractionation sequence from early mafic granodiorite to late leucocratic granite. The great majority of the tin deposits, including greissenized veins and sheets and griessenized granites, occur in biotite muscovite granite. A few appear in porphyritic biotite granite-adamellite.

Analytical data for the two types of granite are summarized in Table 5-II; analyses for some average granitoids are given in Table 5-III. Apart from the high content of Sn in the mineralized granites, other elements and element ratios show very marked differences, notably in F and in the ratios for K:Rb, Rb:Sr, and Mg:Li. By itself F is apparently not diagnostic; sample 1 (barren) has the same F content as sample 9 (mineralized), whereas there is an order of magnitude difference in Rb:Sr and Mg:Li ratios between these two samples. Flinter et al. (1972), on the basis of their work on the New England (Australia) granites, also concluded that F is not indicative of mineralization.

The problem with Groves' data from Tasmania is that all the known mineralized granites have more than 20 ppm Sn, and it is not possible to determine whether other elements would be useful for a low-tin granite that was mineralized. The closest comparison that can be made is between sample 5 (16 ppm Sn, barren) and sample 12 (28 ppm, mineralized). In the mineralized sample the K:Rb and Mg:Li ratios are 2 times and 12.5 times

TABLE 5-III

Content of selected elements, and element ratios for some average granitoids (from S.R. Taylor, 1968, and Turekian and Wedepohl, 1961)

Rock type	K (%)	Mg (%)	Li (ppm)	Ba (ppm)	Rb (ppm)	Sr (ppm)	Sn (ppm)
Average granodiorite (Taylor)	2.5	0.95	25	500	110	440	2
Average granite (Taylor)	3.5	0.33	30	600	145	285	3
Average low-Ca granite (Turekian and Wedepohl)	4.2	0.16	40	840	170	100	3

Rock type	K:Rb	Rb:Sr	Mg:Li	Ba:Rb
Average granodiorite (Taylor)	227	0.25	380	4.5
Average granite (Taylor)	241	0.51	110	4.1
Average low-Ca granite (Turekian and Wedepohl)	247	1.7	40	4.9

lower, respectively, than in the barren granites; the Rb:Sr ratio is 25 times higher than in barren granites. A further problem is that the biotite-muscovite granite is reasonably interpreted as the final differentiation product of the magma that produced the porphyritic biotite granite-adamellite; the latter may, therefore, be expected to show some of the characteristics of a tin-bearing pluton; moreover, samples of the granite-adamellite apparently were taken from the immediate vicinity of mineralization at the Anchor Mine.

A better assessment of the validity of element ratios in identifying mineralized plutons can be made on the basis of the work of Sheraton and Black (1973) on granites in northeast Queensland (Australia). Analytical data and a summary of economic mineralization are given in Table 5-IV. The mean Sn content of granites containing tin is significantly higher than of those that do not contain tin, but the concentration is much lower than in tin granites elsewhere in the world. The highest Sn content (16 ppm) is in the Finlayson granite (a porphyritic biotite-adamellite and granite) which is associated with the Annan River Tinfield. The Mareeba granite (average 9 ppm Sn), which includes porphyritic biotite and biotite-muscovite adamellites, leucogranites, and granodiorites, has tin, tungsten, and copper mineralization; it includes the important Mount Carbine wolfram deposit. The Elizabeth Creek granite, which is mostly leucocractic biotite adamellite, has an average of about

TABLE 5-IV

Content of selected elements, and element ratios of northeast Queensland (Australia) granitic rocks (from Sheraton and Black, 1973)

Granite	Number of samples	Economic mineralization*	K (%)	Mg (%)	Li (ppm)	Rb (ppm)	Sr (ppm)	Sn (ppm)	Cu (ppm)	Pb (ppm)	Zn (ppm)	K:Rb	Rb:Sr	Mg:Li
Elizabeth Creek	73	Sn, Mo, W, Pb, Cu (Au, Bi, Ag, Sb)	3.93	0.07	31	427	26	~5	6	32	33	92	16	22.6
Finlayson	6	Sn	3.98	0.25	94	388	43	16	8	25	32	102	9.1	26.6
Mareeba	23	Sn, W, Cu	3.53	0.26	69	356	88	9	6	26	50	99	4.1	37.6
Esmeralda	30	Au, Sn (Pb, Ag, Cu)	4.16	0.17	39	291	78	~5	7	51	79	143	3.7	43.6
Almaden	15	Cu, Pb, Ag, Zn	2.42	1.44	19	146	196	<4	14	18	45	165	0.74	758
Herbert River	40	(Cu, Pb, W)	3.47	0.25	29	233	101	<4	3	26	28	149	2.3	86.2
Dumbano	19	none	2.78	0.21	6	127	374	<4	3	28	42	220	0.34	350
Dido	12	none	1.25	1.16	7	45	812	<4	26	8	66	277	0.06	1657

* Minor mineralization in brackets

5 ppm Sn, but it includes the Herberton Tinfield, which up to 1968 had produced 15% of Australia's total tin output. The Esmeralda granite also has an average of about 5 ppm Sn and is typically a biotite adamellite grading into granite; economic mineralization is gold and tin, but only gold has been produced in significant quantities.

The other two mineralized granites listed in Table 5-IV — the Herbert River and the Almaden granite — are not tin-bearing; they contain an average of less than 4 ppm Sn. The Almaden granite is typically a hornblende-biotite-granite and has associated contact-replacement lodes of copper, silver, lead, and zinc in limestone or along contact faults (there is some doubt that this mineralization is genetically associated with the granite). The Herbert River granite is predominantly biotite adamellite, but it ranges in composition from biotite granite to hornblende-biotite-granodiorite. Neither the Dumbano granite (dominantly biotite adamellite) nor the Dido granite (hornblende-biotite, tonalite, and diorite) have any known economic minerlization associated with them.

The tin-bearing granites of northeast Queensland are not uniquely distinguished from the other granites by their Sn content, but they are notable for high Rb:Sr and low Mg:Li ratios compared with the barren and non-tin granites; they also have generally lower K:Rb ratios. On the same scale of values, the barren porphyritic biotite granite-adamellite of the Anchor Mine (above and Table 5-II) would be considered anomalous in terms of Rb:Sr ratios and probably also in terms of K:Rb ratios; the Mg:Li ratio is closer to background. The distributions of Cu, Pb, and Zn in the northeast Queensland granites show no discernible patterns that can be related to base metal mineralization.

A limited study of some tin granites of the New England Area of New South Wales (Australia) by Juniper and Kleeman (1979) suggested that, in this area, it is possible to discriminate between tin and barren granites on the basis of seven major elements. The authors stated that the distinction can be made on the basis of SiO_2-CaO + MgO + FeO-Na_2O + K_2O + Al_2O_3, Na + K-Fe-Mg and Ca-Na-K ternary diagrams. Their data indicate that tin granites compared to barren granites are enriched in SiO_2 relative to Al_2O_3 and are enriched in Na + K relative to Fe, Mg, and Ca.

Southwestern England is a well-known and distinct tin province, with tin trade dating from Phoenician times to present. The granitic plutons are clearly enriched in Sn, and copper, tungsten, lead, zinc, and iron have been recovered from the St. Austell granite and from the surrounding Devonian rocks (Fig. 5-6). The St. Austell pluton comprises an early porphyritic biotite-muscovite granite and a porphyritic lithium mica granite; the latter is succeeded by a non-porphyritic lithium mica granite; the final differentiate is a fluorite granite. Most of the tin deposits occur in the Devonian rocks in the south and southeast; within the pluton tin deposits are found chiefly in the porphyritic lithium mica granite and the biotite-muscovite granite. It is

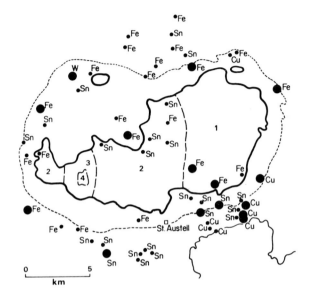

- Deposit with recorded production < 100 tons
● Deposit with recorded production >5000 tons
◯ St. Austell Granite
◌ Metamorphic

PORPHYRITIC GRANITE
1 biotite muscovite granite
2 lithium mica granite

NON-PORPHYRITIC GRANITE
3 lithium mica granite
4 fluorite granite

Fig. 5-6. Distribution of mineral occurrences associated with the St. Austell granite, southwest England (compiled from Edwards, 1976).

interesting that there appear to be no deposits in the final fluorite granite. Analyses of these various granite phases are given in Table 5-V. By any of the various criteria discussed so far, the element ratios K:Rb, Rb:Sr, and Ba:Rb indicate that the St. Austell pluton is potentially tin bearing; the Mg:Li ratio in the biotite-muscovite granite (and possibly in the porphyritic lithium mica granite) would indicate a barren pluton. Data for the Carnmenellis granite — which includes Sn — are also shown in Table 5-V; it is clearly indicated as a tin-bearing granite.

Analyses for examples of tin-bearing granites in other parts of the world are given in Table 5-VI. The most important of these is from Erzgebirge, East Germany. Tin occurs exclusively in younger Variscan granites (Eibenstock Massif in Table 5-VI), whereas the older Variscan granites (Kirchberg Massif) are barren. The mineralized granites are clearly discriminated from the barren granites by lower K:Rb ratios (by a factor of 3—4) and lower Mg:Li ratios (by a factor of 10—30). Similarly, the Carboniferous granite porphyry that is associated with the tungsten-molybdenum-tin-bismuth-base metal mineralization at Mt. Pleasant and with the Beech Hill mineralized

TABLE 5-V

Analyses for selected elements and element ratios in the St. Austell (U.K.) granite (data on St. Austell from Exley, 1958; data on the Carnmenellis adamellite from Butler, 1953)

Granite	Number of samples	K (%)	Mg (%)	Li (ppm)	Ba (ppm)	Rb (ppm)	Sr (ppm)	Sn (ppm)	K:Rb	Rb:Sr	Mg:Li	Ba:Rb
St. Austell granite												
biotite-muscovite granite	1	5.0	0.34	20	370	1120	130	—	44.6	8.6	170	0.33
porphyritic Li mica granite	30	3.8	0.19	32	83	813	48	—	46.7	16.9	59	0.10
non-porphyritic Li mica granite	1	3.9	0.04	121	74	1850	73	—	21.1	25.3	3.3	0.04
fluorite granite	3	3.8	0.05	29	43	1213	44	—	31.3	27.6	17	0.04
Carnmenellis-adamellite	1	4.3	0.28	400	110	400	80	45	10.8	5.0	7.0	0.28

TABLE 5-VI

Selected element contents and ratios for granitic rocks of the South Mountain Batholith, Nova Scotia (data from Smith and Turek, 1976); around the Mt. Pleasant W-Mo-Bi-Sn base metal deposit, New Brunswick (data from Dagger, 1972); and from the Erzgebirge (from Tischendorf, 1973)

Granites*	K (%)	Mg (%)	Li (ppm)	Ba (ppm)	Rb (ppm)	Sr (ppm)	Sn (ppm)	K:Rb	Rb:Sr	Mg:Li	Ba:Rb
Nova Scotia granites:											
Halifax pluton, biotite granite (B)	—	—	116	274	188	84	10	175	2.24	—	1.80
Halifax pluton, muscovite-biotite granite (B)	—	—	109	191	252	34	15	169	7.41	—	0.72
Halifax pluton, alaskite (B)	—	—	83	143	292	29	18	125	10.0	—	0.44
West Dalhousie pluton, biotite granite (B)	—	—	69	502	229	74	7	180	3.09	—	2.11
West Dalhousie pluton, alaskite and leucogranite (B)	—	—	70	369	293	64	11	146	4.58	—	1.46
New Ross pluton, biotite granite (M)	—	—	—	201	359	12	18	107	29.9	—	0.56
New Ross pluton, alaskite and leucogranite (M)	—	—	195	83	550	37	28	68	14.9	—	0.15
New Brunswick granites											
St. George granite (B)	3.5	0.80	—	—	228	197	8.7	153	1.5	—	—
Beech Hill adamellite (M)	3.9	0.10	—	—	522	36	9.6	75	14.5	—	—
Mt. Pleasant granite porphyry (M)	5.9	0.31	—	—	742	13	68	80	57.1	—	—
Erzgebirge, Eibenstock Massif:											
Main intrusion (M)	3.9	0.11	370	—	870	—	27	45	—	3.0	—
1st additional intrusion (M)	3.7	0.07	435	—	1220	—	37	30	—	1.6	—
2nd additional intrusion (M)	3.8	0.07	490	—	1220	—	32	31	—	1.4	—
Erzgebirge, Kirchberg Massif:											
Main intrusion (B)	3.7	0.45	110	—	255	—	14	145	—	41	—
1st additional intrusion (B)	3.7	0.29	100	—	340	—	16	109	—	29	—
2nd additional intrusion (B)	3.8	0.21	75	—	400	—	10	95	—	28	—

M = mineralized; B = barren.

adamellite in New Brunswick (Canada) is clearly distinguished from the nearby barren Devonian St. George granite by Rb:Sr ratios. The distinction is far less clear between the New Ross Pluton in Nova Scotia (Canada), which contains a number of tin occurrences, and the barren Halifax and West Dalhousie plutons (also in Nova Scotia); the Sn content and the K:Rb and Rb:Sr ratios, however, do indicate that the New Ross Pluton should be considered as potentially tin-bearing.

DISCUSSION OF AN INTERPRETATIVE PROCEDURE

The distinction between tin-bearing and barren granites *within any particular region* apparently can be quite reliably determined on the basis of one or more of the ratios K:Rb, Mg:Li, and Ba:Rb. Very low ratios for K:Rb, Mg:Li, and Ba:Rb and very high ratios of Rb:Sr are indicators of tin mineralization in all regions — although not *all* tin granites are thus indicated by all ratios.

To better assess the usefulness of these ratios they are summarized in Table 5-VII. The data are divided into three groups for each ratio and are plotted on Fig. 5-7. The three groups are:

— *Tin granites.* These are divided into granites where tin is or has been produced, and granites that have tin occurrences but which have not been exploited.

— *Barren granites.* These are divided into granites that have no mineralization of any kind, and those that have mineralization of other metals.

— *Average granites.* These are divided into global averages (granodiorite, granite, and low-Ca granite) and some of Tauson's (1974) granite types.

Only four of the tin granites overlap with the global average granites for any ratio. Since many of the barren granites fall on the tin-bearing side of the global averages, it is more useful to examine the overlap between the tin granites and the barren granites. Those tin granites that overlap with barren granites are summarized in Table 5-VIII. Two granites would be rejected as tin-bearing on the basis of two ratios; another ten would be rejected on the basis of one ratio (it is unfortunate that complete data are not available for the four ratios on all granites).

From these relations there seems to be a reasonable expectation that tin granites could be discriminated on the basis of combining the variations in K, Mg, Ba, Li, Rb, and Sr through a multivariate statistical technique such as discriminant analysis or by the Soviet technique of multiplicative ratios (see Chapter 3). A multiplicative ratio designed to give large numbers for tin granites and small numbers for barren granites would be:

$$Rb:Sr \times Rb:K \times Li:Mg \times Rb:Ba = (Rb^3 \times Li)/(Mg \times Ba \times Sr \times K)$$

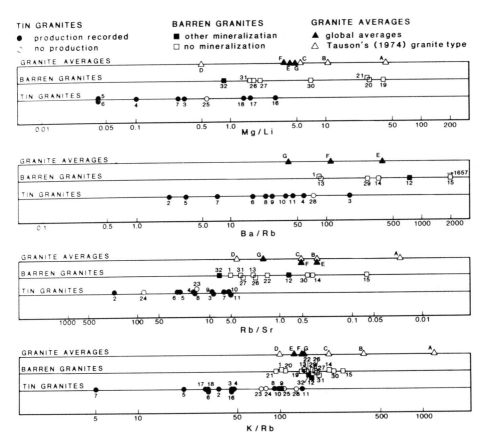

Fig. 5-7. Ratios of Ba:Rb, Mg:Li, Rb:Sr, and K:Rb for mineralized and barren granites. Numbers and letters refer to granites listed in Table 5-VII.

As an illustration, multiplicative ratios for K:Rb, Rb:Sr, and Mg:Li, calculated as $(Rb^2 \times Li)/(K \times Mg \times Sr)$, with the element contents expressed in percent for all of the granites for which there are analyses available are illustrated in Fig. 5-8. The ratios have a range of 7 orders of magnitude. There is only one reversal in the sequence: the Anchor Mine "barren" porphyritic biotite granite-adamellite has a higher ratio than the Esmeralda granite. This is not significant because, as noted above, samples of the "barren" granite were taken from the immediate vicinity of the Anchor Mine and the only significant production from the Esmeralda granite has been gold, not tin. If it is therefore assumed that the "barren" granite-adamellite at the Anchor Mine is a tin-bearing host, a threshold value of the multiplicative ratio $(Rb^2 \times Li)/(K \times Mg \times Sr)$ would be about 500. This type of approach is obviously very promising; given the apparent variation in the Ba:Rb ratio as

TABLE 5-VII

Summary of element ratios and Sn content in granitic rocks

Granites	Number in Figs. 5-7, 5-8	K:Rb	Rb:Sr	Mg:Li	Ba:Rb	Sn (ppm)	Mineral- ization*
Anchor Mine							
porphyritic biotite granite-adamellite	1	107	4.9	48	—	9	B
biotite-muscovite granite	2	37	207	2.2	—	49	A
St. Austell							
biotite-muscovite granite	3	45	8.6	170	0.33	—	A
porphyritic lithium mica granite	4	47	17	59	0.10	—	A
non-porphyritic lithium mica granite	5	21	25	3.3	0.04	—	A
fluorite granite	6	31	28	17	0.04	—	A
Carnmenellis adamellite	7	8	5.0	7.0	0.28	45	A
Elizabeth Creek granite	8	92	16	23	—	5	A
Finlayson granite	9	102	9.1	27	—	16	A
Mareeba granite	10	99	4.1	38	—	9	A
Esmeralda granite	11	143	3.7	44	—	5	A
Almaden granite	12	165	0.74	758	—	<4	b
Herbert River granite	13	149	2.3	86	—	<4	B
Dumbano granite	14	220	0.34	350	—	<4	B
Dido granite	15	227	0.06	1657	—	<4	B
Erzgebirge, Eibenstock,							
main	16	45	—	—	3.0	27	A
1st additional	17	30	—	—	1.6	37	A
2nd additional	18	31	—	—	1.4	32	A
Erzgebirge, Kirchberg,							
main	19	145	—	—	41	14	B
1st additional	20	109	—	—	29	16	B
2nd additional	21	95	—	—	28	10	B
St. George granite	22	153	1.5	—	—	8.7	B
Beech Hill adamellite	23	75	15	—	—	9.6	a
Mt. Pleasant granite porphyry	24	80	57	—	—	68	a
New Ross biotite granite	25	107	30	—	0.56	18	a

Halifax biotite granite	26	175	2.2	—	1.8	10	B
West Dalhousie biotite granite	27	180	3.1	—	2.1	7	B
Granitoids related to Li, Be, Sn, W or Ta	28	130	—	75	—	—	A
Granitoids not related to Li, Be, Sn, W or Ta	29	170	—	270	—	—	B
Lettermullan, Carna granite	30	240	0.4	—	7.2	—	B
Lettermullan, Murvey granite	31	189	3.5	—	1.6	—	B
Lettermullan, garnetiferous Murvey granite	32	160	7.1	—	0.83	—	b
Average background granite							
plagiogranite	A	1250	0.02	—	45	2.7	
ultrametamorphic leucogranite	B	386	0.3	—	11	2.6	
palingenic granite	C	236	0.5	—	5.4	5.3	
plumasitic leucogranite	D	100	4.0	—	0.5	6.3	
granodiorite	E	227	0.3	380	4.5	2	
granite	F	241	0.5	110	4.1	3	
low-Ca granite	G	247	1.7	40	4.9	3	

* B = barren of all mineralization; b = no Sn mineralization, but other metals present; A = economic Sn mineralization; a = Sn mineralization; occurrence.

TABLE 5-VIII

Tin-bearing granites that fall into barren classification on the basis of one or more ratios*

Granite	No. on Table 5-VII, Figs. 5-7, 5-8	K:Rb	Rb:Sr	Mg:Li	Ba:Rb
St. Austell, biotite- muscovite granite	3			X	
Carnmenellis adamellite	7		X		
Elizabeth Creek granite	8	X			n.d.
Finlayson granite	9	X			n.d.
Mareeba granite	10	X	X		n.d.
Esmeralda granite	11	X	X		n.d.
Erzgebirge, Eibenstock					
main	16		n.d.	n.d.	X
1st additional	17		n.d.	n.d.	X
2nd additional	18		n.d.	n.d.	X
New Ross biotite granite	25	X		n.d.	
Granitoids related to Li, Be, Sn, W, or Ta	28	X	n.d.		n.d.

* x = classified as barren; n.d. = no data.

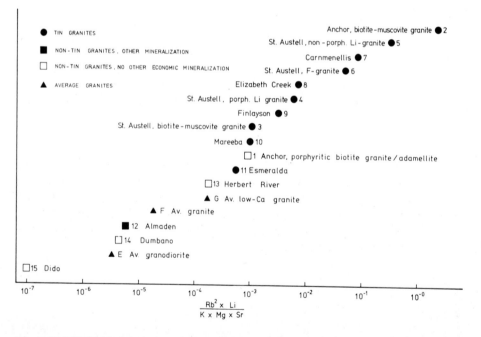

Fig. 5-8. Discrimination between mineralized and barren granites by multiplicative ratios. Numbers and letters refer to granites listed in Table 5-VII.

a function of the tin-bearing capacity of a granite, its introduction into the multiplicative ratio may be expected to improve the discrimination even further. (See also the use of K, Rb, and Sr in Chapter 8.)

CONCLUSIONS

The data presented in this chapter indicate that the development of geochemical relations to unequivocally discriminate between granites that give rise to tin mineralization and those that do not is tantallizingly close. In the majority of case histories examined in this book it is clear that interpretation of rock geochemical data demands a multi-element approach. The tin problem is not an exception. Certainly a high Sn content (greater than 15—20 ppm) and a high variance for the distribution of Sn in granite is indicative of a high probability for the occurrence of associated tin mineralization. It should be noted, however, that a low tin content does not exclude the possibility of associated tin mineralization.

Analysis for a number of major and trace elements — specifically Na, K, Mg, Ca, Fe, Rb, Sr, Li, and Ba — and calculation of discriminant functions or multiplicative functions on the basis of all of the elements should provide an unequivocal discriminator. Such a discriminator would be characterized by high ratios of Rb:Sr, (Na + K):(Ca + Mg), and (Na + K):Fe, and by low ratios for K:Rb, Mg:Li, and Ba:Rb. It is unfortunate that comprehensive data are not available in the published literature on a sufficient number of examples to demonstrate this conclusion. The extremely good discrimination between tin granites and barren granites based on a combination of only K:Rb, Rb:Sr, and Mg:Li ratios (see Fig. 5-8) is sufficient proof of the contention that it is probable that completely reliable discrimination would be possible with a wider range of elements.

The number of samples from granite plutons for the case histories described in this chapter varies from 1 to 73. The number of samples that should be collected is subject to the same constraints as those discussed in the Conclusions of Chapter 4, and should be calculated from tables similar to that given in Appendix 3 after an orientation survey. In the absence of orientation data, it is recommended that a minimum of 30—60 samples from an unzoned granite should be collected for the purpose of determining the potential of the granite for tin mineralization.

Chapter 6

REGIONAL SCALE EXPLORATION FOR VEIN AND REPLACEMENT DEPOSITS

INTRODUCTION

Vein deposits represent one of the smallest targets for exploration. Since they owe their origin to the introduction of material, they are commonly associated with strong wallrock alteration. On the other hand, because of their generally narrow dimensions, chemical halos are of small extent, although generally they are quite well defined. Regional scale anomalies may be expected where veins occur in areas of widespread alteration or are associated with intrusive rocks (see Chapter 5 on tin). Similarly, replacement deposits — if formed through widespread movement of material in solution — may also be expected to have regional scale anomalies associated with them. For example, the important lead-zinc deposits and the less important copper deposits in the southwest of Sardinia are clearly indicated by the distribution of Pb and Cu in the granites of the island based on a reconnaissance survey of 1 sample per 10 km^2 by Hall (1975); regional scale anomalous zones of 50—100 km for Pb and Zn correspond to the location of the main areas of mineralization. A different kind of regional anomaly occurs associated with copper mineralization within basaltic flows and in veins and fractures that cut the flows of the Coppermine River Group in the Northwest Territories in Canada. Cameron and Baragar (1971) showed that the form of the frequency distribution of Cu in the mineralized Coppermine River Group was fundamentally different from the frequency distribution of Cu in the barren but otherwise similar Yellowknife Series; although the latter rocks are inherently richer in Cu, the Coppermine River Group displays a long tail of high values indicative of sulphide segregation and hence mineralized conditions.

Because of the wide variation in form and possible genesis of vein and replacement deposits, consideration of regional exploration rock geochemical applications discussed in this chapter are based on the dimension of the anomalies, rather than on the type of mineralization, as follows: (1) anomalies greater than 6 km in extent; (2) anomalies of 3—6 km in extent; and (3) anomalies of 1—3 km in extent. The last category should strictly be

114

included in the chapter on local and mine scale exploration (Chapter 9) according to the definition of the scale of responses adopted in this book. The local anomalies around veins, however, are generally only a few tens of metres in extent, and the anomalies of 1—3 km are therefore included with the regional data to maintain a similar *relative* distinction of response to other types of deposits.

ANOMALIES OF MORE THAN 6 KM

There are a number of examples of anomalous patterns of greater than 6 km in extent. One of the largest is in the Coeur d'Alene district of Idaho, the most important silver mining area in the U.S. Mineralization occurs as replacement lead-zinc-silver veins in strongly folded quartzites and argillites of the Precambrian Belt Supergroup. The ore minerals are galena, sphalerite, and tetrahedrite; small amounts of chalcopyrite are present in most deposits. The most abundant gangue minerals are siderite and pyrite, and there are lesser amounts of magnetite, pyrrhotite, and arsenopyrite. Gott and Botbol (1973) collected 8000 soil and 4000 rock samples over an area of about 5000 km^2 (Fig. 6-1). The samples were analyzed for Cu, Ag, Te, Cd, and Zn by AAS, for Sb by wet chemistry, and for Mn, Pb, Hg, and As by a semi-quantitative spectrographic method.

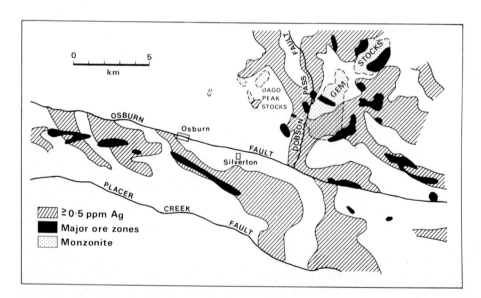

Fig. 6-1. Main ore zones and distribution of Ag in rocks, Coeur d'Alene district, Idaho, U.S. (Redrawn with permission from Gott and Botbol, 1973, *Geochemical Exploration 1972*. Institution of Mining and Metallury, 1973, fig. 3, p.4.)

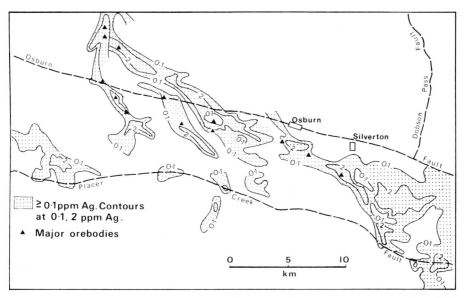

Fig.6-2. Main ore bodies and distribution of Ag in rocks prior to movement along Osburn, Dobson Pass, and Placer Creek faults, Coeur d'Alene district, Idaho, U.S. (Redrawn with permission from Gott and Botbol, 1973, *Geochemical Exploration 1972*. Institution of Mining and Metallurgy, 1973, fig. 4, p. 4.)

The data were used by Botbol and his co-workers (Botbol et al., 1977) to demonstrate the use of characteristic analysis as an effective interpretative technique for multi-element spatial data. Even single-element data — such as the distribution of Ag — clearly defines the location of the mineral deposits and quite remarkably corresponds with the postulated 26 km post-mineralization eastward movement north of the Osborn fault (see Fig. 6-2).

Watterson et al. (1977) observed that prior to faulting the largest ore deposits in the district lay in an arc around the southeastern perimeter of the Cretaceous quartz monzonite Gem-Dago Peak stocks, and that very large halos of Te, Sb, As, Pb, S, and Cd surround the pre-faulting position of these stocks. A reconstruction of the Te halo around the Gem and Dago Peak stocks (Fig. 6-3) shows an anomalous area of some 8 km in diameter; 85% of the ore mined in the Coeur d'Alene district has come from the area of the halos. Watterson and his co-workers found that a low-level Te halo surrounds the entire 780 km² area of the mining district. According to the authors the Te pattern is not a direct reflection of mineralization; the intrusion of the monzonite stocks *post-dates* mineralization (although pre-dating faulting), and the heat source from the intrusions remobilized Te and other volatile elements, redistributing them as halos around the stocks.

Watterson et al. (1977) also investigated the distribution of Te around silver-lead-zinc vein deposits in the Montezuma district, the Cripple Creek

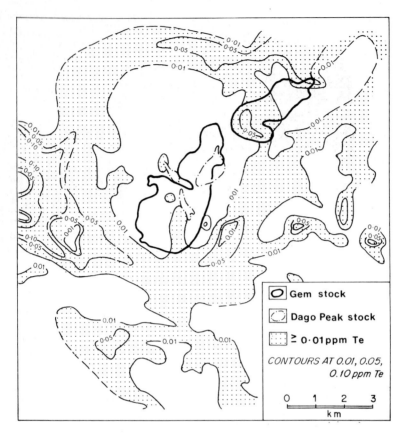

Fig. 6-3. Distribution of Te in rocks around reconstructed (pre-faulting, post-mineralization) Gem-Dago Peak stocks, Coeur d'Alene mining district, Idaho, U.S. (from Watterson et al., 1977).

gold-silver fissure vein deposits, and the Crater Creek copper-gold-lead-silver-zinc veins and molybdenum stockwork veins of Colorado (they also investigated around the copper porphyry deposits in Nevada discussed in Chapter 8). In all cases regional scale anomalies of 6—10 km in extent were found. The data for Crater Creek are shown in Fig. 6—4; within the large regional anomaly, the areas of mineralization generally coincide with higher Te contents, and the authors concluded that the Te content diminishes exponentially from the centre of the mineralization.

ANOMALIES OF 3—6 KM

The Sam Goosly and Bradina deposits in British Columbia have been described as replacement and vein deposits (Church et al., 1976), although

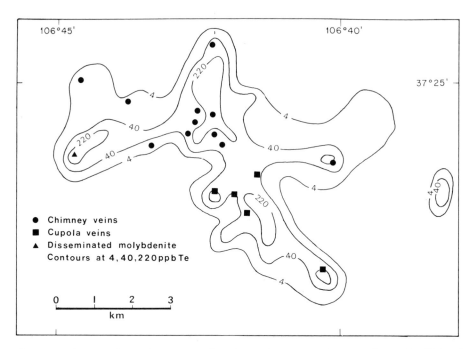

Fig. 6-4. Distribution of Te in rocks of the Crater Creek area, Colorado, U.S. (from Watterson et al., 1977).

Ney et al. (1972) have suggested that the deposits have some similarities to volcanic massive sulphide deposits. The Sam Goosly deposit was discovered by a stream sediment survey and follow-up soil surveys. The deposit is in a Mesozoic acid volcanic sequence adjacent to a Tertiary syenomonzonite intrusion (Fig. 6-5); the mineralization is massive and disseminated pyrite with chalcopyrite, tetrahedrite, pyrrhotite, and minor sphalerite and magnetite. The mineralized zone is surrounded by aluminous alteration, with scorzalites, andalusite, pyrophyllite, and corundum. At Bradina the mineralization is chiefly veins of pyrite and sphalerite, lesser chalcopyrite, and minor galena and tennantite. The host rock at Bradina is an Upper Cretaceous andesitic volcanic sequence intruded by large dykes that appear to be contemporaneous with the veins; the host rocks are altered and have wide halos of disseminated pyrite around the veins.

Church et al. (1976) suggested that the Bradina deposit is a high-level expression of mineralization related to a deeply buried intrusive stock similar to that at Sam Goosly. Their conclusions derive largely from the results of a rock geochemical survey covering about 780 km². A total of 853 samples were analyzed by AAS after a $HClO_4$-HNO_3 digestion for Cu, Pb, Zn, Co, Ni, Ag, Cd, As, Hg, Fe, Mo, and Mn. As a result of factor and cluster analysis they concluded that the intrusive rocks probably were

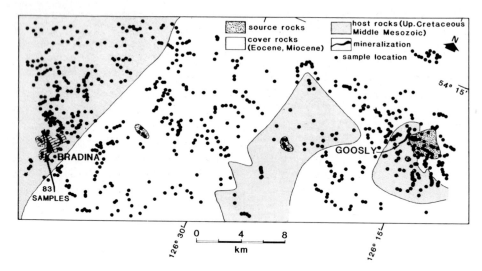

Fig. 6-5. Simplified geology, location of Bradina and Sam Goosly mineral deposits, and location of rock samples, Goosly-Owen Lake area, British Columbia, Canada (compiled from Church et al., 1976).

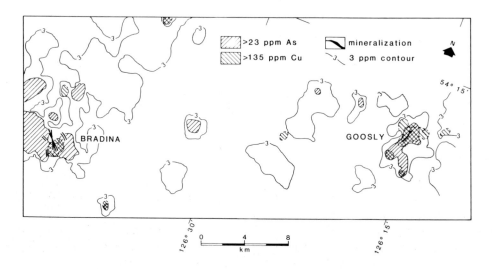

Fig. 6-6. Distribution of As and Cu in rocks around the Bradina and Sam Goosly deposits, British Columbia, Canada. (Redrawn with permission from Church et al., 1976, *Canadian Institute of Mining and Metallurgy Bulletin*, vol. 69, 1976, fig. 5, p. 94).

the source of at least the Cu in the mineralized zones (although they conceded that the intrusions may have been merely a source of heat to mobilize mineral-bearing solutions). The Cu:Pb:Zn ratios in the host rock around mineralization are also similar to the ratios in the ore.

The areal distributions of As and Cu are shown in Fig. 6-6. Comparison of Fig. 6-6 with Fig. 6-5 above shows that, broadly, the host rock formations have more than 3 ppm As, whereas the so-called cover rocks have less than 3 ppm As. The Sam Goosly and Bradina mineralized areas are characterized by extensive (4--6 km) As anomalies — the As content is more than 23 ppm. Copper anomalies of more than 135 ppm that are coincident with more than 23 ppm As are largely confined to the vicinity of the Sam Goosly and Bradina mineralized zones (there is one very small coincident anomaly northeast of Bradina and another southeast of Bradina). The Cu-As relation appears to be a useful exploration guide in this area. If it is assumed, as is generally the case, that As is much more mobile than Cu, then it is reasonable to speculate that the extensive As anomalies of more than 23 ppm that do not have associated Cu anomalies in the vicinity of Bradina may be indicative of significant deeply buried mineralization.

Somewhat smaller regional anomalies of up to 3 km in extent occur in the Taylor mining district in Nevada. Silver (with lesser amounts of antimony, gold, copper, lead, and zinc) are the main metals that have been produced in the district; the ore has come almost entirely from disseminations in jasperoid bodies. The jasperoid bodies and rhyolite porphyry dykes occur within Devonian to Permian limestone, dolerite, and shale beds; they appear to be controlled by probable Tertiary age faults and fracture zones. Lovering and Heyl (1974) postulated a buried Tertiary intrusive body as a source of both silicifying fluids and possibly of the metals.

A total of 95 samples of jasperoid (fine-grained siliceous replacement of pre-existing rock) from 23 locations in the Taylor silver mining district show geochemical halos related to the main area of mineralization (Lovering and Heyl, 1974). Anomalous concentrations were defined as an "enrichment factor" calculated as:

$$\sum x_i / nm$$

where $x_i, \ldots, x_n$ are the individual values for a particular element in n samples at a specific location, and m is the median value for the element for the entire district. Anomalous values for many elements occur within the central Taylor mining district. Enrichment factors of greater than 5 occur *only* in the vicinity of the Taylor mining district for Ag, Hg, Te, Pb, and Zn (Fig. 6-7); the distribution of Ag most clearly defines the district.

The distribution of some of the elements show strong north-south-trending anomalies (e.g., Au and Sb, Fig. 6-8), presumably controlled by the major Taylor fault system. High values of Au and Sb (accompanied by an enrichment of greater than 1.0 for Te and Hg) to the north of the main mining district and high values of Cu, Pb, and Sb (accompanied by an enrichment of greater than 1.0 for Au and Hg) to the south of the main mining district suggest two obvious targets for exploration. The eastward extension of anomalous Pb, Zn, and Au also suggests that the area to the east of the main district is prospective.

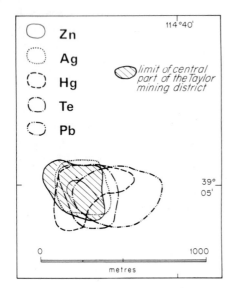

Fig. 6-7. Distribution of Zn, Ag, Hg, Te, and Pb contents in jasperoid ⩾ 5 times median values, Taylor mining district, Nevada, U.S. (compiled from data in Lovering and Heyl, 1974).

A similar study was conducted by Lovering and McCarthy (1978) in the western part of the Detroit mining district in west-central Utah. Prior to World War I, this district produced copper, gold, and silver from jasperoid bodies in fissures and fractures within Cambrian carbonate rocks. Molybdenum has been produced intermittently since 1924 from the eastern part of the district from carbonate and oxide replacement bodies in the basal carbonate beds of early Cambrian age. Lovering and McCarthy collected 660 samples from 150 localities. The samples were largely of jasperioid, although 80 samples were of limestone and dolerite host rock; some samples of float, dump material, gossans, and fracture fillings were also taken. The authors found zoned halos (up to 3 km) for Pb, Hg, Cu, Au, Bi, As, Mo, and Sn which they suggested were related in part to a shallow, buried pluton.

An unusual example of a regional scale anomaly around a vein-type deposit on the island of Mykonos in Greece was attributed to secondary processes by Lahti and Govett (1981). Silver-rich base metal sulphides occur in well defined and steeply dipping barite veins in the basement porphyritic granitoid of monzonitic to granodioritic composition (Fig. 6-9). The veins horsetail into a remnant cover that occurs on higher ground of thin andesitic-basaltic lavas and felsic pyroclastic rocks (up to 30 m thick) and heterogeneous sedimentary rocks — conglomerates, sandstones, marls, and carbonaceous calcareous rocks of probable Tertiary age. Oxidation of the veins occurs down to at least 250 m, and primary sulphides appear to be essentially restricted to zones beneath the volcanic-sedimentary capping. The only

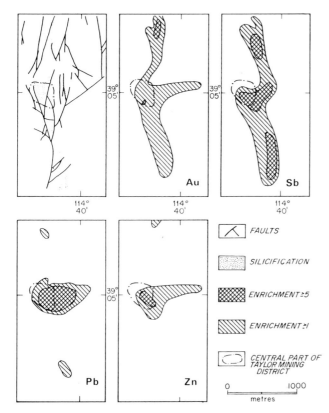

Fig. 6-8. Distribution of Au, Sb, Pb, and Zn in jasperoid showing enrichment relative to median values, Taylor mining district, Nevada, U.S. (Redrawn with permission from Lovering and Heyl, 1974, *Economic Geology*, vol. 69, 1974, fig. 4, p. 54.)

recent mining has been for barite; there is, however, evidence of small-scale mining in the 19th and early 20th century for silver-rich galena, cerussite, and anglesite, as well as iron from limonite-hematite veins.

The most extensive area of sulphide mineralization is on the eastern part of the island at Mavro Vouno where sulphides occur within a barite vein and as disseminations and stringers in the overlying volcanic-sedimentary sequence. An extensive multi-element anomaly 3 km long and up to about 1800 m wide occurs within both the volcanic-sedimentary rocks and the basement granitic rocks along a ridge (Fig. 6-10). The anomaly is crudely zoned with a central coincident anomaly of Cu, Pb, Zn, Ag, Sb, and Hg surrounded by an anomalous zone of Cu, Pb, Zn, Ag, As, and Sb; this is succeeded outwards by anomalous zones of Pb and Ag, accompanied by Cu and Zn in some cases. Drilling data indicate that the anomalous element contents away from mineralization do not persist for depths below about 50 m. The anomaly was

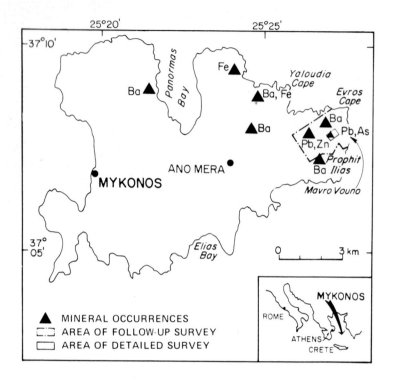

Fig. 6-9. Location of chief mineralized zones and geochemical rock surveys, Mykonos, Greece (from Lahti and Govett, 1981).

interpreted as being due to secondary processes. The outliers of volcanic-sedimentary rocks protect the underlying sulphides from oxidation and leaching, but severe oxidative weathering leaches metals from the sulphides at the outcropping contact between the basement and the volcanic-sedimentary cover along the slopes of the ridge. The clay micas and hydrous Fe and Mn oxides produced by weathering of the granitic basement allow surface fixation of the metals travelling downslope in solution.

A clear example of primary dispersion related to epigenetic copper mineralization around the Gortdrum deposit in Ireland has been described by Steed and Tyler (1979). They analyzed samples from drill core for Cu, Pb, Zn, S, and As by XRF and Hg by AAS. Mineralization occurs as disseminations and fissure fillings in highly deformed limestones and calcareous shales and sandstones at the base of the Carboniferous limestone adjacent to a major strike-slip fault. The deposit originally had 4×10^6 tonnes of 1.19% Cu and 23.5 g/tonne Ag. The principal ore minerals are bornite, chalcocite, chalcopyrite, and tennantite. A pronounced mineral zoning beyond the ore zone is shown by minor disseminations and veins; chalcopyrite becomes the dominant copper mineral, and at a distance of 2.5—

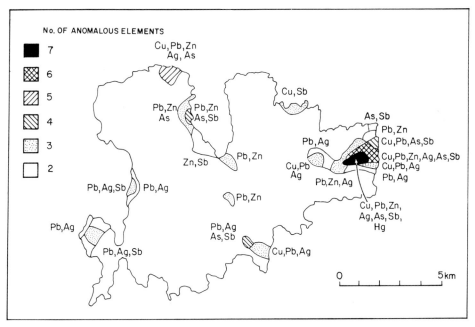

No. OF ANOMALOUS ELEMENTS

- ■ 7
- ⊠ 6
- ▧ 5
- ⧅ 4
- ▦ 3
- ☐ 2

Cu,Pb,Zn
Ag, As

Cu,Sb

Pb,Zn
As

Pb,Zn
As,Sb

As,Sb

Pb,Zn

Cu,Pb,As,Sb

Pb,Ag

Cu,Pb,Zn,Ag,As,Sb

Cu,Pb,Ag

Pb,Zn

Cu,Pb
Ag

Pb,Zn,Ag

Pb,Ag

Zn,Sb

Pb,Ag,Sb Pb,Ag

Pb,Zn

Cu,Pb,Zn,
Ag,As,Sb,
Hg

Pb,Ag

Pb,Ag
As,Sb

Pb,Ag,Sb

Cu,Pb,Ag

0   5 km

Fig. 6-10. Regional geochemical anomalies in rock, Mykonos, Greece (from Lahti and Govett, 1981).

4.5 km from the deposit there is a peripheral zone of pyrite, sphalerite, and galena with only traces of copper sulphides. The various mineral zones are not stratigraphically controlled (since in places they persist through 1000 m of varied lithologies of different ages); particular lithologies favour particular mineral assemblages, and there is a marked variation in trace element abundance (except for Hg) in different rock types. In one particular lithology (a dark, bioclastic limestone) linear anomalies of up to 5 km occur along the fault for Cu, Hg, and As. The Cu anomaly is only 400 m wide adjacent to the orebody and at right angles to the fault, whereas the Hg anomaly is 750 m wide and the As anomaly is about 1000 m wide. In the underlying Old Red Sandstone there are broad anomalies of approximately 1 by 2 km surrounding the ore zone. Anomalies for Pb and Zn are weak but conform to the mineral zoning pattern in being displaced along the major fault for about 2.5 km from the Cu·anomalies.

The anomalies are quite extensive, but care had to be taken to ensure that results from comparable stratigraphic lithological units were considered. This limited the areal coverage since not all units (except the Old Red Sandstone) are present in the entire area. This may be a case where sampling of vein material would yield even larger halos, give better areal coverage, and simplify the interpretation.

ANOMALIES OF 1—3 KM

There are numerous examples of regional geochemical anomalies of 1—3 km associated with vein deposits. Drewes (1973) deliberately collected samples from mineralized or altered rocks in the Santa Rita Mountains of Arizona, and demonstrated that anomalous concentrations of Au, Te, Hg, Bi, Sb, As, Cu, Pb, Zn, and Ag occurred spatially related to known base and precious metals occurrences. Doraibabu and Proctor (1973), on the other hand, collected representative non-mineralized rock samples at 60 sites on a regular grid over the 26 km^2 of the Tres Hermanas granite-quartz monzonite stock in New Mexico and showed regular anomalous patterns of Cu, Pb, and Zn related to vein and manto-like replacement bodies in the adjacent country rock.

Mohsen and Brownlow (1971) showed that the distribution of Mn in biotite from the granodiorite-quartz monzonite Philipsberg Batholith (in Montana) is related to the location of manganese ore deposits in the adjacent

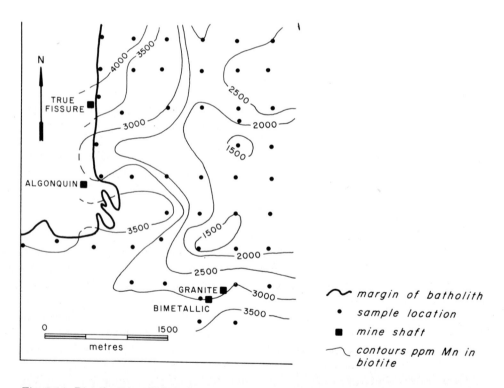

Fig. 6-11. Distribution of Mn in biotite from the western part of the Philipsberg batholith, Montana, U.S. (Redrawn with permission from Mohsen and Brownlow, 1971, *Economic Geology*, vol. 66, 1971, fig. 3, p. 614.)

sedimentary rocks. The manganese ores are replacement deposits of rhodo-
chrosite and Mn-rich dolomite in relatively pure carbonate rocks. The distribution of Mn (determined by XRF) in 64 samples of biotite from the western
edge of the batholith is shown in Fig. 6-11 (this represents only a very small
part of the batholith which has an area of about $117\,km^2$). The average Mn
content is 3194 ppm, with a range of 1400—6172 ppm. The concentration
of Mn generally increases towards the contact, although the values near the
True Fissure and Algonquin manganese deposits are generally lower than
marginal values elsewhere around the batholith (which are up to 6172 ppm).
An exception to the increase towards the margin is the old Granite-Bi-
metallic rich silver mine which lies within the intrusion; the content of Mn
appears to increase towards this also. The proportion of mafic minerals
also increases substantially towards the margin; in the area of the batholith
where detailed samples were taken the proportion of mafic minerals at the
margin near the manganese mine is twice that in the interior of the batholith,
whereas the increase towards the margin remote from the manganese mine
is much less (of the order of 13 times). There is a linear relation between Mn
in biotite and Mn in co-existing hornblende; despite this, the trend of the
distribution of Mn in whole rock is remarkably similar to that shown by Mn
in biotite (Fig. 6-12).

There are inadequate data to assess whether the location of manganese
mineralization can be unequivocably determined by the distribution of Mn
in biotite. The distribution trends are certainly significant, but they could be
related to primary features of the batholith that are not apparent from the
amount of sample coverage. The work of Tolstoi and Ostafiichuk (1963) in
the northern Caucasus (U.S.S.R.) is relevant to the problem. They showed
that in the Ashchikol'sk Massif granodiorites — which have associated copper
mineralization — that biotite and hornblende are principal carriers of Mn;
these minerals are uniformly distributed in unaltered rocks of the core. In

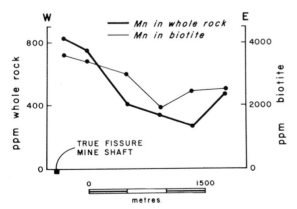

Fig. 6-12. Distribution of Mn in biotite and whole rock along an E—W traverse through
True Fissure Mine (see Fig. 6-11), western part of the Philipsberg batholith, Montana,
U.S. (compiled from Mohsen and Brownlow, 1971).

126

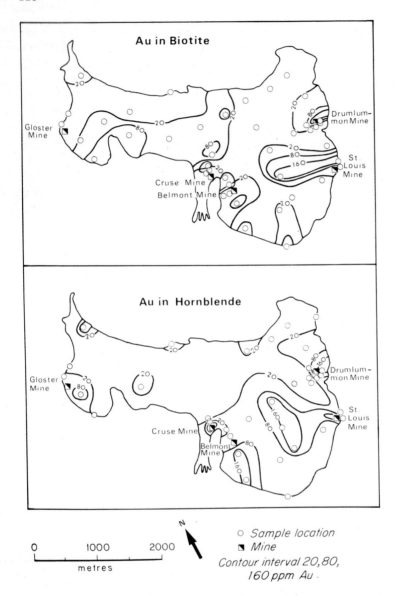

Fig. 6-13. Distribution of Au in biotite and hornblende, Marysville granodiorite stock, Montana, U.S. (Redrawn with permission from Mantei and Brownlow, 1967, *Geochimica et Cosmochimica Acta*, vol. 31, 1967, fig. 2, p. 230 and fig. 3, p. 231.)

altered rocks near the margin of the intrusions, the content of dark minerals is higher, they are irregularly distributed, the content of Mn increases, and the distribution of Mn is positively skewed.

Mantei and his co-workers (Mantei and Brownlow, 1967; Mantei et al., 1970) investigated the distribution of Au, Ag, and base metals in rocks and mineral separates from the Marysville stock in Montana. The granodiorite-quartz diorite intrusion has a surface area of about $8 \, km^2$; it is coarse-grained and shows a decrease in grain size towards the centre. It is intruded into limestone, argillites, and shales of the Belt Series, which are metamorphosed to hornfels for a distance of 0.8—3.3 km from the margin of the intrusion. Steeply dipping fissure gold and silver veins occur in the marginal parts of the intrusion and in the surrounding metamorphic rock. The Drumlummon vein (Fig. 6-13) is by far the richest gold vein; finely dispersed gold is accompanied by tetrahedrite and chalcopyrite. At other mines the gold is accompanied by pyrite, sphalerite, and galena (although at some mines there are no sulphides present).

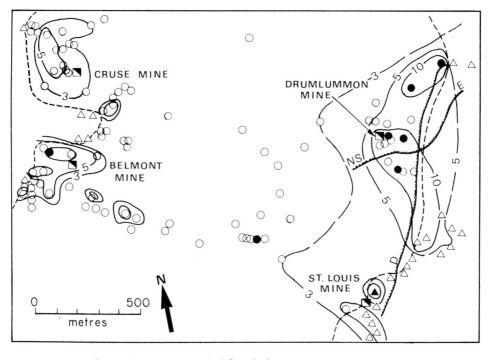

○ ● Sample in stock, > 20 ppb Au
△ ▲ Sample in country rock, > 20 ppb Au
◨ Mine
Veins. D=Drumlummon, NS=North Star, E=Empire.
Margin of stock
Contours at 3,5,10 ppb Au

Fig. 6-14. Distribution of Au in Marysville stock, Montana, U.S. (compiled from Mantei et al., 1970).

The distribution of Au (determined by neutron activation) in 44 biotite samples and 37 hornblende samples separated from samples over the stock are shown in Fig. 6-13. The highest values occur towards the margin of the stock (especially around the mines), although there are also some low values close to the mines and some high values where there is no known mineralization. The distribution of Au in magnetite is less distinctly related to mineralization; high values occur near the Belmont and Cruse mines, but the closest samples had low values. There are no high values in the vicinity of the rich Drumlummon mine.

The distribution of Au and Ag in whole rock (also determined by neutron activation) from 125 samples collected from the central portion of the stock are shown in Figs. 6-14 and 6-15. Mantei et al. (1970) reported background values of 1—2 ppb for Au an 12—18 ppb for Ag. Rocks from near known

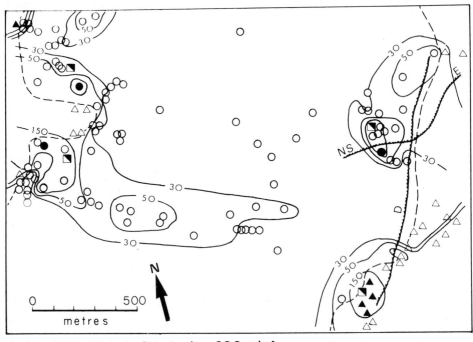

○ ●   Sample in stock, >200ppb Ag
△ ▲   Sample in country rock, >200ppb Ag
◣   Mine
◸   Veins. D=Drumlummon,N.S.=North Star,E=Empire
- - -   Margin of stock
       Contours at 30,50,150ppb Ag

Fig. 6-15. Distribution of Ag in Marysville stock, Montana, U.S. (compiled from Mantei et al., 1970).

mines and veins showed above-background values for both Au and Ag, with values increasing towards the mines by several orders of magnitude. The isolated very high value of Au (118 ppb) at the centre of the area corresponds to a highly altered zone. Copper, Pb, and Zn were also determined (by AAS); Cu shows a small increase towards the mines, but Pb and Zn show no trend.

The results of these investigations clearly indicate a local enrichment of ore elements in whole rock and minerals that is spatially related to mineralization. Moreover, whole rock analyses appear to be quite adequate to define the location of mineralization without resorting to the time-consuming process of mineral separation. It appears that a sample interval of 200 m would have located all known mineralization, although the sampling density was not adequate to be precise about this conclusion.

Mantei and Brownlow (1967) attributed the enrichment of Au to magmatic processes causing concentrations in the last crystallized outer margin of the stock, but in a later study (Mantei et al., 1970) it is suggested that the enrichment is essentially a hydrothermal process related to the veins. This latter conclusion is consistent with the results of investigations by Tilling et al. (1973) who concluded that the background concentration of Au in unaltered igneous rocks was 1—2 ppb, that Au is not concentrated by magmatic differentiation (it is generally higher in mafic than in intermediate rocks and tends to be concentrated in early crystallized mafic minerals), and that investigations at Butte, Marysville, and other regions in the U.S. show that the Au content of fresh igneous rocks does not indicate proximity to mineralization; anomalous Au values occur only in hydrothermally altered rocks. Kwong and Crocket (1978) arrived at similar conclusions based on neutron activation analysis for Au on 168 samples from mafic and felsic volcanic rocks, metasedimentary rocks, mafic-ultramafic intrusions, and felsic plutonic intrusions in the Kakagi Lake area of northwestern Ontario in Canada. They found an average background level of 1—2 ppb Au with no correlation with lithology. Anomalous (greater than 10 ppb) Au contents were encountered in rocks associated with structural features (shears, quartz veins) that permitted the transport of fluids and hydrothermal alteration.

The discovery in 1968 of the Carlin-type Cortez gold deposit in Nevada — micron-sized particles of native gold disseminated in silty dolomitic limestone and dolomitic siltstone of the Silurian Roberts Mountain Limestone — was an early triumph for exploration rock geochemistry. Erickson et al. (1964a) reported anomalous amounts of As, Sb, and W in grab samples of fracture fillings and jasperoid occurring in the limestone. In the same year Erickson and his co-workers (Erickson et al., 1964b) recognized the association of As, Sb, W, and Hg with the disseminated gold deposits of Carlin, Gold Acres, Bootstrap, and Getchell (these geochemical associations were later confirmed with detailed work at the Cortez deposit by Wells et al., 1969). Erickson then analyzed the samples from Cortez for Hg and Au; finding Au to be present in anomalous amounts, he then had additional samples collected.

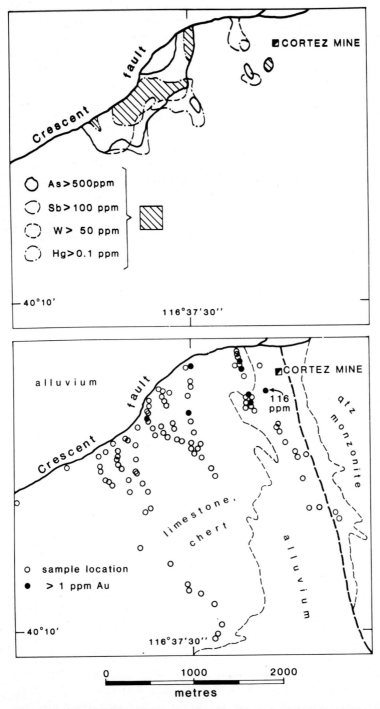

Fig. 6-16. Distribution of Au, As, Sb, W, and Hg in samples of jasperoid, fracture fillings, and shear zones around the Cortez Mine, Nevada, U.S. (compiled from Erickson et al., 1966).

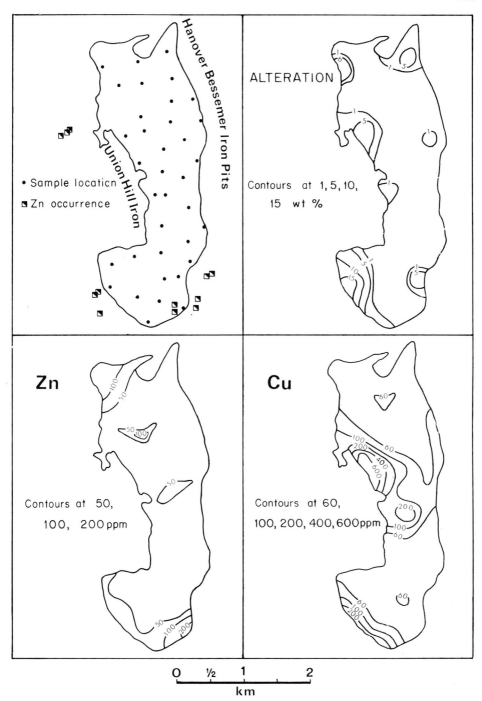

Fig. 6-17. Location of mineral occurrences, sample stations, distribution of alteration minerals (wt. % chlorite, pyroxene, epidote), and Cu and Zn in rocks of the Hanover-Fierro intrusion, New Mexico, U.S. (Redrawn with permission from Belt, 1960, *Economic Geology*, vol. 55, 1960, fig. 1, p. 1248, figs. 4 and 5, p. 1253, and fig. 6, p. 1255.) Threshold values for Cu and Zn are 64 ppm and 48 ppm, respectively.

The results from these samples — which showed Au to be present in amounts up to 116 ppm — were published in 1966 (Erickson et al., 1966) and are summarized in Fig. 6-16.

There are several important aspects of this work that must be emphasized. Firstly, the samples were not representative of the country rock nor were they intended to be; rather, they were deliberately selected samples of rock that appeared to show evidence of the possibility of passage of metal-bearing hydrothermal solutions, i.e., jasperoid, fracture fillings, and altered rock. This was, in fact a geochemical survey of the "plumbing system" of the region. The second very important point is that in no way do the geochemical anomalies outline the Cortez mine (this may, in part, be a function of the lack of outcrop in the alluvium-covered area). The initial discovery of As-Sb-W anomalies, however, provided the focus on the northern part of the area; the subsequent work indicated the presence of significant contents of Au. In fact, the content of Au generally increases to the northeast, i.e., in the direction of the Cortez mine.

The distribution of Cu and Zn in felsic intrusions that have associated mineral deposits has been described by Belt (1960) for three mining districts in New Mexico — Hanover-Fierro, Granite Mountain, and Lordsburg. Blocks of rock weighing about 2–3 kg were collected from surface outcrops, and a sample of about 0.5 kg was taken from the core of the blocks for analysis. Metal determination was by dithizone, but no details were given about the digestion technique. Data for the Hanover-Fierro intrusion are shown in Fig. 6-17. The intrusion ranges from an equigranular medium-grained tonalite restricted to the southern end of the intrusion and along the eastern and western margins, to a porphyritic, fine-grained tonalite, to a calc-alkali granite that accounts for most of the intrusion. The associated mineral deposits are magnetite replacement deposits in dolerite (on the west and east margins) and nearly pure sphalerite deposits replacing limestone around and beneath the southern lobe of the intrusion and along the northwestern margin. The distribution of Cu content increases markedly towards the south-western and western margin where the largest zinc deposits occur; the distribution of Zn shows a similar, although less distinct, pattern. The distribution of alteration minerals is also shown in Fig. 6-17 as contours for total weight per cent for chlorite and epidote and pyroxene. All the zones with a high proportion of alteration minerals are near the intrusive contact where the mineral deposits occur.

Belt (1960) stated that there is no relation between Cu and Zn distribution and primary compositional features of the intrusion; he also concluded that the distribution of Cu and Zn is not significantly correlated with the alteration as measured by the distribution of alteration minerals. Whereas there is not a complete coincidence of high Cu, Zn, and proportion of alteration minerals, it is nevertheless significant that anomalies for these three parameters generally occur close to base metal mineralization. It is concluded

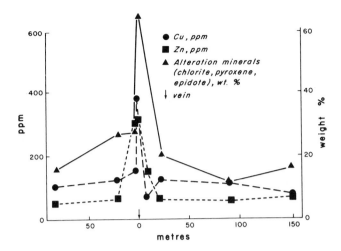

Fig. 6-18. Distribution of Cu, Zn, and alteration minerals in granodiorite intrusion adjacent to Cu-Zn vein, Lordsburg, New Mexico, U.S. (Redrawn with permission from Belt, 1960, *Economic Geology*, vol. 55, 1960, fig. 15, p. 1266.)

that the base metal anomalies are a secondary feature related to hydrothermal mineralization and do not reflect primary magmatic features.

In this particular study, in fact, it appears that alteration minerals may be a better guide to mineralization than the distribution of Cu and Zn. This is reinforced by some more detailed data from around a copper-zinc vein at Lordsburg; the distribution of Cu, Zn, and weight percent alteration minerals (Fig. 6-18) clearly indicate that visible alteration is probably more widespread than the distribution of Cu and Zn.

CONCLUSIONS

Regional scale geochemical responses associated with vein deposits are difficult to appraise in terms of their general applicability because of doubts concerning the genesis of both the veins and the geochemical halos. The very large scale responses (e.g., Coeur d'Alene) appear to be related to regional scale mineralization and probably to intrusive activity; indeed it was suggested that the Te halo at Coeur d'Alene is related to *post-mineralization* intrusion of monzonite stocks. The mineralization and geochemical halos in the Taylor and Detroit mining districts have been ascribed to buried intrusions. The geochemical dispersion patterns in the Philipsberg, Marysville, and Hanover-Fierro stocks are spatially related to vein-type deposits at or beyond the margins of the stocks. In the case of the Marysville gold and silver deposits the dispersion patterns appear to be related to hydrothermal activity that presumably gave rise to mineralization. The same is probably true of the Hanover-Fierro deposits, but the Mn anomalies in the Philipsberg stock may be related to magmatic processes (which, of course, may be genetically related to mineralization).

The available evidence indicates that vein-type and replacement deposits associated directly or indirectly with igneous intrusions or large-scale hydrothermal alteration are likely to have associated regional geochemical anomalies. The dimensions of such anomalies are broadly commensurate with the size of the mining district — large districts have anomalies of large extent, while smaller districts have smaller anomalies.

An important approach to rock geochemical surveys for vein and replacement deposits is demonstrated in a number of the studies described above — anomalies are developed strongly in vein and fracture material and altered country rock. Insofar as such rocks are samples of the "plumbing system" of an area, obviously the potential metal concentrations will be indicated by anomalous contents of ore or associated elements in the system. Given the geological nature of vein deposits (and also their small size) measurable dispersion into pre-ore country rock is small (see Chapter 9). Regional anomalies are not likely to be detected in country rock (as distinct from associated large igneous intrusions).

In terms of practical operations two situations exist:

— Deposits within or adjacent to large igneous intrusions may be expected to have extensive geochemical halos within the intrusions (which may extend into the country rock). Such halos may be detected by widely spaced representative samples of the intrusion (about 1 sample per km^2 for large mineralized districts and up to about 10 samples per km^2 for small mineralized districts).

— Deposits that are less obviously related to igneous intrusions, or where intrusions are not exposed or directly related to the mineralization, may have regional scale anomalies that can be detected in samples of fracture fillings and vein material that are not representative of the country rock but that are part of the hydrothermal plumbing system. The sample density, in part, will be dictated by the distribution and density of fractures and, therefore, the availability of samples; this is likely to be of the order of 10—50 samples per km^2.

REGIONAL SCALE EXPLORATION FOR STRATIFORM DEPOSITS OF VOLCANIC AND SEDIMENTARY ASSOCIATION

INTRODUCTION

Following Stanton's (1972) classification, the types of deposits considered in this chapter are *stratiform* sulphides — they are essentially layered and concordant with the stratification although late deformation may obscure these relations in some cases; they are also *stratabound* in the sense that they occur at some particular stratigraphic horizon. The deposits occur within rocks that range from almost wholly volcanic to sedimentary sequences with little or no evidence of volcanic activity. Specifically excluded from this classification are the stratabound sedimentary copper deposits of the Zambian Copperbelt, limestone-lead-zinc, sandstone-uranium-vanadium-copper, and conglomerate-orthoquartzite-gold-uranium deposits.

The stratiform sulphide deposits are frequently referred to as "volcanogenic" and most commonly as "massive" sulphides (a term that has no textural connotation). To avoid any possible genetic implications (and the tedious qualifiers "volcanic" and "sedimentary"), the deposits will be referred to throughout this chapter as "massive sulphides".

Exploration geochemical studies on massive sulphides are numerous and are the most successful of the rock geochemical work. Apart from the economic importance of the deposits, they have been intensively studied since the late 1960s because of the obvious close genetic association between the stratiform sulphides and their host rocks which gives the intuitive expectation that chemical differences in the host rocks should be widespread. Almost without exception the initial regional scale exploration studies have been made by professional exploration geochemists — results are not due to petrological, mineralogical, nor economic geology studies. The great advances made in understanding the genesis and geological setting of these deposits, however, has undoubtedly been a major factor in stimulating recent exploration geochemical studies, and these factors are important in focusing attention on an appropriate geological terrain and in designing rock geochemical procedures and interpretation methods. It is therefore appropriate to review the geological and geochemical features of massive sulphide deposits before considering the exploration aspects which are discussed in the remainder of this chapter.

GEOLOGY AND GENESIS OF MASSIVE SULPHIDE DEPOSITS

Massive sulphides are generally lens-shaped bodies that lie along volcanic or sedimentary stratigraphic planes with lateral dimensions up to ten times their thickness; this ideal shape may be obscured in areas of deformation. In examples of dominantly sedimentary environments (McArthur River, Northern Territory, Australia) and in some volcanic environments (Heath Steele B-zone, New Brunswick, Canada) there is no obvious "feeder" zone beneath the sulphide bodies. In the typical massive sulphide bodies of the

ARCHAEAN CANADIAN SHIELD

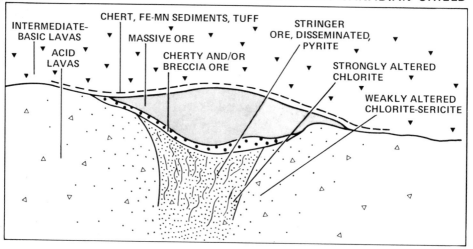

BATHURST NEW BRUNSWICK

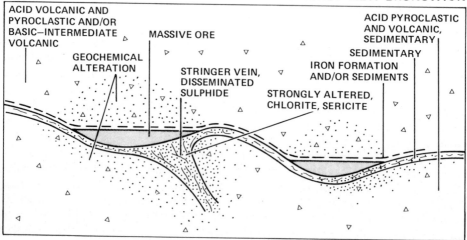

KUROKO

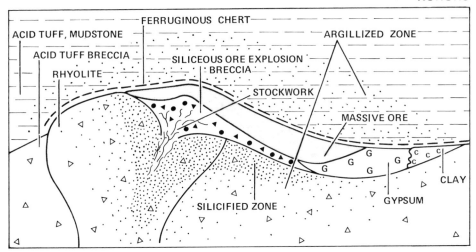

CYPRUS

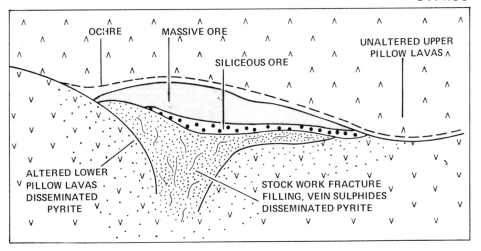

Fig. 7-1. Schematic diagram showing geological relations of major types of massive sulphides.

Canadian Shield, the Palaeozoic deposits of New Brunswick, the Kuroko deposits of Japan, and the Cyprus deposits, the massive ore is typically underlain by stringer or stockwork mineralization (Fig. 7-1).

Although their age, host-rocks, and composition differ, these four types of deposits share the common characteristic of a massive ore zone underlain by a pipe-like zone of stringer mineralization (in many cases highly siliceous);

the foot-wall rocks are altered (chloritic and sericitic) and commonly contain disseminated pyrite. Chemical sediments termed "exhalite" by Ridler and Shilts (1973) — chert and iron sulphides, oxides, and carbonates — commonly occur at the same horizons as the massive sulphides and extend for distances of many kilometers in some cases. Classification of these numerous deposits into different sub-types is difficult because of their gradational character, both in composition and geological environment.

All the deposits are dominantly iron sulphides and differ in terms of relative proportions of Cu, Pb, and Zn and also in terms of precious metal assemblages, relative proportions of volcanic and sedimentary host rocks, petrology of the associated volcanic rocks, age of deposit, and tectonic environment. There are essentially three compositional types: (1) Zn-Cu that occur in differentiated mafic to felsic volcanic rocks and are characteristic of the Archaean; (2) Pb-Zn-Cu in felsic volcanic rocks and also in largely sedimentary rocks, typically of Proterozoic and Phanerozoic age; and (3) Cu (with Zn in some examples) in mafic ophiolite rocks of Phanerozoic age. This classification, with examples, is given in Table 7-1.

Sillitoe (1972, 1973a) has derived a plate tectonic model for the environment of deposition of the Phanerozoic deposits of dominant volcanic origin (Fig. 7-2). He proposed that the copper massive sulphides of ophiolite association were generated at sites of sea-floor spreading — the ocean rises. They were then transported laterally; during subduction small slabs of the oceanic crust containing the sulphide deposits may be tectonically emplaced behind or beneath trenches. Possibly some deposits owe their present location to obduction — overthrusting onto the continental crust on island arcs. The Pb-Zn-Cu type of massive sulphides were generated by calc-alkaline volcanic activity along continental margins or island arcs underlain by subduction zones.

It is now generally accepted by most geologists that massive sulphides are related to submarine volcanism on or very close to the ocean floor. The source of the metals is ascribed to submarine hot-spring or fumarolic activity

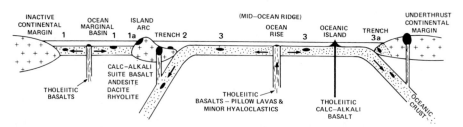

Fig. 7-2. Schematic diagram of genetic model for massive sulphides (compiled from Sillitoe, 1972, 1973a and Hutchinson, 1973). Examples of Cu-ophiolite deposits: *3a* = Cyprus, Turkey, southwest Japan, Philippines, north Italy; *1a—3a* = Notre Dame Bay (?). Examples of Zn-Cu-Pb deposits: *2* = Kuroko, Spain, Portugal, New Brunswick, Mt. Lyell, Captain's Flat.

139

TABLE 7-1

Type examples of massive sulphides and their age range* (from Lambert, 1976; Hutchinson, 1973; Sillitoe, 1972, 1973a)

Ore type	Main Rock Associations	Examples	1	2	3	4	5
Zn-Cu (Pb, Au, Ag) "Superior type"	andesite-rhyolite lavas and pyroclastics calc-alkaline	Noranda, Timmins (Canada); West Shasta (U.S.A.)	?▮---	▮?	▮		
Zn-Pb-Cu (Au, Ag) "Kuroko type"	acid tuffs and lavas mudstones, calc-alkaline	Kuroko (Japan); Bathurst (Canada); Woodlawn, Roseberry, Captains Flat (Australia); Rio Tinto (Spain)	? ▮	¦		▮	
Zn-Pb (Cu, Ag) "MacArthur type"	tuffaceous sandstones, siltstones	McArthur, Mount Isa, Hilton, Broken Hill (Australia); Sullivan (Canada); Rammelsberg (Germany)		▮	▮		
Cu (Zn, Au) "Cyprus type"	basalts, tholeiite	Skouriotissa (Cyprus); Ergani-Maden (Turkey); Notre Dame Bay, Betts Cove (Canada); Balabac Island (Philippines)		---	▮		▮

*1 = Archaean; 2 = Proterozoic; 3 = Palaeozoic; 4 = Mesozoic; 5 = Cenozoic

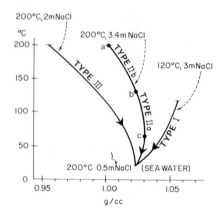

Fig. 7-3. Change in density on mixing NaCl solutions with seawater. (Redrawn with permission from Sato, 1972, *Mining Geology*, vol. 22, 1972, fig. 3, p. 34.)

of saline solutions analogous to the Red Sea brines. A useful model of sulphide deposition is from Sato (1972, 1973) who suggested that many of the variations between deposits can be accounted for by variations in temperature and salinity of ascending metalliferous brines and their behaviour on mixing with normal seawater. The possible theoretical relations calculated by Sato (1972) are illustrated in Fig. 7-3. The two end members, type I and type III, represent ore solutions that are, respectively, more and less dense than seawater. Mixing with seawater causes type I ore solution to steadily decrease in density until it is the same temperature and density as seawater; type III steadily increases in density until it is the same as seawater.

A more complicated situation is represented by type II ore solutions. They are less dense than seawater, but increase in density as they mix with seawater to a maximum (that exceeds the density of seawater at point *c*, Fig. 7-3) with continued mixing the density then decreases until it reaches the same density as seawater at infinite dilution. There are therefore two possible situations: (1) type IIa, an ore solution that has a density greater than seawater (between point *b* and *c* in Fig. 7-3) that first increases and then decreases in density on mixing with seawater; and (2) type IIb, an ore solution that has an initial density less than seawater (between point *a* and *b*, Fig. 7-3) that increases in density on mixing with seawater to the same density as seawater and thereafter follows the same course as type IIa.

Sulphide deposits precipitated from the different types of ore solutions described above will have different physical forms and also varying geographic locations relative to the source area. Similarly, the chemical and mineralogical composition is likely to differ by virtue of the early precipitation of chalcopyrite and the later precipitation of galena and sphalerite consequent upon mixing of the ore solution with seawater (Sato, 1973). The possible variations are shown in Fig. 7-4. They are:

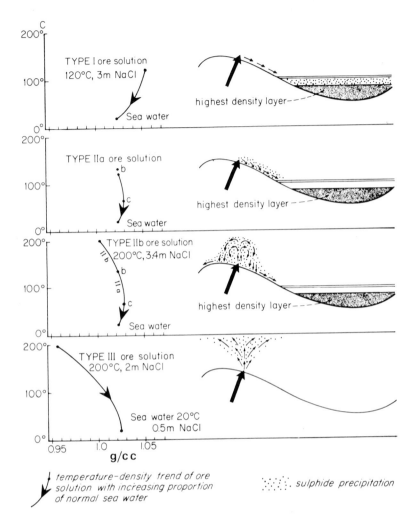

Fig. 7-4. Possible behaviour of ascending ore solutions at the sea floor as a function of initial temperature and solubility of brine. (Redrawn with permission from Sato, 1972, *Mining Geology*, vol. 22, 1972, fig. 3, p. 34 and fig. 5, p. 36.)

Type I ore solution. The ore solution has a greater density than seawater and little mixing will occur when it emerges on the sea floor. The brine will flow downslope and accumulate in topographically low areas; slow mixing will occur. The exact form of the deposit will be controlled by the sea floor topography, but generally such deposits will be thin over a large area. It is probable that the deposit would not show strong vertical zoning (Cu-rich bottom, Pb-Zn-rich top).

Type IIa ore solution. Initially the ore solution, being more dense than sea-water, will flow downslope as in the type I solution. Mixing with seawater increases the density, and new solutions will tend to rise above the mixed seawater/ore solution; therefore mixing will continue until all the brine reaches the maximum density. Sulphides will be deposited near to, and downslope from, the vent. The deposit will be thick and of limited lateral extent.

Type IIb ore solution. The ascending ore solution is less dense than seawater and will consequently rise and mix rapidly. Since the density will increase, the mixed ore solution/seawater will sink and flow downslope. Deposits are likely to be thick and of limited areal extent and could occur anywhere around the vent. It seems probable that deposits close to the vent will be relatively Cu-rich, whereas downslope deposits will be dominantly of the Pb-Zn type.

Type III ore solution. The ore solution is lighter than seawater; as mixing occurs it will steadily increase in density. Since the mixed solution can never exceed the density of seawater, the ore solution will continue to rise until it either reaches the surface or infinite dilution will take place. The sulphides will be precipitated and dispersed in the seawater, eventually settling to the sea floor. Deposits should typically be thin, of great extent, and laminated. Again, a lateral zoning should be expected, with Pb-Zn, Cu-poor deposits at the greatest distance from the vent. It is quite possible, however, to conceive of situations where, because of restricted basin configuration or current patterns, deposits of greater thickness and limited extent could occur.

Many of the local variations in composition and physical form of massive sulphides can be explained by Sato's model. Colley (1976) defined five possible types of Kuroko-type deposits based on occurrences in Fiji and Sato's model. The classification is illustrated in Fig. 7-5; although in many instances it may not be possible to place a particular deposit exactly in the classification and details of the genesis may differ to a greater or lesser extent from Sato's model, it is useful for the interpretation of geochemical data. The five types are:

Type I deposits within the feeder zone of the ore solutions characterized by stringer and disseminated sulphides with extensive alteration (sericite, chlorite, silica) of the surrounding rocks. There will be no significant massive ore below this type of mineralization because it represents the root zone of a massive ore deposit now removed by erosion. Failure to recognize this relation has resulted in futile drilling in this type of mineralization in Cyprus (Constantinou and Govett, 1973). Geochemical major element distribution patterns (described in detail in Chapters 10—12) are likely to be closely related to alteration patterns — characteristically an enrichment of Mg, K, Fe, and possibly SiO_2, and a depletion of Ca and Na. Trace element patterns will vary with the type of deposit; most deposits are likely to have an enhanced Cu content in the surrounding rocks, and the Zn-Pb (Cu) may also have enriched Zn.

Type II characteristic Archaean massive sulphide deposits that immediately

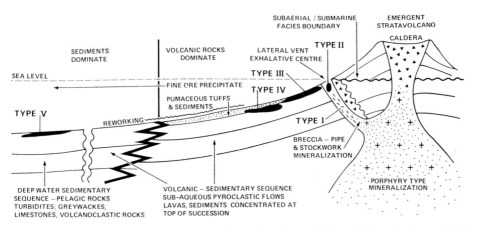

Fig. 7-5. Schematic model showing positions of various varieties of Kuroko-type sulphide deposits within a volcanic-sedimentary succession and their relation to island arc porphyry copper deposits. (Redrawn with permission from Colley, 1972, *Institution of Mining and Metallury, Transactions, Section B*, vol. 85, 1976, fig. 2, p. 192.)

overlie the exhalative centre and zone of stringer and disseminated ore. The geochemical character of the footwall rocks around the feeder zone is, of course, the same as around a type I deposit, but additional broad halos can be expected in footwall rocks lateral to the deposit, and also in the hanging-wall rocks; these halos are especially well developed in pyroclastic and sedimentary rocks. Broad regional halos of Mn and some trace elements are possible in sedimentary hanging-wall rocks from waning exhalations.

Type III which is an extremely common type of deposit (characteristic of the Kuroko deposits in Japan) that lies a little downslope from the exhalative centre. It is essentially similar to type II except for its more lenticular and bedded nature.

Type IV is completely removed from the exhalative centre. It represents deposits formed by considerable downslope migration of dense brine. This type of deposit is quite common and is being increasingly recognized (e.g., Heath Steele B zone deposit in New Brunswick, Canada; see Chapter 11). Since there is no underlying feeder zone, footwall alteration is weak and limited to the effects of downward percolating brine; similarly, hanging-wall alteration is limited to the effects of upward moving waters of compaction. Geochemical halos are correspondingly weaker, although they are of the same character as type III. As pointed out by Colley (1976), this type of deposit could also arise from weak and diffuse fumarolic activity.

Type V is a hypothetical deposit in Colley's classification that is an extreme case of type IV deposits where the distance from the exhalative source is so great that a volcanic association is ill-defined. Deposits of this type would arise from a concentration by sulphide precipitates from Sato's type III brine (see above), and also, according to Colley, from reworking and slumping of deposits from the edges of marine basins into deeper waters. The McArthur River deposit in Australia has some characteristics that fit this type of deposit. Associated geochemical halos are likely to be very broad and diffuse.

In practice it may be difficult to classify a particular deposit into Colley's (1976) or Sato's (1972) scheme unless there are very detailed geological data. Such data are not available prior to discovery but awareness of the possibilities is vital to exploration. The purpose of the above review of the genesis of massive sulphides is to emphasize the variability of their geological relations and hence the variability of their geochemical response. This can be important in regional exploration and is of particular concern in local and mine scale exploration for massive sulphide deposits (see Chapters 10—12).

ARCHAEAN MASSIVE SULPHIDES OF THE CANADIAN SHIELD

General geochemical character

The massive sulphide deposits are the most typical form of mineral deposit of the Archaean rocks of the Canadian Shield. There have been a number of studies designed to develop regional scale rock exploration techniques to find these deposits. A number are discussed here.

The first summary of data (Sakrison, 1971) indicated only that regional Zn enrichment may be a diagnostic feature of productive volcanic belts. This has, to a large extent, been substantiated as a generalization not only for Shield deposits but for massive sulphides of other ages elsewhere (e.g., Cyprus, see below).

Descarreaux's (1973) study of the Abitibi volcanic belt, the largest in the Superior Province of the Canadian Shield, is of fundamental importance. On the basis of samples collected along a section from Matagami in the north to Noranda in the south, Descarreaux concluded that Zn-Cu massive sulphide deposits occur exclusively in association with calc-alkaline rocks. This conclusion, together with the characteristic occurrence of the massive sulphides towards the top of acid flows that mark the end of a differentiated sequence, would be of importance in regional assessment of potentially productive ground if it could be shown to be generally applicable.

Descarreaux calculated that in the Abitibi belt 44% of the volcanic rocks are basalt (about one-half of which are calc-alkaline), 26% are andesite (about three-quarters of which are calc-alkaline), and the remaining more acidic rocks are predominantly calc-alkaline. He also indicated that volcanic and pyroclastic rocks most directly associated with the sulphide bodies showed depletion of Na and enrichment of Mg (see Chapter 10). Bennett and Rose (1973) concluded that these are characteristic features of Archaean massive sulphides of the Canadian Shield. Productive volcanic cycles in the

Archaean are, in fact, broadly distinguishable from non-productive cycles by the same geochemical indicators as in the Phanerozoic, i.e. enhanced contents of Fe, Mg, and Zn and depletion of Na and Ca.

Cameron (1975) compared data from barren and mineralized Archaean volcanic rocks from the Slave Province (Northwest Territories) and from Noranda, Quebec. The characteristics of the sampled areas are summarized in Table 7-II. Major element data, shown as AFM diagrams in Fig. 7-6, clearly support Descarreaux's contention, based on his studies at Abitibi, that massive sulphides on the Canadian Shield occur in rocks that are predominantly calc-alkaline.

A tentative contrary opinion has been stated by Fox (1979) who suggested that massive sulphide deposits of the Zn-Cu (Pb) type are not confined to calc-alkaline sequences. Other workers have stated that Zn-Cu Archaean deposits occur in tholeiitic and not in calc-alkaline rocks based on work at Noranda (Spence and De Rosen-Spence, 1975) and Matagami (MacGeehan, 1977). Sopuck et al. (1980) state that felsic rocks from Zn-Cu productive areas (including Uchi Lake, Sturgeon Lake, Noranda, and Normetal) show a tholeiitic trend, whereas felsic to intermediate rocks from non-productive areas are generally calc-alkaline. Sopuck and his co-workers speculated that the distinction between tholeiitic and calc-alkaline rocks on the basis of AFM diagrams may be misleading — especially in Archaean rocks of the Canadian Shield — because of widespread alkali and iron metasomatism;

TABLE 7-II

Archaean volcanic belts sampled by Cameron (1975)

Area name, location (number of samples)	Mineralization
Indin Lake N.W.T. (109)	no massive sulphides; abandoned Au workings
Yellowknife, N.W.T. (194)	no massive sulphides; important Au
High Lake, N.W.T. (94)	4.7×10^6 tonnes 3.5% Cu, 2.5% Zn
Agricola Lake N.W.T. (40)	drill data: 37 m massive sulphide 3.7% Zn, 0.7% Pb, 1.1% Cu, 68 g Ag/tonne 1.03 g Au/tonne
Sanctuary Lake, N.W.T., 20 km SW of Agricola Lake (117)	no known mineralization; highly anomalous lake sediments
Hackett River, N.W.T., 40 km NNW of Agricola Lake (37)	9×10^6 tonnes 8% Pb-Zn, 154 g Ag/tonne; other prospects
Noranda, Quebec (701)	includes Vauze, Norbec, Waite, Amulet, Millenbach, Delbridge, Quemont, Horne Mines (see Fig. 7-9 for locations)

146

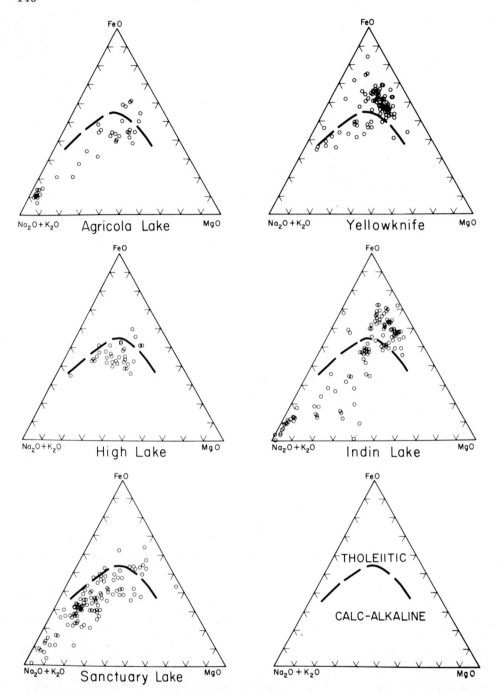

Fig. 7-6. AFM diagrams, Slave Province, Yellowknife, Canada and theoretical tholeiitic—calc-alkaline relation (from Cameron, 1975).

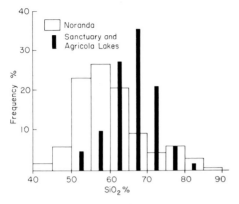

Fig. 7-7. Frequency distribution of SiO_2 in host volcanic rocks of Pb-poor deposits (Noranda, Canada) and Pb-rich deposits (Sanctuary and Agricola Lakes, Canada) (from Cameron, 1975).

also, the felsic rocks that are host to mineralization may be Fe-rich units within a general calc-alkaline succession. Clearly, the nature of the magma that is host to Archaean massive sulphide deposits on the Canadian Shield remains a matter of dispute.

There are two main types of massive sulphide deposits on the Canadian Shield: Zn-Cu; and Pb-Zn (Cu). The Noranda and High Lake deposits are Pb-poor of the Zn-Cu type, whereas the Hackett River, Agricola Lake, and Sanctuary Lake areas are associated with Pb-rich deposits of the Pb-Zn (Cu) type; the latter are generally in much more siliceous rocks than the Noranda type (this difference is illustrated in Fig. 7-7).

Cameron, in his work on the Slave Province and Noranda (see above) noted a tendency for volcanic rocks associated with the Zn-Cu type of massive sulphides to be enriched in Zn compared to barren volcanic rocks but did not find a comparable difference in rocks associated with the Pb-Zn (Cu) type of massive sulphide. He also found strong regional anomalous Zn contents in interflow shales over an area of $5180 \, km^2$ around Timmins compared with similar rocks in other areas of the southern Canadian Shield; there are four massive sulphide deposits, including the Kidd Creek deposit, in the Timmins area.

TABLE 7-III

Zinc content (ppm) of shales from the southern Canadian Shield* (from Cameron, 1975)

Sample location	Number of samples	am	s	gm	Median	90 percentile	95 percentile
Timmins	141	632	3700	165	110	500	1276
Other areas	265	123	177	104	98	154	233

* am = arithmetic mean; s = standard deviation; gm = geometric mean.

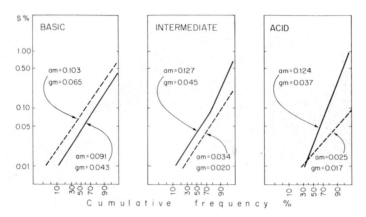

Fig. 7-8. Cumulative frequency (%) of sulphur in basic, intermediate, and acid volcanic rocks from Indin Lake (broken line) and Noranda (solid line) area, Canada (from Cameron, 1975). *am* = arithmetic mean; *gm* = geometric mean.

The results of Cameron's (1975) survey are given in Table 7-III. The distribution of Zn in areas other than Timmins is relatively uniform, with a mean value of 123 ppm compared with a mean of 637 ppm in shales from the Timmins area. The distribution of Zn at Timmins is clearly erratic. The median values for Timmins and the rest of the samples are similar, and there is not a large difference between the geometric means — the high arithmetic mean of Timmins arises from a number of samples with highly anomalous Zn, as shown by the differences in values for the 90 and 95 percentiles.

Cameron also suggested that sulphur is a useful discriminator for mineralized volcanic sequences. The distribution of S in basic, intermediate, and acidic rocks is shown in Fig. 7-8 for Indin Lake and the Noranda area. Whereas there is a decrease in S content from basic to acid rocks in the barren Indin Lake example, the reverse is true for Noranda; the main distinction is that, although there is little difference in S content between basic rocks in the two areas, there is a considerable increase in S in the acid ore-bearing rocks at Noranda. (The general validity of this difference can probably be accepted, but it should be noted that the Indin Lake samples are a mixed population of tholeiite and calc-alkaline types.)

Regional survey in the Noranda district

The rhyolite horizons in the Noranda district have been classified by Spence (1967) into five zones stratigraphically upwards and from west to east. Most of the sulphides occur in zone III, and a few in zones IV and V (see Fig. 7-9). According to J.A. Coope (personal communication, 1981), however, the interpretation of the volcanic stratigraphy is being modified: zone II is disappearing, and zone I probably hosts the Corbet deposit and the new Ansil discovery (Newmont Exploration of Canada has also found a small deposit in zone I).

149

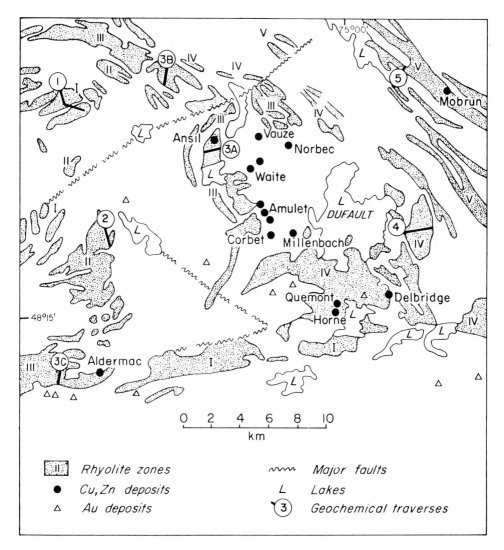

Fig. 7-9. Location of rock geochemical traverses in the Noranda area, Canada. (Redrawn with permission from Larson and Webber, 1977, *Canadian Institute of Mining and Metallurgy Bulletin*, vol. 70, 1977, fig. 1, p. 81.)

A series of seven traverses across the rhyolite zones were sampled at 15- to 30-m intervals by Larson and Webber (1977). The samples were analyzed for Si, Al, Na, K, Ca, Mg, Fe, Ti, P, Mn, Zn, Cu, S, Rb, Sr, and Zr to determine whether rhyolite zones associated with mineralization are distinguishable chemically from barren zones. All the traverses are calc-alkali rhyolites, but Larson and Webber concluded that there was no obvious chemical variation that could be used to discriminate ore-bearing rhyolitic zones from barren zones. Examination of Fig. 7-9 shows that their traverse 3A is across a

150

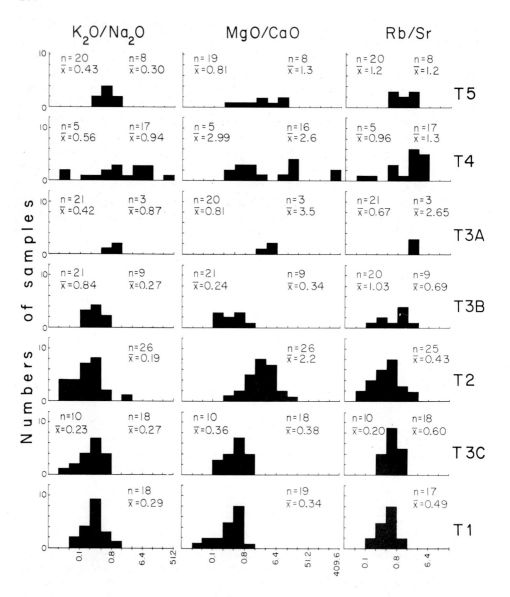

Fig. 7-10. Frequency distribution and mean values (figures on the right) of $K_2O:Na_2O$, $MgO:CaO$, and Rb:Sr ratios in rhyolite volcanic rocks, and mean values (figures on the left) for ratios in pyroclastic rocks in the Noranda area, Canada (data from R. Webber, personal communication, 1977).

zone associated with most of the sulphide deposits; traverse 4 has a number of associated sulphide deposits if those around Noranda are included; traverse 3C and 5 each have one sulphide deposit; traverses 1, 2, and 3B have no sulphide deposits in the vicinity.

Using additional data generously provided by Dr. Webber, frequency distributions for the ratios $K_2O:Na_2O$, $MgO:CaO$, and $Rb:Sr$ in rhyolite are shown in Fig. 7-10. There is a clear regional variation in these ratios that generally increases from west to east and that is generally highest in areas with the greatest number of massive sulphides. Traverses 3A and 4 (those most closely associated with the majority of the sulphide deposits) clearly have anomalously high ratios for $K_2O:Na_2O$, $MgO:CaO$, and $Rb:Sr$. Traverse 5 is also anomalous in $Rb:Sr$. In terms of decreasing anomalous character (skewness, mode, and mean) based on the three ratios, the traverses are ranked as follows: $4 \gg 5 > 3B > 2 > 3C > 1$.

Traverse 3A is omitted from the series because unfortunately there are only three samples of rhyolite; most of the traverse is in pyroclastic material. Accordingly, mean values for pyroclastics are also given in Fig. 7-10. Two of the traverses that have no associated massive sulphides — traverses 1 and 2 — also have no pyroclastic material, which is a significant exploration guide (Larson and Webber, 1977). The pattern is not nearly as clear as for rhyolite (which may be expected since pyroclastics are much more prone to reflect secondary processes than rhyolite). The ranking, in decreasing order of anomalous character, is: $4 > 5 > 3B > 3A > 3C$.

Combining the results of the three ratios for both rhyolites and pyroclastics, the anomalous ranking is: $4 > 5 > 3B > 3A > 2 > 3C > 1$. Although this pattern could be the result of the sample distribution (R.G. Webber, personal communication, 1977), it is significant that proximity to sulphide mineralization is marked by increases in $K_2O:Na_2O$ and $MgO:CaO$ ratios — a trend that is consistent with relations elsewhere. The increase in the $Rb:Sr$ ratio is particularly interesting and suggests that it may also be useful in regional exploration in acid volcanic rocks (see data and discussion of its use in Chapters 8 and 10).

Webber has re-examined his data by correspondence analysis (Dumitriu et al., 1979). The main conclusion from this study revealed that traverses 4 and 3B appear to be of the greatest potential interest; traverse 1 is of the least potential interest. This ranking is similar to the order derived from the histograms and means shown above and differs only in placing traverse 3B rather than traverse 5 as the second priority.

Discrimination between barren and productive volcanic cycles

Davenport and Nichol (1973) produced significant diagnostic data on discrimination between barren and productive volcanic cycles from their work in the Birch-Uchi volcanic-sedimentary belt that covers an area of

152

about $780\,km^2$ in the Superior Province northwest of Lake Ontario. Drill core and 300 1-kg chip surface rock samples from an area of $186\,km^2$ at a density of about 1.5 samples/km^2 around the South Bay Mines Zn-Cu-Ag massive sulphide deposit (0.6×10^6 tonnes 10.02% Zn, 1.66% Cu, and 65 g/ tonne Ag) were collected and analyzed for a variety of elements to determine whether the Confederation Lake volcanic cycle that is host to the ore body could be distinguished geochemically from the underlying Woman Lake volcanic cycle. Both volcanic cycles commence with basic rocks that are successively succeeded upwards by intermediate and acid rocks. The South Bay Mines ore body — and some sub-economic occurrences in the Confederation Lake cycle — are associated with acid tuffs, quartz feldspar porphyry, and argillite, whereas sulphide occurrences (essentially barren pyrite and pyrrhotite) in the Woman Lake cycle are associated with acid to intermediate tuffs and clastic sediments.

The broad differences in element contents of the two volcanic cycles are summarized in Fig. 7-11. Both Fe_2O_3 and Zn are higher in intermediate rocks of the mineralized Confederation Lake cycle than in the barren Women Lake cycle. There is, however, an almost complete overlap in the

ELEMENT	<56% (BASIC) W.L.	C.L.	57–64% (INTERMEDIATE) W.L.	C.L.	>66% (ACID) W.L.	C.L.
SiO2						
Fe2O3				◉		
K2O			●		●	
MgO			○			
Mn					○	
V	○		●		●	
Cr			●		●	
Co			●		◉	
Ni			●		●	
Cu			○		○	
Zn				●		●
Mean Zn, ppm	79	105	74	134	63	128

○ Higher at 95% confidence level
◉ Higher at 99% confidence level
● Higher at 99.8% confidence level
W.L. Woman Lake Cycle
C.L. Confederation Lake Cycle

Fig. 7-11. Comparison of distribution of elements in corresponding rock types between the ore-bearing Confederation Lake cycle (*C.L.*) and the barren Woman Lake cycle (*W.L.*) (compiled from Davenport and Nichol, 1973).

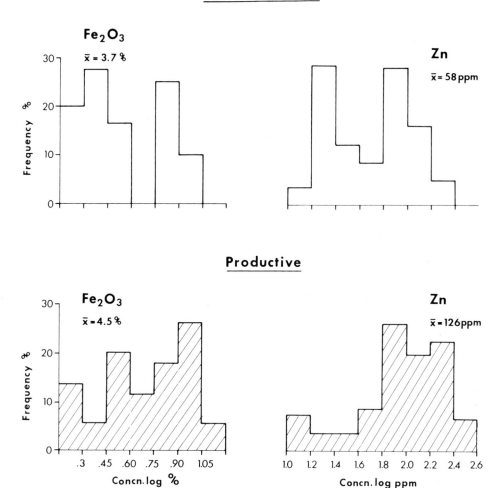

Fig. 7-12. Frequency distribution of Fe_2O_3 and Zn in productive and non-productive cycles of volcanism, Uchi Lake, Canada. (Reproduced with permission from Govett and Nichol, 1979, *Geophysics and Geochemistry in the Search for Metallic Ores*, *Geological Survey of Canada*, *Economic Geology Report*, 31, 1979, fig. 15-10, p. 350.)

two populations (Fig. 7-12) although the form of the frequency distributions is quite different. The significant differences between the two cycles is better seen if the degree of fractionation (as represented by the SiO_2 content) is taken into account. The Fe_2O_3 (Fig. 7-13) and Zn (Fig. 7-14) contents of rocks from productive cycles are clearly higher than those from the non-productive cycle.

154

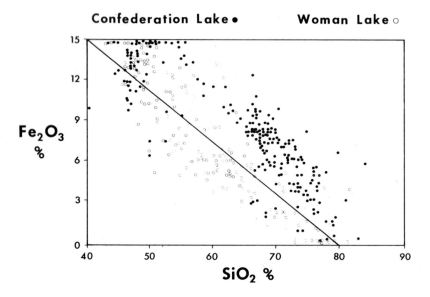

Fig. 7-13. Relation between Fe_2O_3 and SiO_2 contents of productive and non-productive volcanic cycles, Uchi Lake, Canada. (Reproduced with permission from Govett and Nichol, 1979, *Geophysics and Geochemistry in the Search for Metallic Ores*, *Geological Survey of Canada*, *Economic Geology Report*, 31, 1979, fig. 15-11, p. 351; original data from Nichol, 1975.)

Variation in SiO_2 content

Although Zn shows only moderate differences in concentration due to petrological variations (as pointed out by Cameron, 1975), Wolfe (1975) assembled an impressive amount of data to show that Zn is a much more useful discriminator of mineralized volcanic rocks if account is taken of the degree of fractionation; it should be remembered that in background rocks Zn, unlike Cu — which occurs mostly as chalcopyrite — occurs mainly as a substitute for Fe in ferromagnesian silicates and oxides.

Fig. 7-15 shows the distribution of Zn in 1575 Archaean rocks as a function of SiO_2. The values for Zn vary between 50 and 200 ppm and generally show a positive correlation with Fe; in intermediate rocks (52—60% SiO_2) Zn remains constant at about 85 ppm; in rocks with more than 60% SiO_2 the Zn decreases linearly with increases in SiO_2 to about 35 ppm in rhyolites containing about 75% SiO_2. Wolfe (1975) pointed out that for all practical purposes rocks that contain more than 10% total Fe as FeO are tholeiitic and barren of sulphides (however, see discussion above); he concluded that it is only necessary to consider samples with less than 10% Fe. The variation of Zn as a function of SiO_2 for such rocks is shown in Fig.

Confederation Lake •   **Woman Lake ○**

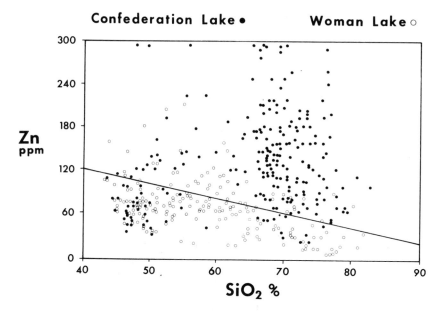

Fig. 7-14. Relation between Zn and SiO$_2$ contents of productive and non-productive volcanic cycles, Uchi Lake, Canada. (Reproduced with permission from Govett and Nichol, 1979, *Geophysics and Geochemistry in the Search for Metallic Ores, Geological Survey of Canada, Economic Geology Report*, 31, 1979, fig. 15-12, p. 351; original data from Nichol, 1975.)

7-16 which shows that Zn remains essentially constant at 85 ppm up to 60% SiO$_2$ (with a maximum upper range of about 125 ppm) and thereafter declines with increasing SiO$_2$. It may be noted that Sopuck et al. (1980) showed that Zn-Cu massive sulphide deposits of the Canadian Shield are essentially restricted to volcanic sequences with more than 68% SiO$_2$. The distribution of Zn and SiO$_2$ in samples from the Upper and Lower Volcanic Cycles at Uchi Lake (presumed to be equivalent to Davenport's and Nichol's (1973) Confederation and Woman Lake Cycles, respectively) is shown in Fig. 7-17. All of the samples from the Upper Cycle (associated with the South Bay Mines deposit) plot in the anomalous zone; seven of the 21 samples from the Lower Cycle (which contains uneconomic sulphide deposits) are also anomalously high in Zn.

Mean SiO$_2$ and Zn contents (from Nichol et al., 1977) of the acid volcanic rocks of the five cycles of volcanism at Sturgeon Lake are also plotted in Fig. 7-17. Each cycle of volcanism begins with mafic to intermediate massive flows, pillow lavas, and lesser tuffs; these rocks grade up into less abundant dacite and rhyolite tuffs. The top of the cycle is commonly marked by minor amounts of sedimentary material. The Mattabi ore body (5.9 × 10^6 tonnes of 6.89% Zn, 0.7% Cu, 0.68% Pb, and 83.4 g/tonne Ag) is within rhyolite of cycle 1. Three smaller deposits lie at the contact of cycle 2

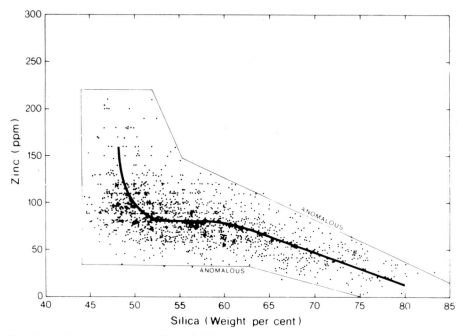

Fig. 7-15. Relation between Zn and SiO$_2$ content in 1575 samples of Archaean volcanic rocks (from Wolfe, 1975).

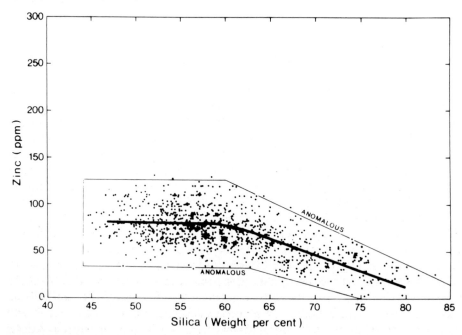

Fig. 7-16. Relation between Zn and SiO$_2$ content in Archaean rocks containing less than 10% FeO (as FeO + 0.9 Fe$_2$O$_3$) (from Wolfe, 1975).

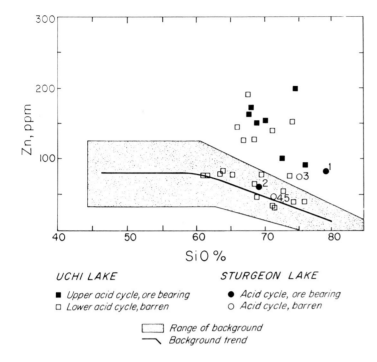

Fig. 7-17. Variation in Zn and SiO_2 in Archaean volcanic rocks containing less than 10% total Fe as FeO (compiled from Wolfe, 1975 and Nichol et al., 1977).

rhyolites and overlying cycle 3 andesites. The largest of these is the Sturgeon Lake orebody, estimated to have 1.2×10^6 tonnes of 9.06% Zn, 3.01% Cu, 1.1% Pb, 170.9 g/tonne Ag, and 0.63 g/tonne Au. On the basis of Wolfe's $SiO_2 : Zn$ discriminator discussed above, cycle 1 is clearly anomalous. Cycle 2, however, is similar to the non-productive cycles 4 and 5, whereas cycle 3 is anomalous.

This classification is based on arithmetic mean analyses and gives no indication of the proportion of samples in each cycle that might be anomalous; hence Nichol et al., (1977) calculated the residual values of Zn, Fe_2O_3, and Na_2O in samples after regression against SiO_2 as an indicator of the differentiation index (the residuals are the difference between the actual analyses and the calculated amount of the component attributable to the normal petrology as measured by the SiO_2 content). Residual threshold values were selected to provide the optimum discrimination between productive and non-productive cycles; the results are shown in Fig. 7-18. In terms of Zn, Fe_2O_3, and Na_2O residuals, cycle 1 is clearly the most anomalous, whereas cycles 4 and 5 are essentially background. Both cycle 2 and cycle 3 are also anomalous and, interestingly, cycle 3 is more anomalous than cycle 2.

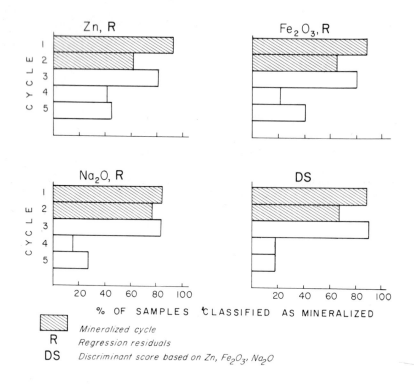

Fig. 7-18. Distribution of element residuals and discriminant scores in felsic rocks for mineralized and unmineralized volcanic cycles at Sturgeon Lake, Canada (compiled from Nichol et al., 1977).

The combined effect of the Zn, Fe_2O_3, and Na_2O residuals was estimated by the discriminant analysis score technique used by Pantazis and Govett (1973) described in Chapter 12. The results of the discriminant analysis are also shown in Fig. 7-18 and indicate even more decisively that cycle 1 is anomalous and cycles 4 and 5 are not. Whereas 65% of the samples from productive cycle 2 are anomalous, 88% of the samples from cycle 3 are classified as anomalous (compared to 87% of the samples from productive cycle 1). On the basis of these data, cycle 3 is clearly a very promising target, although exploration has not so far revealed any economic mineralization. Nichol and his co-workers (Sopuck et al., 1980) extended the SiO_2 normalizations and discriminant analysis data treatment to other localities in the Canadian Shield with similar favourable results.

Textural variation

The distinction of textural type can be extremely important (Govett and Nichol, 1979). For example, at Noranda relatively high Zn contents are

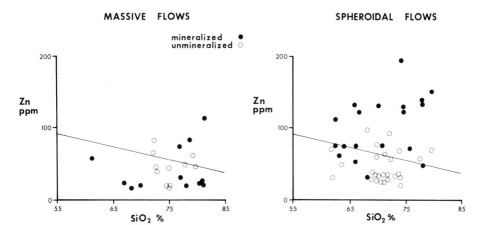

Fig. 7-19. Relation between the composition and texture of productive and non-productive cycles, Noranda area, Canada. (Reproduced with permission from Govett and Nichol, 1979, *Geophysics and Geochemistry in the Search for Metallic Ores, Geological Survey of Canada, Economic Geology Report*, 31, 1979, fig. 15-13, p. 352; original data from Nichol, 1975.)

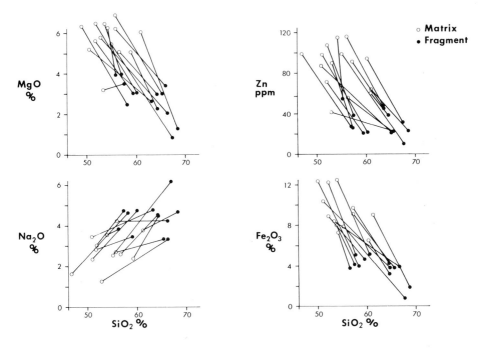

Fig. 7-20. Relation in composition of fragment and matrix of pyroclastics at Kakagi Lake, Canada. (Reproduced with permission from Govett and Nichol, 1979, *Geophysics and Geochemistry in the Search for Metallic Ores, Geological Survey of Canada, Economic Geology Report*, 31, 1979, fig. 15-14, p. 352; original data from Sopuck, 1977.)

160

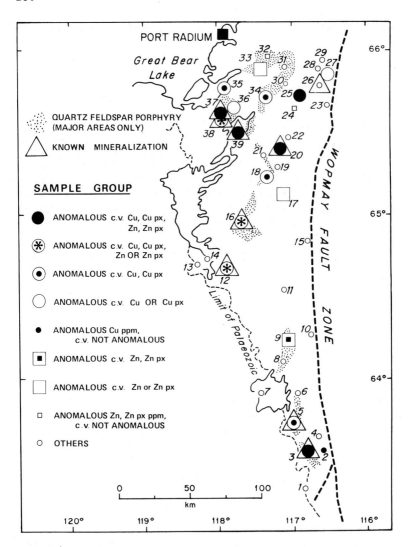

Fig. 7-21. Location of sample groups and summary geochemical results for acid volcanic rock geochemical sampling (Great Bear Province, Canada; compiled from Garrett, 1975). Numbers refer to locations mentioned in the text and in Fig. 7-22.

apparent only in spheroidal flows of the productive cycle and do not appear to exist in the presumably less permeable massive flows (Fig. 7-19). Similarly, variations in chemical composition with textural type give rise to significant variations in composition. In pyroclastic units of the non-productive Kakagi Lake area the fragments are more siliceous and soda-rich than the matrix, whereas the matrix is enriched in MgO, Fe_2O_3 (3-fold), and Zn (5-fold); see Fig. 7-20. Since these are some of the useful elements indicative of mineralization, the nature of the sample needs to be considered carefully during interpretation.

PROTEROZOIC MASSIVE SULPHIDES OF THE CANADIAN SHIELD

Regional scale sampling of acid porphyry rocks of the Proterozoic Bear Province of the Canadian Shield by Garrett (1975) also show that the distribution of Cu and Zn indicate mineralized districts, although the interpretative procedures are more complex due to a less straightforward relation between the distribution of trace elements and mineral deposits. The acid porphyry rocks — mapped as quartz-feldspar porphyries — are both extrusive and intrusive and form about 30% of the Great Bear Batholith. The acid flows and pyroclastics vary from dacite to rhyolite in composition and are locally associated with sedimentary rocks.

The most important mineralization is in the Rainy Lake area (37 and 38 in Figs. 7-21 and 7-22) where silver and copper are produced from the Terra Mine; copper occurs in stratabound bodies within the volcanic-sedimentary pile, and silver occurs in cross-cutting veins. Chalcopyrite associated with magnetite, chert, and jaspilite horizons and in quartz veins occurs at Tommie Lake (20); similar occurrences with Co and As minerals occur at Lou Lake (3). A porphyry copper-type deposit is known at Bode Lake (16). Other mineral occurrences have been noted near Mazenod (5), De Vries (10), Hardisty (12), and Blackjack (26) Lakes. It was noted by Garrett that this field area is not typical of the Archaean Shield areas discussed above where most of the massive sulphides occur and where the dominant volcanic rocks are of basic and intermediate composition with lesser acidic rocks (the reverse is the case in the Bear Province).

Duplicate samples from an average of 17 sites remote from known mineralization were collected from the locations shown in Fig. 7-21. The samples were analyzed for total Cu and Zn after $HF-HClO_4$ digestion, and for sulphide Cu and Zn (Cu px, Zn px) after a H_2O_2/ascorbic acid digestion (Lynch, 1971). Sampling and analytical variability is significantly less than regional variability.

About two-thirds of the Cu is in the sulphide form, probably in pyrite, chalcopyrite, and pyrrhotite (when it is present). There is a positive correlation between Cu and Cu px, indicating that an increase in the total Cu in the rock is accommodated in sulphides. Only about 20% of the Zn is in a sulphide form, the remainder presumably being in magnetite and ferromagnesian minerals; there is no correlation between total Zn and Zn px. There is a general increase in Zn content (total and sulphide) from south to north; this may be a feature of primary geological variation or an indication of better prospects for stratabound zinc deposits in the north.

As a measure of the geochemical relief in each designated area Garrett

Locality name	No.	Cu gm,ppm	Cu cv %	Cu px gm,ppm	Cu px cv %	Zn gm,ppm	Zn cv %	Zn px gm,ppm	Zn px cv %	Symbol on map
	1	4–8	100–150	<4	150–200	40–60	50–100	15–20	<50	○
Hump L.	2	20–40	50–100	12–16	50–100	40–60	<50	10–15	50–100	●
Lou L.	3	8–12	>250	4–8	>300	<20	100–150	5–10	100–150	▲
	4	8–12	<50	<4	<50	<20	<50	5–10	<50	○
Mazenod L.	5	8–12	>250	4–8	>300	40–60	50–100	10–15	<50	◉▲
	6	4–8	50–100	<4	100–150	20–40	<50	5–10	<50	○
	7	<4	<50	<4	50–100	60–80	<50	10–15	<50	○
	8	8–12	50–100	4–8	100–150	40–60	50–100	10–15	50–100	○
Rae L.	9	12–16	50–100	8–12	50–100	60–80	100–150	15–20	150–200	■
	10	4–8	50–100	<4	50–100	20–40	<50	5–10	50–100	○
	11	4–8	100–150	4–8	100–150	40–60	<50	10–15	<50	○
Hardisty L.	12	8–12	150–200	4–8	200–250	80–100	50–100	15–20	150–200	✳▲
	13	<4	50–100	<4	50–100	20–40	50–100	5–10	<50	○
	14	<4	<50	<4	<50	20–40	50–100	5–10	<50	○
	15	8–12	100–150	4–8	100–150	40–60	50–100	10–15	50–100	○
Bode L.	16	12–16	>250	8–12	>300	60–80	50–100	15–20	100–150	✳▲
Angle L.	17	8–12	100–150	4–8	100–150	40–60	50–100	10–15	150–200	□
	18	8–12	200–250	4–8	250–300	80–100	50–100	15–20	50–100	◉
	19	8–12	100–150	4–8	100–150	80–100	50–100	15–20	50–100	○
Tommie L.	20	>24	>250	>16	200–250	80–100	>200	>30	>200	▲
	21	4–8	50–100	<4	100–150	80–100	50–100	10–15	50–100	○
	22	4–8	100–150	<4	100–150	40–60	50–100	10–15	<50	○
	23	8–12	50–100	8–12	50–100	80–100	<50	10–15	<50	○
Wylie L.	24	12–16	50–100	8–12	50–100	100–120	<50	>30	50–100	▫
Hansen L.	25	12–16	150–200	4–8	200–250	80–100	150–200	25–30	>200	●
Blackjack L.	26	20–24	100–150	12–16	100–150	60–80	<50	10–15	50–100	△ (in circle)
	27	12–16	150–200	8–12	150–200	80–100	<50	10–15	<50	○
	28	8–12	50–100	<4	100–150	40–60	50–100	10–15	50–100	○
	29	4–8	<50	<4	50–100	20–40	<50	<5	<50	○
	30	8–12	50–100	4–8	100–150	80–100	<50	25–30	50–100	○
	31	4–8	50–100	<4	50–100	60–80	<50	15–20	50–100	○
Cruikshank L. {	32	8–12	100–150	4–8	150–200	100–120	<50	20–25	<50	▫
	33	8–12	100–150	4–8	100–150	100–120	<50	25–30	100–150	□
Calder R.	34	4–8	200–250	<4	250–300	60–80	<50	15–20	50–100	◉
Conjuror Bay	35	8–12	200–250	<4	250–300	140–160	<50	20–25	<50	◉
	36	12–16	150–200	8–12	150–200	>160	<50	>30	<50	○
Rainy L. {	37	20–24	>250	8–12	>300	120–140	>200	25–30	>200	▲
	38	16–20	200–250	12–16	200–250	80–100	50–100	15–20	100–150	✳▲
Clut L.	39	12–16	>250	4–8	>300	120–140	>200	20–25	>200	▲

△ KNOWN MINERALIZATION

■ ANOMALOUS cv. Zn, Znpx

□ ANOMALOUS cv. Zn or Znpx

▫ ANOMALOUS Zn,Znpx ppm,cv. NOT ANOMALOUS

○ OTHERS

● ANOMALOUS cv.Cu,Cupx,Zn,Znpx

✳ ANOMALOUS cv.Cu,Cupx,Zn,or Znpx

◉ ANOMALOUS cv.Cu,Cupx

○ ANOMALOUS Cu,or Cupx

● ANOMALOUS Cu ppm,cv NOT ANOMALOUS

Fig. 7-22. Summary interpretation of rock geochemical data from acid volcanic rocks from Great Bear Province, Canada (compiled from Garrett, 1975).

(1975) chose the arithmetic coefficient of variation (cv = the standard deviation expressed as a percentage of the mean). Thus, a high cv indicates a high geochemical relief that reflects erratic high values (positively skewed distributions) and which may indicate local metal concentration processes. The abundance is given by the geometric mean (gm). Thus a high cv with a low gm is considered to be possibly more important than the reverse situation (i.e., high abundance and low cv). The analytical data are summarized in Fig. 7-22. All known occurrences of mineralization, except Blackjack Lake, are reflected in an anomalously high cv for Cu and Cu px; all except Blackjack Lake (26) and Mazenod Lake (5) also have anomalously high cv's for Zn or Zn px or for both. Blackjack Lake is characterized by very high Cu and Cu px, and Mazenod Lake has a very high cv for Cu and Cu px. Only one area, Hansen Lake (25) has similar characteristics, but it has no known mineralization; it is clearly an extremely interesting exploration target.

Interpretation of the remainder of the data must necessarily be speculative because other anomaly combinations in Fig. 7-22 are assumed to be of lesser importance in the absence of known associated mineralization. Possible areas of interest are:

Calder River (34) which has high Cu and Cu px cv's with low means;

Conjuror Bay (35) which has high Cu and Cu px cv's, a low Cu px gm, and high Zn;

Rae Lake (9) which is moderately anomalous in Zn;

Location (18) which is moderately anomalous in Cu.

The results from this survey clearly demonstrate that the broad areas that contain known types and occurrences of mineralization could have been found by this type of rock geochemical survey. It is presumed that it would be useful for similar types of mineralization in similar geological conditions.

PALAEOZOIC MASSIVE SULPHIDES IN NEW BRUNSWICK

In the Bathurst district of New Brunswick there are more than 50 known occurrences of base metal sulphide mineralization within an area of Palaeozoic volcanic-sedimentary rocks. Only four of these occurrences are, or have been, producing mines; 19 are classed as "major occurrences". Most of these 23 deposits are typical massive sulphides of the New Brunswick type (see above); their locations are given in Fig. 7-23 and the grade and tonnage are given in Table 7-IV (the geology and the main deposits are described in Chapter 11).

To determine whether rock geochemistry could be used on a regional reconnaissance scale, Govett and Pwa (1981) collected 419 samples of rhyolite from an area of $2000 \, km^2$. Each sample consisted of about 1 kg of chip samples and was crushed, split, and ground to pass through an 80-mesh

TABLE 7-IV

Grade and tonnage of some massive sulphide occurrences in northern New Brunswick, Canada* (from Govett and Pwa, 1981)

No.	Name	Reserves (10^6 tonnes)	Grade (%)			Zn:Pb
			Cu	Pb	Zn	
1	Brunswick 12	>100	0.28	3.83	9.30	2.4
2	Brunswick 6	>2.4	0.42	1.77	4.91	2.8
3A	Heath Steele A, C, D	34	1.16	1.63	4.55	2.8
3B	Heath Steele B					
4	Wedge	3	3.0	—	1.75	
5	Armstrong	3.5	0.3	0.4	2.3	5.8
6	Rocky Turn	0.2	0.3	1.5	7.0	4.7
7	Orvan Brook	0.2	0.3	2.2	4.1	1.9
8	McMaster	0.2	0.6	—	—	
9	Caribou	49	0.47	1.7	4.48	2.6
10	Devils Elbow	1.0	0.7	—	—	
11	Indian Falls	?	—	Pb + Zn = 3—4		
12	California L.	?	—	Pb + Zn = 9—12		
13	Canoe Landing	3.5	0.5	0.5	1.5	3.0
14	Que. S.R.	probably similar to No. 15				
15	Headway	0.4	1.36	Pb + Zn = 8.2		
16	Austin Brook	1.0	—	1.86	2.93	1.6
17	9-mile Brook	0.15	0.4	1.2	1.0	0.8
18	Nepisiquit	3.9	0.3	0.5	2.4	4.4
19	Stratmat	0.45	—	Pb + Zn = 10.0		
20	Half Mile Lake	1.6	0.3	0.8	6.0	7.5
21		3—4	<0.5	Pb + Zn = 9.0		
22	Captain	0.8	1.5	—	—	
	Captain North	0.2	—	Pb + Zn = 11.5		
23	Key Anacon	1.9	0.24	2.2	5.9	2.7

*Data on non-producing deposits are the writers' best estimates from published and unpublished sources. Differences between tonnage and grades given here and in text (Chapter 11) are due to different cut-off grades.

screen. The samples were analyzed by AAS for total Cu, Pb, Zn, Na, K, Ca, Mg, Fe, and Mn after HF-HNO_3-HCl digestion. The data were processed by calculating the geometric mean of all samples in cells of approximately 10 km² (3.1 km east-west and 3.8 km north-south); there are 15 such cells to each map sheet (Fig. 7-23).

The regional variations in major elements (presumably related to a number of volcanic centres) limits their usefulness in defining the location of the mineral occurrences. The ore elements, however, quite clearly indicate the areas of interest, although there is not a simple relation between the ore elements and the location of the deposits. Only Cu shows a direct variation in concentration with the mineralized areas; all of the producing zinc-lead

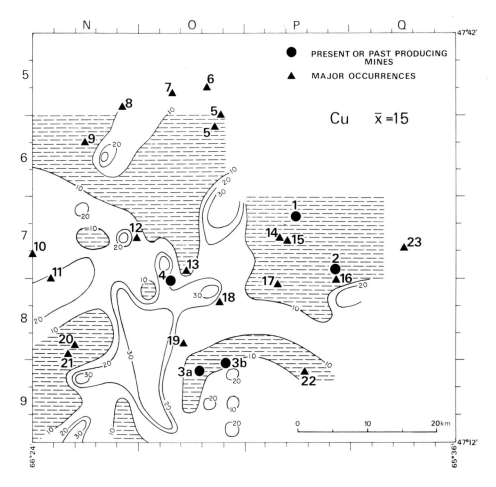

Fig. 7-23. Distribution of Cu in rhyolite, Bathurst area, Canada. (from Govett and Pwa, 1981). Numbers refer to deposits listed in Table 7-IV; N5, O7 etc. are map sheets.

mines and many of the larger deposits fall within areas of less than 10 ppm Cu. Ratios of the ore elements similarly define the location of the major deposits; Zn:Pb ratios in the restricted range of 2.0—2.4 (Fig. 7-24), Zn:Cu ratios of more than 7.0 (Fig. 7-25), and Pb:Cu ratios of more than 3.0 (Fig. 7-26) all define the location of the producing zinc-lead mines.

To refine the interpretation the deposits are grouped into five categories in Table 7-V and are summarized as follows:

Group A: ⩾ 3.0% combined Pb and Zn; ⩾ 1.0 million tonnes reserves.
Group B: ⩾ 3.0% combined Pb and Zn; < 1.0 million tonnes reserves.
Group C: < 3.0% combined Pb and Zn; ⩾ 1.0 million tonnes reserves.
Group D: < 3.0% combined Pb and Zn; < 1.0 million tonnes reserves.
Group E: dominantly Cu deposits.

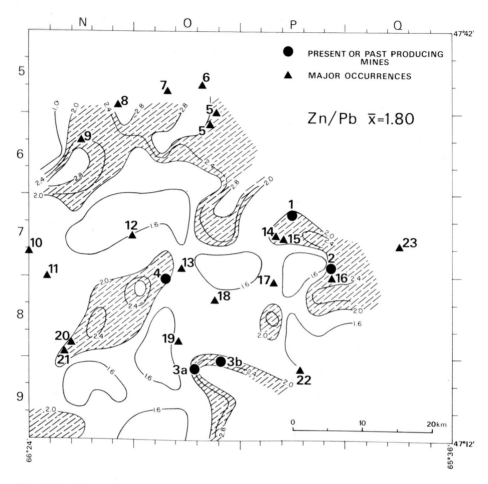

Fig. 7-24. Distribution of Zn:Pb ratios in rhyolite, Bathurst area, Canada (from Govett and Pwa, 1981).

Table 7-V also summarizes the anomalous geochemical responses for the ratios of Pb:Zn, Zn:Cu, Pb:Cu, and the value of Cu.

From Table 7-V it is clear that all deposits in group A fall within zones defined by Zn:Pb ratios of 2.0—2.4, Zn:Cu ratios greater than 7.0, Pb:Cu ratios greater than 3.0, and Cu contents of less than 10 ppm. The only other deposits to fall within this category are group B deposits 14 and 15 (which are near Brunswick No. 12). If the anomalous range of the Zn:Pb ratio is expanded from 2.0—2.4 to 2.0—2.8, deposit 5 of group C is also placed in the same category. (There is very little difference in the total area of anomaly between using 2.4 and 2.8 as the upper limit of the Zn:Pb ratio since the contours are very close in this range.) An interesting relation is that the

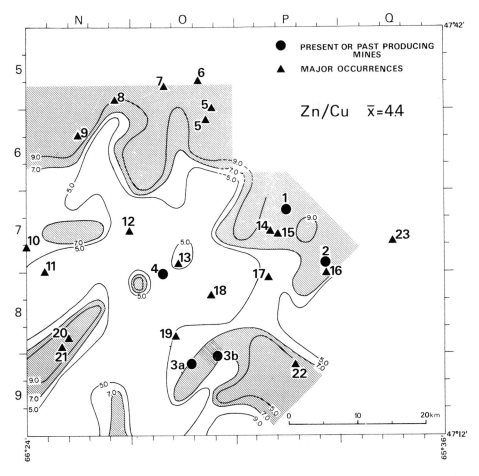

Fig. 7-25. Distribution of Zn:Cu ratios in rhyolite, Bathurst area, Canada (from Govett and Pwa, 1981).

Zn:Pb ratios in the sulphides of the three producing zinc-lead mines (1, 2, 3) and the large Caribou (9) deposit are all in the range of 2.4—2.8.

Deposits 6 and 7 of group B have anomalous Cu and Zn:Cu ratios, and probably have Zn:Pb ratios of 2.4—2.8. The Captain deposit (22) is anomalous for all trace element relations except for the Zn:Pb ratio; this deposit was originally placed in group E as a copper deposit, but was changed consequent upon a report by Gummer et al. (1980) that significant lead and zinc resources have been discovered nearby (Captain North, Table 7-IV). Of the remaining deposits in group B, deposits 11 and 12 have no anomalous trace element relations (there is also no information on grade and tonnage for these deposits), and deposit 19 has only an anomalous Pb:Cu ratio.

TABLE 7-V

Summary of geochemical responses around massive sulphide deposits in northern New Brunswick, Canada (from Govett and Pwa, 1981)

Deposit type	Deposit No.	Zn:Pb =2.0–2.4	Zn:Cu >7.0	Pb:Cu >3.0	Cu <10 ppm
Group A Pb + Zn ≥ 3.0% ≥10⁶ tonnes	1	X	X	X	X
	2	X	X	X	X
	3A,B	X	X	X	X
	9	X	X	X	X
	16	X	X	X	X
	20	X	X	X	X
	21	X	X	X	X
	(23)	X	X	X	X
Group B Pb + Zn ≥ 3.0% <10⁶ tonnes	6	2.4–2.8	X		X
	7	2.4–2.8	X		X
	11				
	12				
	14	X	X	X	X
	15	X	X	X	X
	19			X	
	22		X	X	X
Group C Pb + Zn < 3.0% ≥10⁶ tonnes	5	2.4–2.8	X	X	X
	18				
	13				X
Group D Pb + Zn < 3.0% <10⁶ tonnes	17			X	X
Group E dominantly Cu deposits	4	X			
	8		X	X	X
	10				

Group C deposits (apart from the Armstrong deposits (5) that have trace element relations of group A type) have few trace element anomalies. Deposit 13 has anomalous Cu, but is probably not volcanogenic in origin. Deposit 18 (Nepisiquit) has no anomalous trace element relations, but it has 3.5 million tonnes of 2.9% combined Pb and Zn.

There is only one deposit in group D — deposit 17, the smallest of the deposits considered — and it has anomalous Cu and Pb:Cu ratios. Group E consists of copper deposits. Deposit 4 (Wedge) was once mined for copper and has only an anomalous Zn:Pb ratio; deposit 8 has anomalous Cu and Zn:Cu ratios; and deposit 10 (Devils Elbow) has only an anomalous Pb:Cu ratio.

Within the constraints of available knowledge about known deposits and the distribution of samples, the following priority ratings can be assigned to trace element relations:

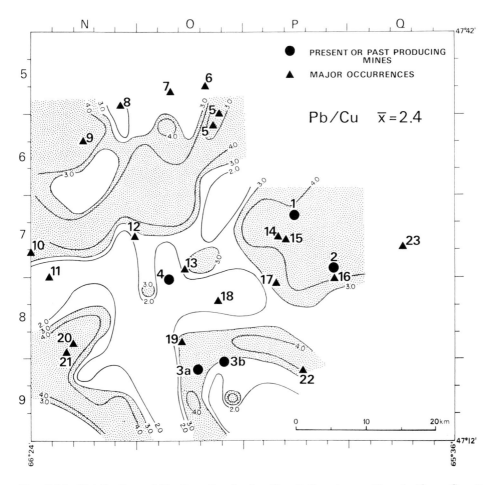

Fig. 7-26. Distribution of Pb:Cu ratios in rhyolite, Bathurst area, Canada (from Govett and Pwa, 1981).

Priority 1 (includes deposits 1, 2, 3, 5, 6, 9, 14, 15, 16, 20 21, and probably 23). Zn:Pb = 2.0—2.8; Zn:Cu > 7.0; Pb:Cu > 3.0; Cu < 10 ppm.

Priority 2 (includes deposits 7, 8, and 22). A minimum of Zn:Cu > 7.0 and Cu < 10 ppm.

Priority 3 (includes deposits 10, 13, 17, and 19). Pb:Cu > 3.0 and/or Cu < 10 ppm.

Deposits 11, 12, and 18 (with no trace element anomalies) and deposit 4 (a Zn:Pb ratio anomaly only) are excluded on the basis of the above classification. These conclusions are shown in Fig. 7-27 which shows priority zones 1 and 2.

Govett and Pwa (1981) point out the limitations of the data — lack of complete sample coverage, interpretative contouring assuming continuous

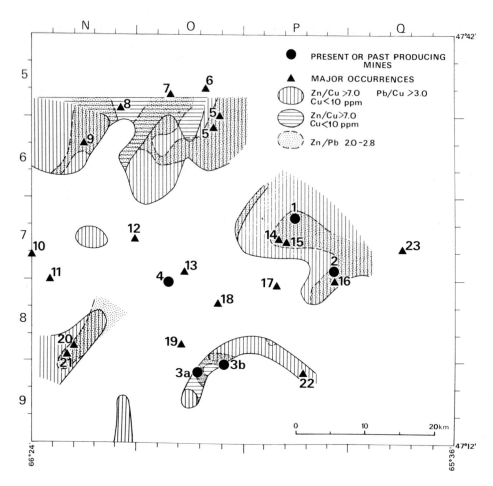

Fig. 7-27. Composite Zn:Cu, Pb:Cu, Zn:Pb, and Cu anomaly map for rhyolite, Bathurst area, Canada (from Govett and Pwa, 1981).

variation between adjacent sample points, lack of geological data on many of the deposits, the probability that geochemical responses vary with variations in deposits, and contouring across structural units. Nevertheless, it is possible to define priority areas for exploration. Broadly these are: (a) a northern zone that includes the Caribou (9), Rocky Turn (6), Orvan Brook (7), and Armstrong deposits (5); (b) an eastern zone including the Brunswick deposits (1, 2) and extending towards Key Anacon (23); (c) an arcuate southeastern zone that includes Heath Steele (3A, 3B); and (d) a southwestern zone that includes the Half Mile Lake deposits (20, 21).

In terms of seeking deposits comparable to Brunswick No. 12 (1) and Heath Steele (3A, 3B), zones of Zn:Pb ratios of 2.0—2.4 within the broader target areas are obviously of prime importance. Although no analytical data are available from around the Key Anacon deposit (23), it is encouraging that contours for all trace element anomalies are open to the east of the Brunswick deposits — which suggests that the ground between Key Anacon and Brunswick No. 6 and Brunswick No. 12 is prospective. The discovery of lead-zinc mineralization near the old Captain copper deposit (noted above) within prospective zone 3 subsequent to the work by Govett and Pwa being completed is encouraging.

The most important conclusion from this investigation is the demonstration that regional geochemical patterns related to the occurrence of sulphide deposits can be obtained from a relatively sparse sample coverage of an average of one sample/5 km^2 over an area of 2000 km^2. This is, therefore, one of the few reconnaissance surveys in which the sampling is on a regional scale.

MASSIVE SULPHIDE DEPOSITS IN CYPRUS

One of the first regional rock geochemical surveys for sulphides of volcanic-sedimentary association to be published outside the U.S.S.R. was that of Govett and Pantazis (1971) who investigated the trace element distribution in pillow lavas over the entire Troodos volcanic series in Cyprus. The work showed that the trace element concentration varied geographically in relation to the location of cupriferous pyrite deposits.

The numerous small deposits lie towards the top of the volcanic sequence which consists of basaltic, largely pillowed, lavas and dyke intrusions that increase in intensity stratigraphically downwards. (The geology is discussed in more detail in Chapter 12.) An outcrop length of about 550 km was sampled along 20 traverses crossing the strike more or less at right angles and located to give approximate equal geographic distribution (Fig. 7-28). Samples of the centres of pillow lavas were taken at approximately 200-m intervals over a total traverse length of 93 km.

The results of the survey showed that the cupriferous pyrite deposits on the island are confined to particular stratigraphic horizons of the tholeiitic basalts that could be defined by the distribution of Cu (depleted), Zn (enriched), Ni (depleted), and Co (enriched). The results for Cu and Zn are shown in Fig. 7-29. No sulphide deposits occur in a region where the Cu:Zn ratio exceeds 1.2; all but two of the deposits occur in rocks where the ratio is less than 1.0, and 71% of the deposits are in rocks where the ratio is less than 0.8.

The variations in Cu:Zn ratios results from both depletion of Cu and enrichment of Zn. Similar relations have been noted elsewhere; the host metavolcanic rocks of Ingladhal (India) cupriferous pyrite-pyrrhotite deposit in an Archaean greenstone belt are depleted in Cu and enriched in Co compared with similar rocks elsewhere (Mookherjee and Philip, 1979), and

172

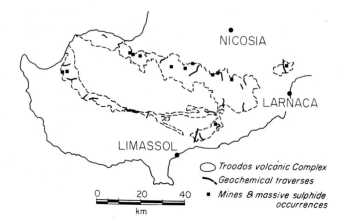

Fig. 7-28. Location of massive sulphide occurrences and geochemical traverses, Cyprus.

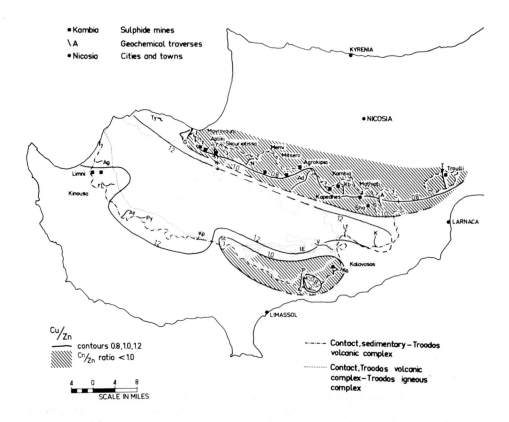

Fig. 7-29. Distribution of Cu:Zn ratios in basaltic pillow lavas, Cyprus (from Govett, 1976b).

TABLE 7-VI

Average content of some elements in the rocks of eastern Tuva* (from Berman and Agentov, 1965).

		Lower Cambrian Tuva	Lower Cambrian eastern Tuva	Mineralized areas
Cu	mafic	66	10	64
	intermediate	240	18	67
	felsic	26	32	24
Pb	mafic	44	0.4	126
	intermediate	23	2.8	79
	felsic	16	1.0	154
Zn	mafic	110	8.8	161
	intermediate	100	28	318
	felsic	100	17	321
Mn	mafic	570	670	1100
	intermediate	400	576	2020
	felsic	300	485	1314
Co	mafic	77	1.9	20
	intermediate	51	8.0	13
	felsic	7.9	1.0	4
Ni	mafic	90	6.8	40
	intermediate	71	12	77
	felsic	13	1.8	5.4
Ba	mafic	31	264	620
	intermediate	32	497	618
	felsic	73	1780	397

*Numbers of spectrographic analyses: volcanic rocks of Tuva, mafic 54, intermediate 60, felsic 42; volcanic rocks of eastern Tuva, mafic 580, intermediate 232, felsic 450; mineralized areas, mafic 119, intermediate 79, felsic 191.

depletion of Cu in andesite underlying massive sulphide deposits in the Lake Dufault area in Canada compared to the stratigraphically higher dacite was noted by Roscoe (1965).

OTHER EXAMPLES OF VOLCANIC-HOSTED MASSIVE SULPHIDES

Berman and Agentov (1965) described "pyritic-polymetallic mineralization" in the Lower Cambrian spilite-keratophyre in eastern Tuva (U.S.S.R.) that are presumed to be massive sulphides. The grade appears to be low (0.03—0.3% Cu, 0.08% Pb, and 0.6—0.9% Zn). The deposits occur in lavas associated with volanic vents, and sulphide disseminations also occur in marine siliceous shales and tuffites. The analytical data for the Tuva volcanic rocks (presumed to be background), the eastern Tuva volcanic rocks where mineralization occurs, and wallrocks from the mineralized areas are shown in Table 7-VI. The volcanic rocks of eastern Tuva (slightly differentiated calc-alkaline suite) are generally depleted in Cu, Pb, and Zn and are enriched in Mn and Ba compared to background. The Cu content in rocks in the vicinity of mineralization is at background levels in mafic and felsic rocks but is depleted in intermediate rocks. Compared to the eastern Tuva region the mineralized rocks are enriched in Cu, Pb, Zn, Mn, and Ba (except in felsic rocks). The authors gave no information on locations and distances of sampling relative to deposits, but there are clear regional scale differences in element concentrations that have similarities and differences compared with other examples described above; regional increases in Mn are common, as is the regional decrease in Cu; the regional decrease in Zn appears to be unique to the region.

The Buchans massive sulphide deposits of central Newfoundland (Canada) occur at the northern end of the Appalachian structural province. Since 1928 some 14.5×10^6 tonnes of ore of an average grade of 14.88% Zn, 7.7% Pb, 1.36% Cu, 116.6 g/tonne Ag, and 1.4 g/tonne Au have been mined (Thurlow et al., 1975). The deposits lie within a volcanic sedimentary sequence of Ordovican-Silurian age; four cycles of generally mafic to felsic volcanic rocks with associated sediments have been recognized. The sulphide bodies are confined to the two lower cycles, and the more important deposits occur within calc-alkaline dacitic pyroclastic flows. Thurlow and his co-workers could not identify significant local geochemical patterns beyond about 30 m from mineralization, but did document regional scale (3 km) enrichment (98—99.99% significant by the Mann-Whitney U test) of Ba, Zn, and Pb in siltstone, andesite, and dacite units host to mineralization compared with barren units of similar lithologies. In some instances Cu, Ag, and Hg are also enriched in lithologic units associated with mineralization.

MASSIVE SULPHIDES OF SEDIMENTARY ASSOCIATION

There are particularly good opportunities — as pointed out by Coope (1977) and Coope and Davidson (1979) — to use rock geochemistry on a regional scale in situations where there are extensive sediments or exhalites close to the ore horizons. This is demonstrated by the work of Bignall et al. (1976) on recent sediments in the Red Sea. They have shown that anomalous contents of Hg, Cu, Zn, and Mn in sediments extend 10 km from the Atlantis II Deep (Fig. 7-30), and anomalous contents of Cu, Zn, and Mn extend 9 km from the Nereus Deep (Hg was not determined). In both cases the anomalous elements show a gradient of increasing concentration towards the metalliferous zones in the Deeps. The authors concluded that Mn, especially, probably had an even wider anomalous dispersion than that reported.

In the Lower Carboniferous Waulsortian limestone that is host to the Tynagh (Ireland) massive sulphide deposit (Pb-Zn-Cu-Ag, $BaSO_4$) a Mn halo was reported by Russell (1974) to be detectable in drill core and surface

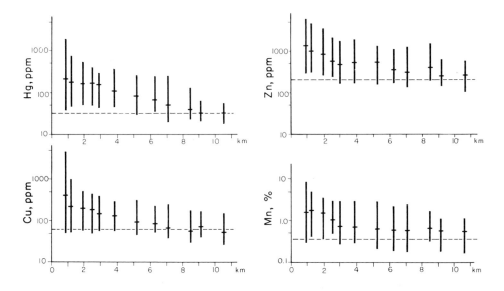

Fig. 7-30. Range (vertical bars) and average (horizontal bars) of Cu, Hg, Mn, and Zn in sediment cores from around Atlantis II Deep with distance from major metalliferous sediments. (Redrawn with permission from Bignall et al., 1976, *Institution of Mining and Metallurgy, Transactions, Section B*, vol. 85, 1976, fig. 4, p. 277.) Broken lines represent background values in sediments from axial valley sides.

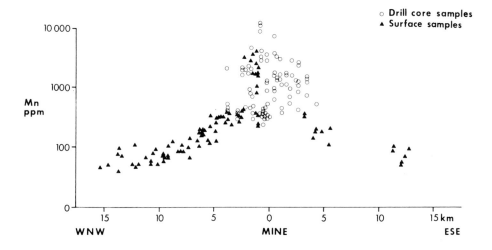

Fig. 7-31. Distribution of 0.2 *m* acetic acid-soluble Mn in limestone that is host to the Tynagh deposit (Pb-Zn-Cu-Ag-BaSO$_4$), Ireland. (Reproduced with permission from Govett and Nichol, 1979, *Geophysics and Geochemistry in the Search for Metallic Ores, Geological Survey of Canada, Economic Geology Report*, 31, 1979, fig. 15-7, p. 349; original data from Russell, 1974.)

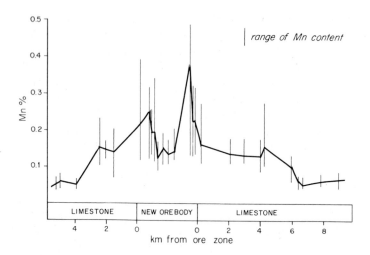

Fig. 7-32. Distribution of total Mn in limestone around the Meggan Pb-Zn-BaSO$_4$ deposit, Germany, (Redrawn with permission from Gwosdz and Krebs, 1977, *Institution of Mining and Metallurgy, Transactions, Section B*, vol. 86, 1977, fig. 1, p. 74.)

samples for a radius of 7 km from the deposit (Fig. 7-31). Beyond 7 km from the deposit Mn contents rarely exceed 100 ppm, whereas closer than 7 km Mn contents are never less than 100 ppm; within 3 km of the deposit the Mn content exceeds 200 ppm, and the variance characteristically becomes large. Zinc is anomalously high for about 1 km from the deposit. Although the Tynagh deposit is apparently epigenetic in relation to its host rock, the dispersion is believed to be syngenetic with the ore and related to a submarine exhalative source and represents "spent" mineralizing solutions subsequent to the main phase of sulphide deposition. An iron formation associated with the sulphide deposit has anomalously high contents of Ba, Pb, and Zn (Russell, 1975).

The Meggan (Germany) pyrite-sphalerite-barite ore body lies within Middle Devonian micritic limestone 2—4 m thick that can be traced for a distance of 25 km along the southeast flank of the Elspe syncline. A total of 342 samples from underground workings, drill holes, and outcrops over a distance of about 18 km were analyzed for Mn by Gwosdz and Krebs (1977). A Mn halo of about 7-fold contrast extends at least 5 km from the deposit (Fig. 7-32). The range of values at any particular location is 2—4 fold (being greater nearer to the deposit). Peak Mn values lie lateral to, and not immediately over, the deposit which is consistent with the model for the dispersion of Mn in an aqueous environment around an active brine source proposed by Whitehead (1973a).

An apparent example of a sedimentary halo occurs around the McArthur River massive sulphide deposit in the Northern Territory, Australia (180 × 10^6 tonnes of 10% Zn, 4% Pb, 0.2% Cu, and 47 g/tonne Ag). Mineralization is in shales of Middle Proterozoic age, and its volcanic association is limited to tuff in the sediments. Lambert and Scott (1973) analyzed 160 samples from eight drill holes (two through the deposit, two within a few hundred metres of mineralization, and four holes at about 2.5 km, 7 km, 17 km, and 23 km from the deposit). They found that Fe and Mn contents are anomalous in stratigraphic footwall and hanging-wall dolomites extending 15–20 km from a major fault zone lying immediately to the east of the deposit. Anomalously high contents of Pb and Zn occur at the same stratigraphic horizon as the sulphides for a distance of 23 km from the deposit.

CONCLUSIONS

The great majority of massive sulphides occur in two major geological environments: Cu deposits in tholeiitic ophiolite sequences; and Zn-Cu and Zn-Pb (Cu) deposits typically in an island arc-type environment (the Pb-rich deposits are generally associated with the more felsic volcanic rocks). A preliminary geological and petrochemical investigation will determine the broad potential of a region.

Despite the successful exploration application of studies restricted to trace element distributions in some cases, it is obvious that petrological variation — and especially the place in a fractionation series — is just as important in interpreting trace element distributions in volcanic rocks as it is in plutonic rocks. Ideally SiO_2 should be determined on all rock samples for regional exploration to allow trace and major element data to be normalized to SiO_2 variation. This poses a practical problem due to the analytical difficulty — and hence the high cost — of determining SiO_2 compared to the simplicity of measuring Cu, Pb, or Zn by AAS. The high analytical costs are nevertheless clearly justified to identify the prospective part of a volcanic sequence which thereafter allows a compensating increase in the probability of exploration success.

Only three of the investigations described in this chapter — the Cyprus case (Govett and Pantazis, 1971), the Canadian Great Bear Province study (Garrett, 1975), and the New Brunswick Bathurst study (Govett and Pwa, 1981) — really qualify as regional surveys in the sense that small numbers of samples more or less evenly distributed geographically were used to identify a regional scale distribution of elements that could be used to identify targets. In these three contrasting environments the trends were defined on the basis of only the ore elements, Cu, Pb, and Zn. All of the other studies described above were based upon a comparison of relatively large numbers of samples in isolated productive and non-productive areas or from productive and non-productive volcanic cycles in one area. Nevertheless, even from these studies it is reasonable to conclude that regional geochemical patterns exist that can be used in regional exploration, and that productive volcanic cycles can be distinguished from non-productive cycles.

Zinc is apparently the most useful indicator among the trace elements; it is generally enriched in volcanic and sedimentary rocks associated with massive sulphide deposits of all types except possibly the Archaean Pb-Zn-(Cu) type (according to data in Cameron, 1975). The behaviour of Cu seems to depend on the type of deposit; it is regionally depleted around the copper deposits in the basalts of Cyprus and in the rhyolites of New Brunswick, but it is enriched in the quartz-feldspar porphyry volcanic rocks that have copper mineralization in the Northwest Territories of Canada. The distribution of Pb is generally not a useful indicator in regional exploration, although the Zn:Pb ratio is important in New Brunswick. There are few data on major elements: productive acidic volcanic sequences seem to be generally enriched in Mg and depleted in Ca and Na; in some cases Fe and K are also regionally enhanced.

The *form* of the frequency distributions of trace elements gives significant information — as was the case with igneous rocks. Rock sequences within which the frequency distribution of trace elements have a strong positive skew are more likely to be ore-bearing than those which have distributions that are normal or which show only small degrees of positive skewness.

Regional geochemical anomalies associated with massive sulphides are of two distinct genetic types that also differ, to some extent, in their anomalous element associations. One is related to the primary character of the volcanic rocks and may be revealed in major and trace element differences (e.g., enhanced SiO_2, Mg, and Zn, or depleted Na and Cu). The second type of geochemical anomaly is directly related to actual ore-forming and sedimentary processes on the sea bed; it is most likely to be detected in the associated sedimentary or exhalative horizons by enhanced contents of minor and trace elements (e.g., Mn, Zn, and Cu) and generally may be expected to give the most extensive anomaly.

A characteristic of regional scale rock geochemical data is the generally high variance of the element distribution. This indicates the wisdom of taking duplicate samples to obtain some measure of the variance. Also, when assessing different rock sequences or the same sequence in different areas, an adequate number of different sites for each sequence or area must be sampled to allow tests of differences to be statistically significant.

PART III. LOCAL AND MINE SCALE EXPLORATION

Chapter 8

LOCAL AND MINE SCALE EXPLORATION FOR PORPHYRY-TYPE DEPOSITS

INTRODUCTION

In recent years there has been considerable progress in understanding the genesis of porphyry-type copper deposits and in defining their characteristic environment of occurrence and associated alteration (Lowell and Guilbert, 1970; Rose, 1970; De Geoffroy and Wignall, 1972; Sillitoe, 1972, 1973b; Guilbert and Lowell, 1974). Given the great importance of these deposits as sources of copper and molybdenum (porphyry deposits supply 45% of the world's copper according to Sutulov, 1974) it is surprising that relatively few detailed rock geochemical studies have been reported.

The characteristics of a "typical" porphyry are shown in schematic form in Fig. 8-1. It consists of a zoned deposit of disseminated copper, copper-molybdenum, or molybdenum sulphides with associated hydrothermally altered host rocks of Jurassic to Tertiary age. The composition is intermediate to felsic (granodiorite to quartz monzonite in Cordilleran Americas and commonly quartz diorite in island arc areas such as the Philippines, Papua New Guinea, the Solomon Islands, and Puerto Rico). The host intrusion may be porphyritic.

The most notable characteristic of porphyry-type deposits is their size; the ore bodies are generally pipe-like with dimensions of 1000 m by 2000 m. The "average" characteristic features listed by Guilbert and Lowell (1974) are:

— 136×10^6 tonnes of 0.8% Cu (0.45% hypogene, 0.35% supergene) and 0.015% Mo;

— 70% of the ore occurs in igneous host rocks, 30% in pre-ore rocks;

— visible alteration extends over some 750 m beyond the Cu zone and is zoned from potassic (K-feldspar, biotite, sericite, quartz, anhydrite) — which is the earliest and at the core — outward through phyllic (quartz-sericite-pyrite), argillic (quartz-kaolin-montmorillonite), and propylitic (epidote-calcite-chlorite) zones;

— over the same interval of alteration sulphide species vary from chalco-pyrite-molybdenite-pyrite through successive assemblages to galena and

182

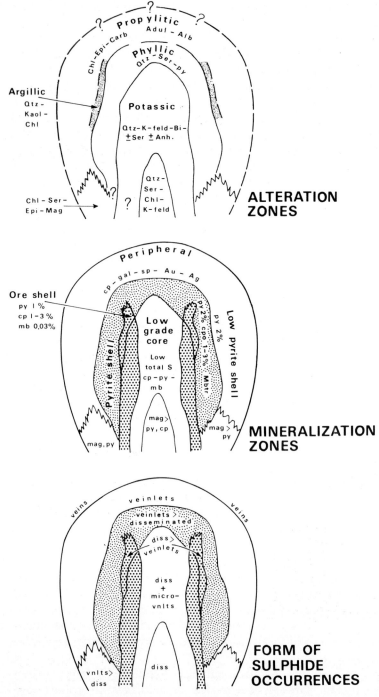

Fig. 8-1. Schematic illustration of zoning around copper porphyry deposits. (Redrawn with permission from Guilbert and Lowell, 1974, *Canadian Institute of Mining and Metallurgy Bulletin*, vol. 68, 1974, figs. 1, 2, and 3, p. 100.)

sphalerite with minor silver and gold in solid solutions in sulphides, as metals, and as sulphosalts;

— the sulphides from the core outwards occur in the form of disseminations — microveinlets, veinlets, and veins — to some peripheral structures that may contain high-grade mineralization.

In terms of general exploration the porphyry deposits offer a large target with characteristic and visible alteration zones. Moreover, their large size generally ensures extensive secondary dispersion trains readily detectable in stream sediment surveys, especially in tropical environments where exposure may be poor. For example, Govett and Hale (1967) recorded a drainage train of more than 16 km in the Philippines, and many of the copper porphyry deposits discovered over the past decade in Central and South America were first detected with stream sediment geochemistry.

Trace element data on host rocks that might be useful in rock geochemical surveys are sparse. Data assembled by Kesler et al. (1975a) suggest that there may be significant regional major element differences between genetically associated host rocks. Specifically, the authors suggested that intrusions in island arc environments (e.g., Caribbean, British Columbia) are enriched in Cu and Zn and depleted in Pb relative to intrusions in a cratonic environment (e.g., southwestern U.S.A.). Given the range of Cu contents (10—65 ppm) and Zn contents (28—74 ppm) in different phases of the Guichon Batholith (Table 8-II below), this conclusion should be treated with caution. Certainly Kessler and his co-workers documented a significant compositional

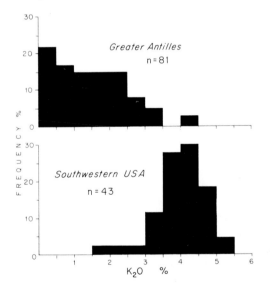

Fig. 8-2. Distribution of K_2O in plutonic rocks associated with copper porphyry mineralization in the northern Caribbean and the southwestern U.S. (modified from Kesler et al., 1975a, by data kindly supplied by S.E. Kesler).

difference in intrusions between the two environments, showing the marked enrichment in K in intrusions in the southwestern U.S. compared with those in the northern Caribbean (Fig. 8-2). Mason (1979) suggests that amphiboles in porphyry copper stocks show Mg-enrichment towards the rims, whereas in barren stocks the rims are enriched in Fe. Divis and Clark (1979) propose that porphyry hydrothermal systems show an increase in the $^{6}Li:^{7}Li$ isotopic ratio as determined by AAS.

A detailed evaluation of the nature of geochemical dispersion around porphyry copper deposits similar to that given for massive sulphides (Chapters 10—12) is not possible due to limited data and to variations in sampling procedures and in the range of elements analyzed by various workers. Case histories and studies are described and evaluated in the remainder of this chapter.

CANADIAN CASE HISTORIES

The Guichon Batholith

The most extensive exploration-oriented rock geochemical work has been done on deposits in the Guichon Batholith; it is the pluton that Warren and Delavault (1960) identified as a copper-positive target in their early work on rock geochemistry (see Chapter 4). The batholith is a concentrically zoned Triassic intrusion that ranges in composition from a diorite at the outer margins (the Hybrid Phase) through the Highland Valley Unit (comprising the Guichon quartz diorite and the Chataway granodiorite), the Intermediate Unit (comprising the Bethlehem and the Skeena granodiorites and the Witches Brook and Bethlehem porphyry dykes), to the central Core Unit (comprising the Bethsaida granodiorite to quartz monzonite and quartz porphyry dykes). These relations are illustrated in Fig. 8-3. Apart from the Craigmont deposit (Fig. 8-4), which is a replacement deposit in Triassic strata, there are no known copper deposits in the Hybrid Phase, and all deposits are restricted to the interior of the batholith.

Brabec and White (1971) measured Cu (342 samples) and Zn (253 samples) on rock samples collected fairly uniformly over the entire batholith; the elements were measured by AAS after an aqua regia digestion. The authors drew attention to significant differences in Cu content of the various phases of the pluton (as did Olade and Fletcher, 1976a; see Table 8-I) and the pronounced concentration of low values in the core of the batholith (Fig. 8-4). Zinc shows only a slight tendency to decrease in concentration towards the core. Cubic trend surfaces for Cu (Fig. 8-5) show a very pronounced regional trend of low values in the Bethsaida Phase and also in the northwestern part of the area. The major porphyry deposits therefore occur in Cu-deficient areas — notwithstanding that the batholith as a whole, with an

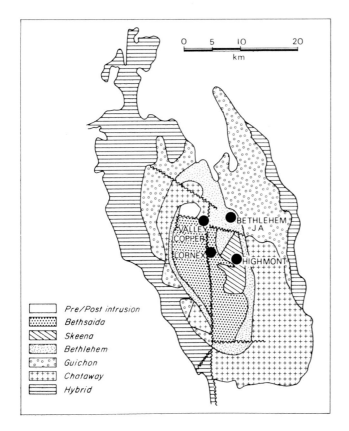

Fig. 8-3. Geology of the Guichon Batholith, British Columbia, Canada. (Redrawn with permission from Olade and Fletcher, 1976a, *Economic Geology*, vol. 71, 1976, fig. 1, p. 734.)

average total Cu content of about 50 ppm, is relatively enriched in Cu compared with a Cu content of 30—35 ppm for an average intermediate granite (Brabec and White, 1971).

The most comprehensive investigations of the batholith are those carried out on the Highland Valley deposits (Olade and Fletcher, 1975, 1976a, 1976b; Olade, 1977, 1979). A total of 1800 samples from mineralized areas and 60 from non-mineralized areas of the batholith were analyzed for 30 elements. Surface samples consisted of 4-kg composite samples of several fist-sized chip samples; drill core samples were composites of 5-cm lengths of core taken over 3 m.

The geometric mean and range for selected elements in unmineralized samples of various phases of the batholith are given in Table 8-I. There is a well defined trend of decreasing concentration of Cu, Zn, Mn, Co, Ni, and V from the marginal Hybrid Phase to the more felsic and generally younger

TABLE 8-I

Geometric mean and range (in ppm) for trace element content of unmineralized samples of the Guichon Creek Batholith (from Olade and Fletcher, 1976a)*

Intrusive unit	n	Cu[1]	Zn[1]	Mn[1]	Co[1]	Ni[1]	V[2]	Rb[3]	Sr[3]	Ba[2]	S[3]
Hybrid	8	51 12—1143	74 65—81	692 570—740	13 11—36	33 16—44	80 50—100	50 (3) 18—82	468 (3) 255—580	500 440—600	380 (3) 310—475
Guichon	8	67 30—98	60 47—77	574 420—720	12 12—14	27 16—38	50 40—60	50 (9) 24—76	735 (9) 610—1000	300 200—500	373 (8) 300—620
Chataway	7	45 17—240	45 37—72	430 350—630	11 9—16	20 10—33	40 20—50	48 (4) 37—74	747 (5) 680—785	500 300—600	394 (5) 290—610
Bethlehem	10	30 9—196	34 22—46	370 250—460	8 6—10	10 6—24	35 30—40	43 (7) 18—82	690 (7) 580—890	560 500—800	390 (7) 290—500
Bethsaida	6	24 4—135	28 15—37	362 250—420	5 3—7	6 5—8	15 10—20	35 (6) 32—75	590 (6) 530—635	520 500—700	395 (6) 270—750

* n = number of samples unless indicated otherwise in parentheses; [1] = HF-HNO₃-HClO₄ digestion/atomic absorption; [2] = emission spectroscopy; [3] = X-ray fluorescence.

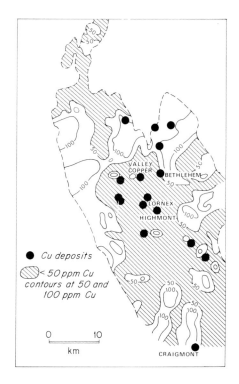

Fig. 8-4. Distribution of Cu in the Guichon Batholith, British Columbia, Canada (compiled from Brabec and White, 1971).

Bethsaida Phase of the core; there is also a decrease in the Ni:Co ratio towards the core. Depletion of Fe and Mg towards the core reflects a decrease in ferromagnesian minerals, and hence a decrease in available sites for trace elements that substitute for Fe and Mg — Zn, Mn, Ni, Co, and V. The fact that Cu, which is presumably present dominantly as a sulphide (see below), also decreases towards the core is taken by Olade and Fletcher (1976a) to indicate that porphyry mineralization is unlikely to have resulted from the differentiation of the batholith towards a Cu-rich magma. The S content is fairly constant throughout all phases, and variations in Rb and Sr are small and reflect variations in K and Ca, respectively (see below).

Olade and Fletcher worked on four porphyry deposits in the Guichon Batholith (see Fig. 8-3 for locations):

Bethlehem JA: 272×10^6 tonnes of 0.45% Cu and 0.017% MoS_2. Principal sulphide minerals are chalcopyrite and bornite in the ratio of 5:1. Other metallic minerals are molybdenite, pyrite, and specularite.

Valley Copper: 907×10^6 tonnes of 0.48% Cu. Principal ore minerals are bornite and chalcopyrite. Other metallic minerals are pyrite, sphalerite, molybdenite, magnetite, and specular hematite.

188

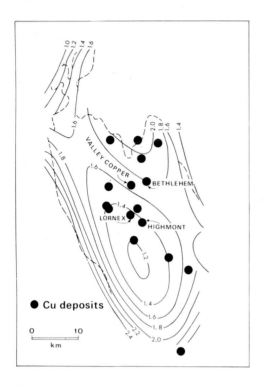

Fig. 8-5. Cubic trend surface (20% fit, logarithmic values) for Cu in the Guichon Batholith, British Columbia, Canada. (Redrawn with permission from Brabec and White, 1971, *Geochemical Exploration, Canadian Institute of Mining and Metallurgy, Special Volume*, 11, 1971, fig. 9, p. 295.)

Lornex: 272×10^6 tonnes of 0.43% Cu and 0.015% MoS_2.
Highmont: 136×10^6 tonnes of 0.3% Cu and 0.051% MoS_2 in 6 mineralized zones.

The dominant influence on the distribution of trace elements is variation in ferromagnesian minerals due to lithological changes. The effects of hydrothermal alteration and mineralization can, however, be detected in the distribution of some elements. The distributions of Zn, Sr, Rb, Mo, S, and Cu across the Valley Copper deposit — which is totally within the Bethsaida Phase — are shown in Fig. 8-6 and illustrate the main points. Zinc, Sr, and Mn (which is not illustrated) are depleted in the zone of intense phyllic and potassic alteration and are enhanced at the periphery of the deposit where argillic alteration is dominant; Ba (also not illustrated in Fig. 8-6) is depleted over the ore zone. The concentration of Rb is above regional background at the limit of sampling and reaches a maximum in the phyllic alteration zone. Both S and Cu are higher than regional background at the limit of sampling; Cu reaches a maximum over the phyllic core, whereas S reaches maxima at the argillic-phyllic boundary. The distribution of Mo is erratic, but it shows

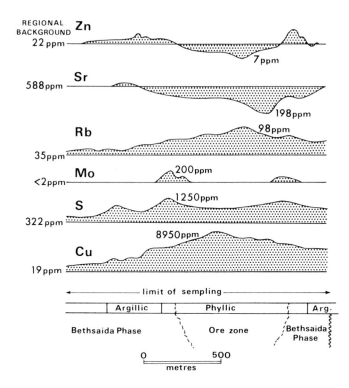

REGIONAL
BACKGROUND **Zn**
22 ppm
7 ppm

Sr
588 ppm
198 ppm

Rb
98 ppm
35 ppm

Mo
<2 ppm
200 ppm

S
322 ppm
1250 ppm

8950 ppm
Cu
19 ppm

limit of sempling

| Argillic | | Phyllic | | Arg. |

Bethsaida Phase   Ore zone   Bethsaida
Phase

0   500
metres

Fig. 8-6. Schematic illustration of element distribution across the Valley Copper porphyry deposit, British Columbia, Canada. (Redrawn with permission from Olade and Fletcher, 1976a, *Economic Geology*, vol. 71, 1976, fig. 9, p. 742.)

strong peaks at the edge of the ore zone. There are no consistent patterns for Hg, Cl, F and B.

At Bethlehem JA the patterns for Mn, Zn, Sr, Rb, Cu, and S are similar to those at Valley Copper. The distribution of Ba shows a pronounced positive anomaly over the ore zone (compared to a negative anomaly at Valley Copper). There is also a pronounced Hg and Cl anomaly associated with mineralization. At Lornex the situation is complicated by the deposit — which is in the Skeena Phase of the Bethlehem granodiorite — being faulted against fresh and unaltered Bethsaida Phase rocks. The distribution of Ba, Sr, Mn, Hg, Zn, B, Mo, and Cu is shown in Fig. 8-7; the fault zone is defined by strong anomalies for Mn, Hg, and Zn, and a minor Mo anomaly. At Highmont mineralization is associated with strong Mo, Cu, and B anomalies; the latter two elements are anomalous considerably beyond the alteration zones.

The extent and contrast of the anomalies for the more useful elements are summarized in Fig. 8-8. Obviously the ore elements Cu and S show the greatest anomalous contrast and extent; at Valley Copper and Bethlehem JA

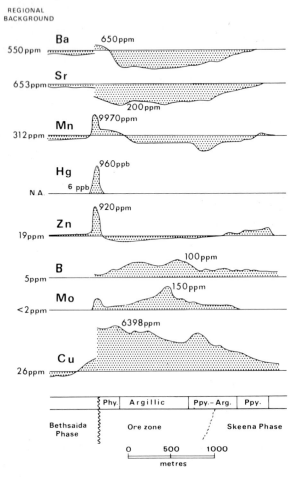

REGIONAL
BACKGROUND

Fig. 8-7. Schematic illustration of element distribution across the Lornex copper porphyry deposit (4,900 level), British Columbia, Canada. (Redrawn with permission from
Olade and Fletcher, 1976a, *Economic Geology*, vol. 71, 1976, fig. 10, p. 743.)

the Cu and S concentrations are considerably above background levels to the
limit of the sampling (i.e., at least 0.5 km from the ore zone).

A study by Olade and Fletcher (1976b) on Bethlehem JA and Valley
Copper, using a sulphide-selective $KClO_4$-HCl acid leach described in Olade
and Fletcher (1974), showed that this procedure extracts virtually all of the
Cu in both mineralized and non-mineralized samples — which indicates
that Cu is present in these rocks largely as a sulphide. The procedure is not
specific for Fe sulphides, and it only extracts 10—20% of the total Fe in unmineralized samples (and this is mostly from silicates); a greater proportion
of total Fe is extracted from mineralized samples.

The geology of the Bethlehem JA deposit is shown in Fig. 8-9 together

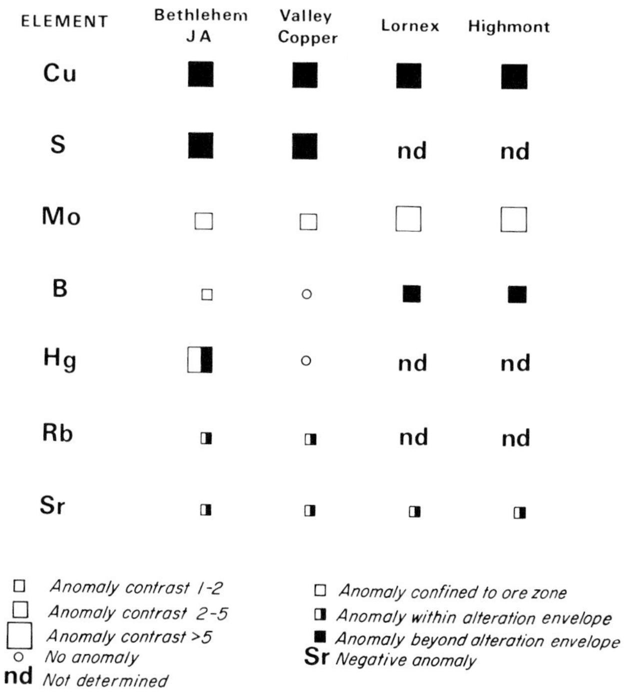

ELEMENT	Bethlehem J A	Valley Copper	Lornex	Highmont
Cu	■	■	■	■
S	■	■	nd	nd
Mo	☐	☐	☐	☐
B	☐	○	■	■
Hg	◨	○	nd	nd
Rb	◪	◪	nd	nd
Sr	◪	◪	◪	◪

☐ Anomaly contrast 1-2
☐ Anomaly contrast 2-5
☐ Anomaly contrast >5
○ No anomaly
nd Not determined

☐ Anomaly confined to ore zone
◪ Anomaly within alteration envelope
■ Anomaly beyond alteration envelope
Sr Negative anomaly

Fig. 8-8. Anomaly extent and contrast in rocks around the copper porphyry deposits in the Highland Valley area of the Guichon Batholith, British Columbia, Canada (compiled from Olade and Fletcher, 1976a).

with the sulphide zoning and the distributions of S, sulphide-Fe, total Fe, and sulphide-Cu. The S content is 0.1% (about 2 times background) at the limit of sampling and increases to more than 1.4% in the mineralized zone; maximum values (more than 4%) occur at the Guichon/Bethlehem JA contacts where chalcopyrite is most abundant. The pattern of sulphide-Fe is similar to that of S and is quite different from total Fe; the highest values generally coincide with the chalcopyrite zone. The highest concentrations of total Fe occur in the more mafic Guichon quartz diorite rather than in the Bethlehem JA granodiorite. Thus, the use of a specific sulphide leach enhances the Fe pattern due to sulphides and suppresses variations related to silicate mineralogy and lithology. The distribution of sulphide-Cu is clearly highly anomalous to the limit of sampling.

Olade (1979) measured the content of Cu and Zn in 26 mineral separates of biotite, magnetite, and quartz feldspar from background samples and samples from around the Highmont and Valley Copper deposits. The results are summarized in Table 8-II. Copper is enriched in all mineral separates and increases in concentration with proximity to mineralization; Zn does not show a similar enrichment and, in fact, shows a tendency to decrease in

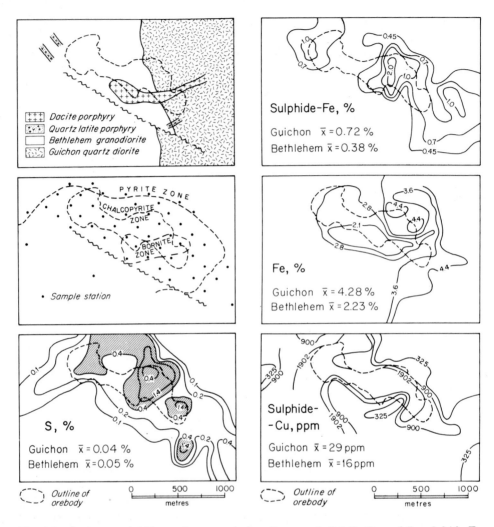

Fig. 8-9. Geology, sulphide zoning, sample locations, and distribution of S, sulphide-Fe, total Fe, and sulphide-Cu at the Bethlehem JA copper porphyry deposit, British Columbia, Canada (from Olade and Fletcher, 1976b).

concentration towards mineralization. On the basis of a sulphide-selective leach the author showed that in anomalous samples Cu in the mineral separates is present dominantly in a sulphide form, whereas Zn is apparently dominantly in a non-sulphide form. In anomalous samples the amount of Cu in whole rock increases as the amount of biotite increases. The chemical composition of biotite also varies with proximity to mineralization; biotite from mineralized areas contains higher Fe and lower Mg than biotite from background samples.

TABLE 8-II

Distribution of Cu and Zn in biotite, magnetite and quartz-feldspar mineral separates, and in whole rock samples, Highland Valley, British Columbia, Canada (from Olade, 1979).

			Biotite	Magnetite	Quartz-feldspar	Whole rock
Mean Cu (ppm)	background, weakly	$n = 10$	98	67	29	27
	anomalous, strongly	$n = 7$	883	251	210	334
	anomalous,	$n = 9$	2549	576	1440	1919
Mean Zn (ppm)	background, weakly	$n = 10$	349	68	17	30
	anomalous, strongly	$n = 7$	292	54	19	30
	anomalous,	$n = 9$	275	59	23	29

Although the variations in Cu in mineral separates are significant (as pointed out by Olade, 1979 and as is evident from the data in Table 8-II), the contrast in whole rock is better. In terms of exploration, there is no advantage in undertaking costly mineral separations.

The characteristic alteration zones associated with porphyry copper deposits are a central K-rich zone (K-feldspar, biotite, sericite) and a marginal Ca-rich zone (epidote, calcite, montmorillonite). It is therefore reasonable to expect an increase in K and a decrease in Ca in rocks towards the centre of a copper porphyry. Since Rb and Sr are geochemically coherent with K and Ca, respectively, these two elements should also vary in concentration with proximity to mineralization (see discussion below). Olade (1977) and Olade and Fletcher (1975) investigated the distribution of these elements on 115 drill hole samples from the Bethlehem JA and the Valley Copper deposits. Analytical results for Rb, Sr, K_2O, Na_2O, CaO, and MgO are given in Table 8-III. The Valley Copper deposit, which lies wholly within the Bethsaida Phase, shows a relative enrichment of Rb, Sr, and K_2O at the periphery of the deposit and a depletion of Na_2O, CaO, and MgO relative to regional background. Within the alteration zones K_2O generally increases from the periphery of the deposit towards the centre, whereas Na_2O, CaO, and MgO show a decrease. Although the Bethlehem JA deposit has two hosts — the Guichon and Bethlehem Phases — which makes comparison with background more difficult, the same trends noted for Valley Copper are discernible. These element patterns are logically attributable to alteration of plagioclase feldspars to K-feldspar, K-clay, and sericite in the central potassic zones (which involved addition of K) and leaching of Na and Ca; Mg is lost through the breakdown of ferromagnesian minerals.

The distribution of alteration zones, K_2O, CaO, Rb, Sr, and Rb:Sr ratios

TABLE 8-III

Distribution of Rb, Sr, K_2O, CaO (from Olade and Fletcher, 1975) and Na_2O and MgO (from Olade, 1977) and some element ratios in background samples from the Bethlehem JA and Valley Copper deposits*

	Geometric mean (ppm)		Arithmetic mean (%)				Ratio		
	Rb	Sr	K_2O	CaO	Na_2O	MgO	Rb:Sr	K:Rb	Ca:Sr
Regional background									
Guichon Phase (10, 7)	50	935	1.86	5.45	4.06	2.31	0.05	309	42
Bethlehem Phase (12, 9)	44	693	1.74	3.96	4.52	1.31	0.06	336	41
Bethsaida Phase (6, 9)	36	588	2.06	2.90	4.83	0.54	0.06	475	35
Bethlehem JA									
Propylitic zone (22, 16)	41	686	1.77	3.17	4.42	1.97	0.06	238	34
Argillic zone (12, 10)	47	617	1.62	2.83	3.92	1.19	0.08	248	27
Potassic zone (8, 6)	81	235	3.84	1.48	2.91	0.74	0.34	388	58
All samples (54, 54)	50	579	1.91	2.82	4.27	1.41	0.09	317	35
Valley Copper									
Argillic zone (18, 11)	57	641	2.44	2.66	3.03	0.42	0.09	303	20
Phyllic-potassium zone (30—)	69	396	3.26	2.08	—	—	0.17	364	36
Phyllic (—12)	—	—	—	—	2.07	0.39	—	—	—
Potassic (—10)	—	—	—	—	2.62	0.50	—	—	—
Quartz-rich (7, 7)	45	529	2.08	1.66	3.16	0.34	0.09	364	38
All samples (61, 61)	59	562	2.78	1.95	2.78	0.42	0.10	391	25

* Numbers in brackets refer to number of samples for Rb, Sr, K_2O, CaO, and Na_2O, MgO respectively.

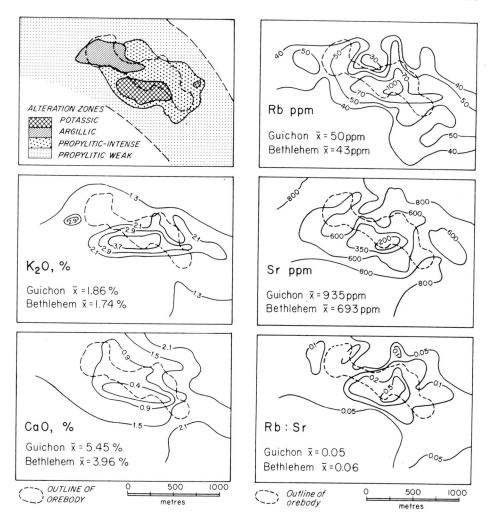

Fig. 8-10. Geology and distribution of K_2O, CaO, Rb, Sr, and Rb:Sr ratio at the Bethlehem JA copper porphyry deposit, British Columbia, Canada, (Redrawn with permission from Olade and Fletcher, 1975, *Economic Geology*, vol. 70, 1975, fig. 2, p. 16 and fig. 5, p. 18 and from Olade, 1977.)

at the Bethlehem JA deposit are shown in Fig. 8-10. The ore zone is clearly defined by positive anomalies for K and Rb and negative anomalies for Ca and Sr. The sympathetic variation between K and Rb and between Ca and Sr is evident. The mineralogical control is also clear; the east-west-trending zone of high K and Rb coincides with pervasive potassic alteration (K-feldspar and sericite); high values of Ca and Sr occur in the outer propylitic zone (epidote, chlorite, zeolites, and carbonate minerals). The distribution pattern of Rb:Sr ratios does not offer any particular advantage over the single element patterns.

In terms of exploration, Cu and S anomalies offer the largest target. Olade and Fletcher (1976a) suggest that comparatively widely spaced samples of 0.5 to 1.0 km should indicate proximity to major mineralized zones.

Other Canadian studies

In the Copper Mountain area of British Columbia porphyry-type mineralization occurs in volcanic rocks of the Triassic Nicola Group where they are intruded by slightly younger Si-poor stocks. All known major mineralization occurs in the Mine Series of the Nicola Group which is bounded on the north by the Lost Horse Stock and on the south by the Copper Mountain Stock (Fig. 8-11). The mineralization (69×10^6 tonnes of 0.56% Cu) occurs as disseminated and fracture-filling chalcopyrite-pyrite (with bornite becoming important in rocks adjacent to the Copper Mountain Stock); only south of the Ingerbelle deposit have extensive pyrite halos been identified. Two types of metasomatic mineral assemblages have been recognized: K-rich in the east of the mineralized belt, and Na-rich in the west. There is, however, no systematic mineralogical zoning that can be used to readily identify mineralization.

Gunton and Nichol (1975) collected 443 samples of an average weight of about 2500 g; each sample comprised about 20 separate chip samples. In the vicinity of the deposits samples were collected (as far as practical) on a 152.4 m × 152.4 m grid. Background material from the Nicola Group was

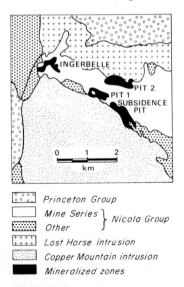

Fig. 8-11. Geology of Copper Mountain area, British Columbia, Canada (modified from Gunton and Nichol, 1975).

TABLE 8-IV

Average of elements in bedrock from the Ingerbelle—Copper Mountain area, British Columbia (from Gunton and Nichol, 1975)[*]

	Nicola Group			Copper Mountain intrusion	
	background	south contact	north contact Mine Series	south	north
	$n = 40$	$n = 51$	$n = 173$	$n = 18$	$n = 38$
CaO (%)	9.3	8.9	10.2	10.3	9.9
K_2O (%)	1.6	2.7	2.1	3.0	2.7
Na_2O (%)	2.8	3.6	4.4	3.1	3.6
P_2O_5 (%)	0.47	0.55	0.59	0.66	0.61
Rb (ppm)	8.3	16.1	18.4	16.5	17.8
Sr (ppm)	399	649	919	1425	1464
Cu_t (ppm)	242	140	1184	299	267
Cu_s (ppm)	55	111	779	264	192
$Cu_s : Cu_t$ (%)	22.7	79.3	65.8	88.3	71.9
K:Rb	1602	1392	947	1509	1259
Ca:Sr	167	98	79	52	48
Rb:Sr	0.021	0.025	0.020	0.012	0.012

[*] All elements except Cu determined by X-ray fluorescence; Cu_t is total Cu measured by AAS after a $4HClO_4/1HNO_3$ digestion; Cu_s is sulphide Cu measured by AAS after extraction with ascorbic acid/hydrogen peroxide.

collected at a distance of up to 24 km from intrusive contacts. The results, summarized in Table 8-IV, show:

— Na, K, P, Rb, Sr, Cu, and sulphide-Cu (determined by the ascorbic acid/hydrogen peroxide technique of Cameron et al., 1971) are enriched in the Nicola formations surrounding the Copper Mountain Stock relative to background areas;

— P, Sr, Cu, and sulphide-Cu are enriched and K and Mn are depleted in Nicola rocks of the mineralized belt relative to the less intensely altered Nicola rocks south of the Copper Mountain Stock;

— the rocks around the Ingerbelle deposit in the west are characterized by higher Na and lower K than the rocks around the Copper Mountain deposits in the east.

The most useful exploration elements are Cu and Sr; they are 8-fold and 1.5-fold higher, respectively, over an area 3 times more extensive than the actual area of the deposits.

Although the Copper Mountain copper deposits are of the porphyry type, the fact that mineralization occurs within the intruded volcanic sequence rather than within the intrusion itself must cause differences in chemical response relative to the deposits within the Guichon Batholith. There is, for example, a considerable difference between the proportion of sulphide to total Cu in background Nicola rocks (23%) compared to that in contact Nicola rocks (66—79%). More importantly, the differences in alteration between Copper Mountain and Guichon are marked by considerable differences in element variations associated with mineralization. Compared with Guichon, the rocks at Copper Mountain directly associated with mineralization (Mine Series) are enriched in Na, Ca, and Sr instead of being depleted; they are depleted in K instead of enriched and are characterized by lower Rb:Sr ratios rather than by higher ratios. It is important to note, however, that the Mine Series *is* enriched in K and Rb compared to background Nicola rocks. The Copper Mountain dioritic intrusion itself is extraordinarily enriched in Cu, with an average of 229 ppm in the south and 267 ppm in the north. Gunton and Nichol (1975) concluded that samples should be analyzed for Cu, P, Sr, Na, and K and collected at a density of about 4 samples/km^2 for reconnaissance and 30 samples/km^2 for detailed exploration.

Wolfe (1974) studied an early Precambrian porphyry-type molybdenum-copper deposit in northwestern Ontario (400 km northeast of Winnipeg). The Setting Net Lake Stock is a porphyritic granodiorite-quartz monzonite intruded into Precambrian mafic volcanic and sedimentary rocks. A 460 m × 2500 m easterly trending mineralized zone occurs in the northern part of the stock in altered quartz monzonite; widespread molybdenite is present in 1- to 2-mm quartz veins and in minor disseminations accompanied by pyrite and chalcopyrite. The average grade in the eastern part of the mineralized zone is 0.067% MoS_2 with minor Cu. Alteration consists of sericitization and albitization of plagioclase, chloritization of biotite, and introduction of pyrite. Wolfe collected duplicate 500-g to 1-kg composite rock chip samples 5—15 m apart from 25 sites over the mineralization and from 25 sites over unmineralized ground. Analytical results for Cu, Mo, Zn, and Mn are given in Table 8-V.

The mineralized zone is characterized by the average contents of Mo and Cu being 15-fold and 7.6-fold higher, respectively, than unmineralized rocks. The distribution of both Mo (Fig. 8-12) and Cu (Fig. 8-13) show a southwest trend that is discordant to the east-west trend of the molybdenite zone as revealed by surface blasting, trenching, and diamond drilling. Notwithstanding the paucity of outcrop in the northwestern part of the stock which influences the geochemical pattern, there are good grounds for suspecting the

TABLE 8-V

Average content of Cu, Mo, Zn and Mn (ppm) in 50 rock samples from 25 sites in each of the mineralized and non-mineralized zones of Precambrian granodiorite—quartz monzonite, Setting Net Lake, Ontario, Canada (from Wolfe, 1974)

	Cu	Mo	Zn	Mn
Non-mineralized	8	1.7	30	277
Mineralized	61	26	30	240

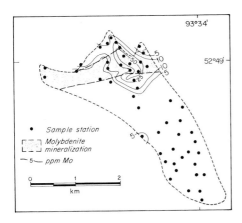

Fig. 8-12. Distribution of total Mo in rocks of the Setting Net Lake stock, Ontario, Canada (from Wolfe, 1974). Contoured values represent an average of two determinations on two separate samples at each site.

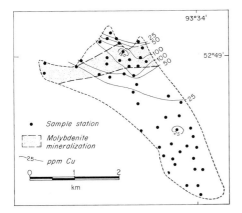

Fig. 8-13. Distribution of hot HNO_3/HCl-soluble Cu in rocks of the Setting Net Lake stock, Ontario, Canada (from Wolfe, 1974). Contoured values represent an average of two determinations on two separate samples at each site.

presence of additional mineralization. The abundances of Mo and Cu in the south part of the stock are consistent with background levels in granitic rocks and give no indication that the northern part of the stock is mineralized.

U.S. CASE HISTORIES

Mineral Butte

The Mineral Butte copper deposit (Pinal County, Arizona) is presumed to be a porphyry-type mineralization; it occurs mostly within a Precambrian

200

TABLE 8-VI

Distribution of elements in bedrock, Mineral Butte, Arizona (from Chaffee, 1976a)*

	Range (ppm)	Median (ppm)	Average low-Ca granites (ppm)
Cu	5—23000	45	10
Co	N(5)—70	L(5)	1.0
F	60—1360	400	850
Au	L(0.02)—10	L(0.02)	0.004
Pb	N(10)—300	20	19
Mo	N(5)—100	L(5)	1.3
Ag	N(0.5)—20	N(0.5)	0.037
Zn	0.5—300	5	39

*L(5) detected at level below detection at 5 ppm; N(5) = not detected at detection limit of 5 ppm. Cu, Au, Ag and Zn measured by AAS; Co, Pb and Mo measured by semi-quantitative emission spectrograph; F by specific ion electrode.

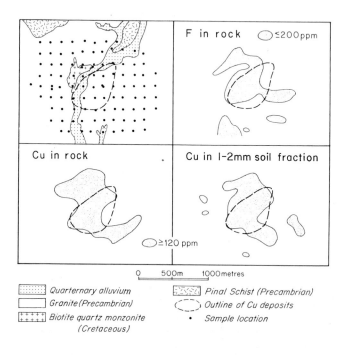

Fig. 8-14. Geology, sample location, distribution of Cu and F in rock, and distribution of Cu in soil at the Mineral Butte deposit, Arizona, U.S. (Redrawn with permission from Chaffee, 1976a, *U.S. Geological Survey Bulletin*, 1278-D, 1976, fig. 2, p. 5; fig. 4, p. 14; and fig. 6, p. 18.)

granite at its contact with an Upper Cretaceous biotite quartz monzonite stock. Only secondary copper minerals occur at the surface. Chaffee (1976a) collected bedrock and residual soil samples on a 150 m grid over the deposit. The rocks were analyzed for 39 elements; only Cu, Co, F, Au, Pb, Mo, Ag, and Zn were sufficiently above the detection limit or showed any variation related to mineralization. No background samples were taken specifically; the median values shown in Table 8-VI are therefore a biased estimate of background. Comparing the median values with average values for a comparable low-Ca granite, Cu is clearly enriched, whereas F is depleted.

The spatial distributions of Cu and F in bedrock and Cu in soil are shown in Fig. 8-14. The distribution of Cu in both bedrock and soils shows a strong positive anomaly that corresponds fairly well with the outline of mineralization, whereas the distribution of F in bedrock shows a similar negative anomaly. Of the other elements considered, the distribution of Mo is anomalous over a wide area and shows little relation to mineralization; Pb has a negative anomaly of about the same dimensions as the Cu anomaly; Ag has a positive anomaly similar to Cu; Zn has a positive anomaly similar to Cu but it is open to the south; Co is anomalous over a more restricted area than Cu; and Au is anomalous at only a few sample sites. The ore element, Cu, gives the best indication of mineralization; there seems to be no advantage in this case in determining other elements. More importantly, the distribution of Cu in soils is at least as indicative of the mineralized zone as the distribution of Cu in rocks; here there would be no advantage in collecting rocks samples at all.

Lights Creek

The Lights Creek Stock (125 km NNW from Reno, Nevada) is a medium- to fine-grained quartz monzonite to granodiorite intruded into Triassic-Jurassic metavolcanic rocks with lesser gabbro, diorite, and an older granodiorite. The surface outcrop covers 18 km^2. There is extensive copper porphyry-type mineralization with variable amounts of disseminated chalcopyrite, pyrite, and bornite throughout the stock. The greatest amount of mineralization (see Fig. 8-15) is at Moonlight Valley, Copper Valley, and Sulphide Ridge. The Superior Mine is a discrete vein and stockwork system that has been worked underground; the Engels Mine (in adjacent diorite-gabbro-metavolcanic rocks) is a more massive type of mineralization in a shear zone that is nevertheless believed to be genetically related to the Lights Creek Stock. Putman (1975) collected 2052 samples from 98 drill core and 31 surface locations as shown in Fig. 8-15. The samples were digested in aqua regia, and Cu, Pb, and Zn were measured by AAS; metal extraction was virtually complete by this technique.

From data given by Putman it can be calculated that less than 1% of the samples have Cu contents lower than 30 ppm (most of the stock has less

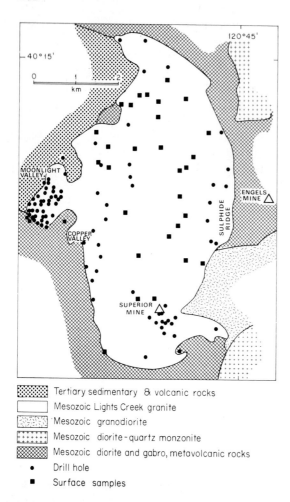

Tertiary sedimentary & volcanic rocks
Mesozoic Lights Creek granite
Mesozoic granodiorite
Mesozoic diorite-quartz monzonite
Mesozoic diorite and gabro, metavolcanic rocks
• Drill hole
■ Surface samples

Fig. 8-15. Geology and sample locations, Lights Creek, California, U.S. (compiled from Putman, 1975).

than 300 ppm Cu), and the stock therefore, is highly enriched in Cu. The trend surface map for Cu (Fig. 8-16) shows a very pronounced increase in Cu content towards the margins of the stock in the vicinity of mineralization. Drill hole data also show that the Cu content increases upwards through the stock in areas of high surface concentrations of Cu, and also that local variance of Cu is much higher in areas of high Cu contents. The variation of Pb and Zn is much less significant, although there is an increase in Pb content towards the west and an increase in Zn content towards the southwest and west.

It appears that the Lights Creek Stock could have been identified as a

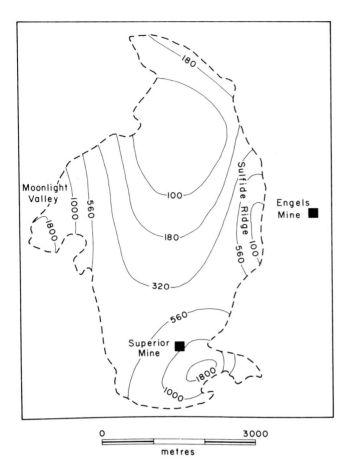

Fig. 8-16. Trend surface distribution of Cu in Lights Creek stock, California, U.S. (Redrawn with permission from Putman, 1975, *Economic Geology*, vol. 70, 1975, fig. 6, p. 1232.)

copper-positive intrusion warranting investigation on the basis of measurement of Cu content in relatively few samples. The trend surface distribution pattern of Cu within the stock strongly suggests that analysis of Cu in rocks from a more detailed survey would serve to delimit the main area of interest. This conclusion, however, should be treated with some caution; the trend surface pattern would be distorted due to the relatively few surface samples, the large number of drill core samples in mineralized areas, and, in the case of Moonlight Valley, the very large number of sample locations relative to the remainder of the intrusion. It should also be noted that although the Guichon Batholith in British Columbia is enriched in Cu as a whole, the porphry deposits occur in Cu-deficient phases of the batholith (see above).

The Kalamazoo deposit

The Kalamazoo and San Manuel copper porphyry deposits (48 km north-east of Tucson, Arizona) are faulted segments of a single deposit formed during the emplacement of a Late Cretaceous monzonite porphyry that was intruded into a Precambrian quartz monzonite (Fig. 8-17). Chaffee's (1976b) study was based on 3-m lengths of drill core at intervals of approximately 15 m along drill holes 23 and 30. Because of rotation of the Kalamazoo deposit from the original vertical position, the two vertical drill holes pass through the side of the deposit and not through the top of the deposit. Hole 23 passed through the Tertiary and Cretaceous sedimentary rocks and the Cretaceous stock, whereas hole 30 passed mostly through the Precambrian stock. The samples were analyzed for 60 elements; 17 of these elements showed some zonal variation in both holes — Au, Zn, Tl, Li, Na, K (determined by AAS); Cu, Mo, Ag, Co, Mn, Rb, Fe, Ba, B (determined by semi-quantitative emission spectroscopy); S (determined titrimetrically); and Se (determined colorimetrically). Background was chosen as the mean concentration in conglomerate in hole 23 on the grounds that these rocks were derived from material away from hydrothermal effects and provide the best estimate of pre-mineralization chemistry. The content of K in the two stocks is quite different, and Chaffee could not reconcile it with the content in conglomerate, he therefore selected separate backgrounds for the two holes.

The author concluded that trace elements show consistently better anomalies than the major elements; with the possible exception of Ba, trace elements do not appear to show significant variations due to differences in

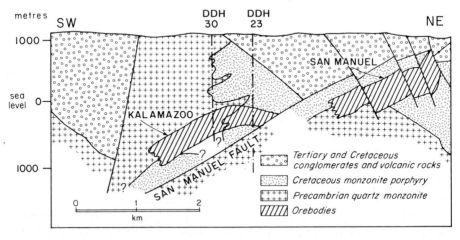

Fig. 8-17. Generalized geology of Kalamzoo and San Manuel copper porphyry deposits (Arizona, U.S.) showing location of drill holes (DDH30, DDH23) through the Kalamazoo deposit (from Chaffee, 1976b).

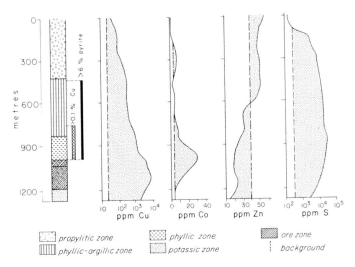

Fig. 8-18. Distribution of Cu, Co, Zn and S in DDH30 through the Kalamazoo copper porphyry deposit, Arizona, U.S. (from Chaffee, 1976b).

the chemistry of the two stocks. Four characteristic patterns are illustrated by Cu, Co, Zn, and S in Fig. 8-18. The main features are summarized below:

— The maximum Cu content occurs in the outer part of the potassic alteration zone, and anomalous values persist for more than 900 m from the ore zone. The patterns for Mo, Au, and Ag are similar, but the anomalies are not as extensive; Mo extends 60—240 m but not beyond the pyrite aureole; Au extends to more than 60 m; Ag does not extend beyond the level of 0.1% Cu.

— The maximum concentration of Co occurs in the inner phyllic zone, with a minor anomaly in the propylitic zone (Au also has a similar anomaly in the propylitic zone). The anomaly extends 150—300 m.

— The distribution of Zn shows a very strong negative anomaly in the potassic-phyllic zone, apparently changing to a positive anomaly in the phyllic zone and continuing in the propylitic zone — giving an anomaly for more than 900 m. The distribution of Mn is similar. Both Tl and Rb show negative anomalies throughout the drill sections.

— The maximum concentrations of S occur several hundreds of metres beyond the ore zone and extend for at least 900 m. The sharp decrease at the top of hole 30 (also shown by Mn and Se) is attributed to weathering. The distribution of Se is similar to that of S.

As was the case in the Highland Valley deposits in the Guichon Batholith, the distribution of Cu and S offers the most direct and extensive positive halos. But, as Chaffee (1976b) rightly points out, the zonal distribution of element patterns offers substantial promise for detecting blind deposits and determining the likely position of the halo relative to an ore zone — hence indicating the direction in which exploration should proceed. In terms of elements that are easily determined by AAS, the measurement of Cu, Co, Zn, and Mn offer the most convincing anomaly pattern in this example.

Copper Canyon

Theodore and his co-workers have described the geology and the results of their geochemical investigations of the Copper Canyon deposit in Nevada in a number of publications (Nash and Theodore, 1971; Theodore and Nash, 1973; Theodore et al., 1973; Theodore and Blake, 1975). The large, low-grade copper and gold deposit at Copper Canyon occurs in fractured and altered Pennsylvanian sedimentary rocks less than 300 m from a Tertiary granodiorite laccolith. Nash and Theodore (1971) point out that although this is strictly a contact metasomatic deposit, it has numerous features that are similar to porphyry copper deposits.

There are two distinct ore bodies (Fig. 8-19): the east ore body occurs

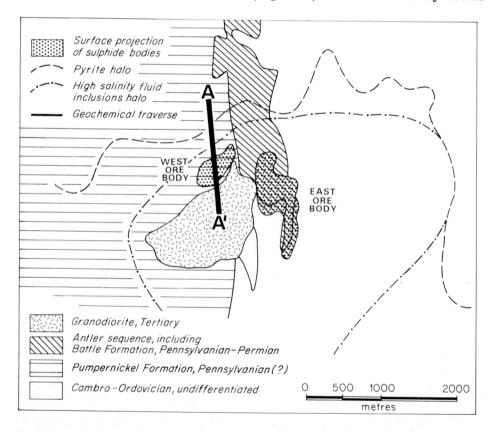

Fig. 8-19. Geology of the Copper Canyon area, Nevada, U.S. (compiled from Nash and Theodore, 1971 and Theodore and Nash, 1973).

in the lower conglomeratic member of the Middle Pennsylvania Battle Formation replacing hematite and calcite zones; the west ore body is a sulphide-rich manto deposit replacing the chert and argillite of the Pumpernickel Formation. Sulphide minerals in both deposits are pyrite, chalcopyrite, pyrrhotite, and marcasite, with lesser amounts of arsenopyrite, native gold and silver, and traces of sphalerite, molybdenite, and galena. Wall rocks are pyritized for 600—3600 m beyond the ore bodies.

Some of the results obtained from 2927 rock samples collected over an area of 16 km^2 are described by Theodore and Nash (1973). Each sample consisted of about 1-kg chips from about 10 m^2 and was analyzed spectrographically for Sb, As, Ba, Bi, B, Cd, Cr, Co, Cu, Pb, Mn, Mo, Ni, Ag, Sr, Sn, V, and Zn; Au and Hg were determined by AAS. A geochemical zonation of 10 elements — Ag, Au, Bi, Co, Cu, Hg, Mo, Pb, Sr, and Zn — was most clearly defined in the Pumpernickel Formation around the west ore body. This zonation is illustrated in Fig. 8-20 (these data specifically exclude dispersion effects due to fracture-controlled base metal vein deposits around the laccolith). In summary the data show:

— Ag is weakly enriched (0.5—10 ppm) throughout the pyritized rocks;
— Bi is weakly enriched (10—200 ppm) around the west ore body;

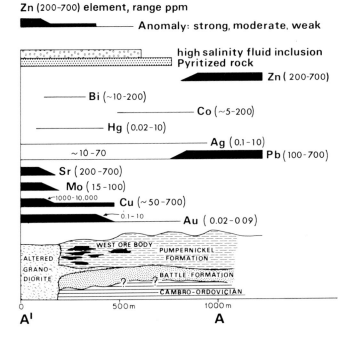

Fig. 8-20. Schematic illustration of element zoning in bedrock at Copper Canyon, Nevada, U.S. (Redrawn with permission from Theodore and Nash, 1973, *Economic Geology*, vol. 68, 1973, fig. 2, p. 567.)

— Pb and Zn are strongly enriched in many samples just beyond the pyritized rock, and erratic high values of Pb extend into the granodiorite;

— Mo and Sr are more abundant in the granodiorite than in wall rocks, although Sr appears to be *relatively* depleted in the intrusion (see below);

— Au has a well-defined but weak anomaly (0.02—0.09 ppm) in an area 300 m from the intrusion and is strongly enriched (0.1—1.0 ppm) over the west ore body and in the granodiorite;

— anomalous Co (5—20 ppm) occurs over the outer 450 m of pyritized rock;

— weak Hg anomalies (0.02—1.0 ppm) occur across the northern part of the intrusion;

— Cu is moderately enriched over the west ore body and is highly anomalous (1000—10,000 ppm) in the granodiorite.

The distribution of high-salinity fluid inclusions forms a broad halo in both intrusive and sedimentary rocks 3—4 km across centred on the east ore body (Fig. 8-19). A high-salinity halo falls within the pyrite halo and is therefore a less extensive target. Its shape, however, is strongly controlled by lithology and intensity of fracturing; the restricted extent of the halo to the west and south is apparently due to the lower permeability of the relatively unfractured argillite of the Pumpernickel Formation.

A comparison of the Copper Canyon granodiorite with other granodioritic intrusions in the Battle Mountain mining district (Table 8-VII) shows that it is mainly distinguished by higher K and lower Ca; this is reflected in higher Rb

TABLE 8-VII

Analyses for selected elements in granodioritic rocks from the Battle Mountain Mining District, Nevada, U.S.

Location[*]	Cu	Pb	Zn	U	Rb	Sr	Ca	K	Rb:Sr	K:Rb
Non-mineralized										
(3)	6.6	14	44	2.3	102	640	3.36	2.64	0.16	259
(4)	120	20	140	3.6	119	590	3.21	2.62	0.20	220
(5)	7.6	20	29	4.7	126	490	2.71	2.70	0.26	215
(6)	4.2	28	50	4.2	111	540	2.93	2.47	0.21	223
(7)	4.8	20	43	2.4	120	570	2.57	2.66	0.21	222
(8)	900	22	160	4.3	147	414	1.93	3.19	0.36	217
Average	174	21	78	3.6	121	541	2.79	2.71	0.22	224
Copper Canyon										
(9)	1500	20	110	9.0	215	319	0.68	5.27	0.68	245
(12)	900	33	245	1.9	140	348	2.50	3.57	0.40	255
(17)	6600	21	31	5.1	274	209	0.21	5.56	1.31	203
(18)	555	18	39	2.1	175	204	0.67	4.57	0.86	261
Average	2411	23	106	4.5	201	274	1.02	4.74	0.73	236

[*] Location numbers (in brackets) as given in source (Theodore et al., 1973).

and lower Sr. There is little difference in the K:Rb ratios between the Copper Canyon and other granodiorites, but the Ca:Sr ratio is considerably lower in the Copper Canyon granodiorite. The greatest difference is shown by the Rb:Sr ratios which at Copper Canyon are more than three times the average of the other intrusions. The difference in trace element contents are less marked between Copper Canyon and non-mineralized intrusions; there is no significant difference in Pb and Zn contents, Cu is very enriched at Copper Canyon (but its distribution is erratic), and Cl is erratically enriched at Copper Canyon.

There is no doubt that the Copper Canyon granodiorite could have been identified as a target area on the basis of K and Ca (or better identified by Rb and Sr) together with the anomalously high Cu contents. There is, however, some truth in the conclusion of Theodore and Nash (1973) that the distribution of Cu and some other elements could lead to attention being directed to the intrusion rather than to the surrounding country rocks; only the high-salinity fluid inclusion halo (and to a lesser extent the pyrite halo) has a locus centred over mineralization.

Breckenridge district

Rock geochemical data from the area of the old Wirepatch mine in the Breckenridge mining district of Colorado was used by Pride et al. (1979) to suggest that a molybdenum deposit may exist at depth. Mineralization at the Wirepatch mine consists of pyrite, sphalerite, galena, and minor gold and silver; it occurs in an intrusive breccia with a rhyodacite porphyry that itself is intruded into monzonite and quartz monzonite porphyry. These intrusive rocks are collectively known as the Wirepatch complex and were intruded during Tertiary times into the Cretaceous Pierre shale. The Wirepatch complex, as well as the Pierre shale, has a central phyllic-argillic alteration zone (sericite, kaolin, with or without quartz and pyrite), succeeded outwards by a chloritic propylitic zone (with or without pyrite, calcite, and epidote).

The alteration zones are well defined as geochemical halos of K_2O, which increases from less than 3% K_2O to more than 6%, and Rb, which increases from less than 100 ppm to more than 400 ppm towards the Wirepatch mine over a distance of 200—900 m. The same pattern is shown by a decrease of CaO from more than 1% to less than 0.1%, a decrease of Na_2O from more than 3% to less than 0.1%, and a decrease of Sr from more than 500 ppm to less than 10 ppm. These elements thus show a zoning characteristic of porphyry deposits.

The elements Cu, Pb, and Zn show a similar anomalous pattern, with the highest concentrations in the most intensely altered rocks, although the pattern is less regular and extensive. The content of Mo is low (less than 5 ppm in much of the area), although there are two small areas of modest enrichment (greater than 10 ppm) southeast of the Wirepatch mine. The authors nevertheless suggested that given the proximity to known Mo occurrences, the age and nature of the intrusions, and the nature and pattern of alteration and element distribution, there is a possibility that molybdenum mineralization may have been deposited at depth in a higher-temperature environment.

Ely district

A belt of copper porphyry mineralization in the Ely mining district of Nevada (extending about 13 km westward from the town of Ely) occurs in Tertiary monzonite stocks intruded into a sequence of Palaeozoic sedimentary rocks. The mineralized belt has a major Te anomaly associated with it; the highest values of Te occur near the outer edge of the mineralized zone (Fig. 8-21). Watterson et al. (1977) stated that there is an element zoning in the district from the centre outward in the order: S, Cu, Fe, Zn, Cd, Te, Sb, As, Pb, Hg, Ag, and Mn. The Te distribution (which Watterson and his co-workers stated may prove to be much more extensive with a better analytical sensitivity) is clearly a regional scale anomaly that could provide the basis for reconnaissance exploration (see also Chapter 4).

The area was reported on in greater detail by McCarthy and Gott (1978).

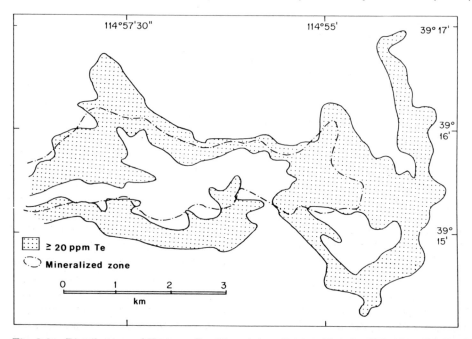

Fig. 8-21. Distribution of Te in rocks, Ely mining district, Nevada, U.S. (from McCarthy and Gott, 1978).

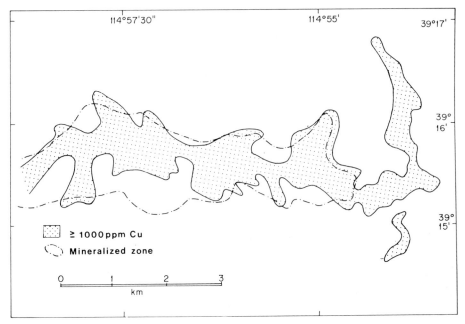

Fig. 8-22. Distribution of Cu in rocks, Ely mining district, Nevada, U.S. (from McCarthy and Gott, 1978).

The main mineralization is chalcopyrite and pyrite, with minor amounts of barite and molybdenite. Galena, sphalerite, silver, and gold occur in small replacement deposits in the peripheral sedimentary rocks. A total of 1466 samples of jasperoid, gossan, chert, limestone, shale, porphyry, and fracture fillings were analyzed by a semi-quantitative spectrographic method for Cu, Mo, Bi, Fe, Pb, Ag, and Mn; by colorimetric methods for Zn and Te; and by flameless AAS for Hg. There is a very pronounced zoning of elements: Cu (Fig. 8-22), S, Fe, and Mo occur in the central part of the district; Zn and Bi overlap and extend beyond the central area and are themselves overlapped by Te, As, and Sb. These latter three elements extend further out than Zn; Pb extends further out than Zn and is closely associated with Ag (which extends even further than Te). The outermost element is Mn. The zoning pattern is illustrated schematically in Fig. 8-23 where it is seen that an anomalous Mn:Cu ratio gives a halo of at least 7 km width compared to a width of the mineralized zone of less than 2 km. Obviously, the zoning pattern and the size of anomalies depend in part upon the background and the defined level of "anomalies". This is not discussed by McCarthy and Gott in their short paper; clearly if the Cu cut-off of 1000 ppm or the Zn cut-off of 5000 ppm were changed, the zoning pattern could also be expected to change. Irrespective of this, the anomalous halos are extensive and very clearly outline the Ely mining district.

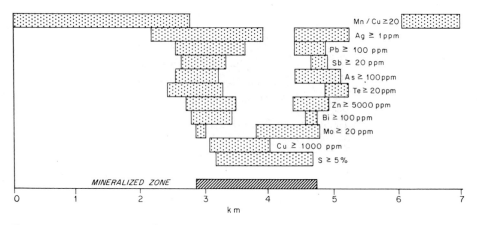

Fig. 8-23. Schematic illustration of metal zoning in rocks, Ely mining district, Nevada, U.S. (compiled from data in McCarthy and Gott, 1978).

Other U.S. studies

A molybdenite prospect near Nojal Peak (Lincoln County, New Mexico) in a Tertiary syenodiorite was investigated by limited rock sampling by Griswold and Missaghi (1964). The syenodiorite is intruded into Tertiary basalts which have suffered considerable alteration (argillic, pyritization) at the contact with the stock; the alteration is not believed to be related to later hydrothermal alteration associated with molybdenite deposition. During the latter part of the 19th century up until 1918 small-scale mining for gold and Pb-Zn-Cu-Ag veins took place in the area. The known molybdenite occurrences are small and have a grade of 0.12% Mo. A total of 110 rock chip samples (about 2.3 kg at each location) were taken and analyzed for Mo. The locations of 75 of these samples are shown in Fig 8-24; the remainder were taken outside the area to better assess background. Apart from the obvious correlation of Mo content with degree of alteration of syenodiorite, even the unaltered syenodiorite carries substantial Mo contents. The basaltic andesite also has a very high Mo content. Although present information suggests that the extent of molybdenite mineralization is not economic, this is an example where the trace element content of an intrusion indicates the presence of concentrations of the element.

The Cu content of biotite from a small number of samples taken from around the Esperanza and Sierrita copper porphyry deposits (Sierrita Mountains, Arizona) shows a trend of increasing concentration towards mineralization (Lovering et al., 1970). The host rock is a composite stock of Palaeocene age that is largely granodiorite with a porphyritic core and quartz monzonite around the deposit (Fig. 8-25). Although high anomalous contents of Cu in biotite extend as far as 4 km from the deposits, there are too few samples to determine unequivocally whether this pattern is due entirely

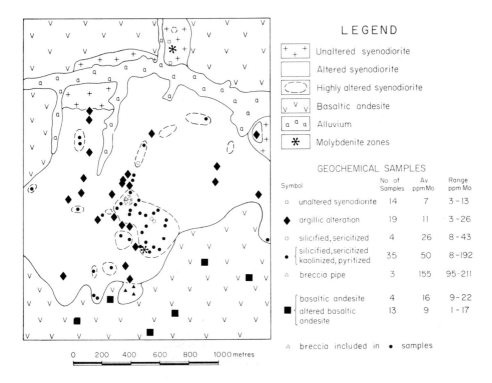

LEGEND

+ + +	Unaltered syenodiorite
☐	Altered syenodiorite
⟨ ⟩	Highly altered syenodiorite
v v v	Basaltic andesite
a a a	Alluvium
*	Molybdenite zones

GEOCHEMICAL SAMPLES

Symbol		No of Samples	Av ppm Mo	Range ppm Mo
a	unaltered syenodiorite	14	7	3 – 13
◆	argillic alteration	19	11	3 – 26
o	silicified, sericitized	4	26	8 – 43
•	silicified, sericitized kaolinized, pyritized	35	50	8 –192
△	breccia pipe	3	155	95 – 211
	basaltic andesite	4	16	9 – 22
■	altered basaltic andesite	13	9	1 – 17

△ breccia included in • samples

Fig. 8-24. Distribution of Mo in rocks of the Rialto stock, New Mexico, U.S. (compiled from Griswold and Missaghi, 1964).

to proximity to mineralization (the regional implications of this study are discussed in Chapter 4).

An operationally useful exploration technique based on radiometric measurement of K was proposed by Davis and Guilbert (1973). They demonstrated that the distribution of K in rocks around porphyry deposits at Morenci, Ajo, and Mineral Park (all in Arizona) and at Santa Rita (New Mexico) reflects alteration patterns, and that the K distribution can be quite adequately determined with a multichannel gamma-ray spectrometer. Potassium was found to be enriched 1.2—3 times over mineralization compared to chemically and temporally similar nearby barren intrusions.

USE OF RUBIDIUM, STRONTIUM AND POTASSIUM

Variations in the concentrations of Rb and Sr and their major element partners K and Ca with proximity to porphyry-type mineralization have been mentioned in the context of several case histories discussed earlier in this chapter. Variations in these elements have also been shown to be useful

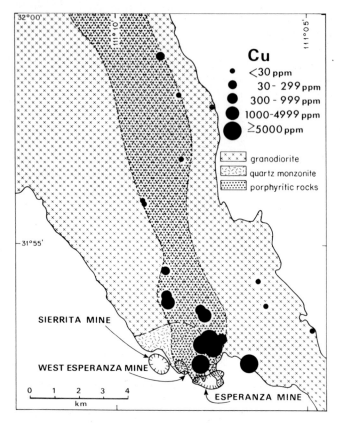

Fig. 8-25. Distribution of Cu in biotite, Pima mining district east of Sierrita Mountains, Arizona, U.S. by spectrographic analyses. (Redrawn with permission from Lovering et al., 1970, *U.S. Geological Survey Professional Paper*, 700-B, 1970, fig. 2, p. 6.)

in regional scale exploration for mineral deposits, especially in the case of tin (Chapter 5). Variations in the content of Rb will also be shown to be useful in exploration for vein-type and massive sulphide deposits in succeeding chapters.

Traditionally Rb and Sr are used as a measure of fractionation of a magma. Thus, Rb is geochemically coherent with K; as K increases with fractionation, so does Rb. Due to the larger size of Rb it tends to become concentrated under conditions of extreme fractionation; an increase in Rb relative to K outside the normal range of K:Rb ratios of 150—300 indicates increased fractionation. The size of Sr^{2+} is intermediate between that of K^+ and Ca^{2+}; it enters plagioclase (but can be captured in K-minerals in early fractions), and the Ca:Sr ratio decreases during fractionation.

Over and above the effects of fractionation, Oyarzun (1975) suggested that in hydrothermally altered rocks (and those affected by low-grade metamorphism) Sr should be depleted due to depletion of Ca; Rb should be relatively (or absolutely) enriched because K is not easily removed due to its strong bonding to secondary micas and clay minerals and because K is actually introduced during some forms of alteration. Thus, the Rb:Sr ratio should increase in rocks that have been hydrothermally altered. Earlier in this chapter this situation was demonstrated by the results obtained by Olade and Fletcher (1975) on the Guichon Batholith. An apparently unique exception to the generalization is provided by the situation at Copper Mountain (described by Gunton and Nichol, 1975) where Sr (and Ca) is enriched rather than depleted.

Lawrence (1975) suggested, on the basis of the empirical relations from the Galway pluton at Lettermullan in Ireland, that high Rb:Sr ratios are indicative of sulphide mineralization as well as high fractionation. At Letter-mullan the pluton is an adamellite — the Carna Granite — that passes towards the margin of the batholith successively through the Transitional Granite, the Errisbeg Townland Granite (porphyritic adamellite), a non-porphyritic phase of the Errisbeg, and into the Murvey Granite (albite—quartz—K-feldspar leucogranite). A garnetiferous, highly siliceous phase of the main Murvey Granite occurs between it and the wall rock. Rare molybdenite, chalcopyrite, and pyrite are restricted to the garnetiferous Murvey Granite and aplite and quartz veins; alteration is restricted to chloritization of biotite and minor sericitization of K-feldspar.

The distribution of Rb:Sr ratios across the granitic sequence is shown in Fig. 8-26. In the Murvey Granite it is 5—8 times higher than in the other phases, and it is 2 times higher again in the garnetiferous phase. Analytical data for Ca, K, Ba, Rb, and Sr for the various granite phases are shown in Table 8-VIII. The K:Rb ratios show a fairly steady decline from the Carna

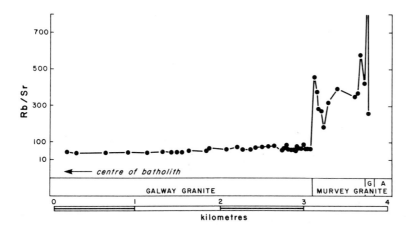

Fig. 8-26. Distribution of Rb:Sr ratios across the Galway granite of Lettermullan, Ireland (from Lawrence, 1975). G = garnetiferous granite; A = aplite.

TABLE 8-VIII

Average contents of selected elements in granites of Lettermullan (from Lawrence, 1975)

	Carna Granite	Transitional Granite	Errisbeg Townland Granite	Non-porphyritic Errisbeg Townland Granite	Murvey Granite	Garnetiferous Murvey Granite	Aplite
n	5	5	19	7	11	2	8
Ca (%)	2.36	2.20	1.74	1.74	0.41	0.33	0.30
K (%)	3.24	3.30	3.40	3.05	4.02	3.76	4.25
Ba (ppm)	971	1035	820	818	329	194	107
Rb (ppm)	135	138	146	149	212	235	262
Sr (ppm)	304	308	242	236	60	33	26
Ca:Sr	77.6	71.4	71.9	73.7	68.3	100	115
K:Rb	240	239	233	205	189	160	162
Ba:Rb	7.19	7.50	5.62	5.49	1.55	0.83	0.41
Rb:Sr	0.444	0.448	0.603	0.631	3.53	7.12	10.08

Granite to the garnetiferous Murvey Granite (which is largely a reflection of increasing fractionation). The Ba:Rb ratio shows an abrupt decline in the Murvey Granite that probably indicates an increase in volatiles; this ratio is halved in the garnetiferous Murvey Granite, and halved again in the aplite (i.e., it shows a similar trend to the Rb:Sr distribution).

The variation in Rb and Sr with proximity to copper porphyry mineralization was illustrated by Oyarzun (1975) and in more detail by Armbrust et al. (1977) with data from El Teniente, Rio Blanco, and Los Bronces (Disputada) mines in Chile.

The El Teniente ore body (5.44×10^6 tonnes of 1.55% Cu) lies mainly within folded andesitic lavas (Cretaceous) and partly within Tertiary intrusive quartz diorite and dacite porphyry; mineralization surrounds a 1200-m-diameter breccia pipe. The sulphide minerals are bornite-chalcopyrite near the pipe, grading outwards to chalcopyrite-pyrite, and finally pyrite with minor chalcopyrite. A leached capping 0.13 m thick overlies oxide and supergene enriched zones. Alteration in the andesite is to biotite adjacent to the pipe, grading outwards to a propylitic assemblage of epidote, chlorite, and magnetite with minor pyrite, tourmaline and biotite; quartz diorite adjacent to andesite has been intensely altered to a quartz-sericite rock which passes outwards to a propylitic assemblage of chlorite, epidote, calcite, quartz, pyrite, and minor sericite; much of the dacite porphyry is relatively unaltered, but locally there is pervasive sericitization-silicification. Anhydrite and tourmaline are abundant in all three rock types.

Mineralization at Rio Blanco, which consists of disseminations and veinlets of pyrite, chalcopyrite, and martite (with minor magnetite, specularite, and molybdenite) is confined to an andesitic roof pendant in a granodiorite quartz monzonite intrusion. Alteration consists of an inner zone (coinciding with the ore zone) of quartz-sericite-pyrite and minor pyrite, and an outer propylitic zone of calcite, chlorite, epidote, and minor sericite.

The Los Bronces ore deposit (32.6×10^6 tonnes of 1.2% Cu) is 1500 m west of Rio Blanco and is confined to a breccia pipe within granodiorite. The host granodiorite is relatively fresh, with chloritization of ferromagnesian minerals and minor sericite; the breccia is intensely sericitized and silicified. There is no significant supergene enrichment at either ore zone.

A total of 140 rock samples were collected at El Teniente from mine workings and from nine drill holes; 39 samples were collected from mine workings at Rio Blanco and from two drill holes in the breccia pipe at Los Bronces. Unaltered rocks from up to 3 km from the mines were collected for estimation of background. Each sample weighed about 500 g and consisted of one or two fragments.

Analytical data for K, Rb, and Sr are given in Table 8-IX for intrusive rocks and in Table 8-X for extrusive rocks. There is a clear increase in Rb content (2—3 times background in quartz diorite, rather less in granodiorite-quartz monzonite, and up to 4 times background in andesite) and a decrease

TABLE 8-IX

Comparison of K, Rb, and Sr contents and K:Rb and Rb:Sr ratios in barren and mineralized intrusive rocks

Sample No.		Rock type, location (number of samples)	K (%)	Rb (ppm)	Sr (ppm)	K:Rb	Rb:Sr
A1		average low-Ca granite	4.2	170	100	247	1.70
A2		average granite	3.47	145	240	285	0.51
A3		average granodiorite	2.55	110	440	230	0.25
B1	b.	QD, Chile (7)	1.19	50	475 (5)	238	0.11
B2	b.	GRD-QM, Chile (9)	2.66	108	550 (7)	246	0.20
B3	m.	El Teniente, QD, propylitic (3)	2.11	113	363 (2)	187	0.31
B4	m.	El Teniente, QD, quartz sericite, hypogene (11)	3.43	157	393	218	0.40
B5	m.	El Teniente, QD, quartz sericite, supergene (4)	3.36	156	109	215	1.43
B6	m.	Rio Blanco, GRD-QM, propylitic (3)	3.05	127	408 (2)	240	0.31
B7	m.	Rio Blanco, GRD-QM, quartz sericite (4)	3.60	144	64	250	2.25
B8	m.	Los Bronces, GRD-QM, quartz sericite (5)	4.42	191	—	231	—
C1	b.	Guichon Creek, QD (10)	1.54	50*	935*	308	0.05
C2	b.	Guichon Creek, GRD (12)	1.44	43*	693*	335	0.06
C3	b.	Guichon Creek, GRD-QM (6)	1.71	36*	588*	475	0.06
C4	m.	Valley Copper, GRD-QM, argillic (18)	2.03	57*	641*	356	0.09
C5	m.	Valley Copper, GRD-QM, phyllic-potassic (30)	2.71	69*	396*	392	0.17
C6	m.	Valley Copper, GRD-QM, quartz-rich (7)	1.73	45*	529*	384	0.09
D1	m.	Copper Mountain, D, south (18)	2.49	16.5	1425	1509	0.01
D2	m.	Copper Mountain, D, north (38)	2.24	17.8	1464	1259	0.01
E1	b.	Galway, Carna G (5)	3.24	135	304	240	0.44
E2	b.	Galway, Transitional G (5)	3.30	138	308	239	0.45
E3	b.	Galway, Errisbeg Townland G (19)	3.40	146	242	233	0.60
E4	b.	Galway, non-porphyritic Errisbeg G (7)	3.05	147	236	205	0.63
E5	b.	Galway, Murvey G (11)	4.02	212	60	189	3.53
E6	(m.)	Galway, garnetiferous Murvey G (2)	3.76	235	33	160	7.12

F1	b.	Battle Mountain, GRD (6)	2.71	121	541	224	0.22
F2	m.	Battle Mountain, Copper Canyon, GRD (4)	4.74	201	274	236	0.73
G1	b.	Anchor mine, GA (6)	3.9	365	75	107	4.9
G2	m.	Anchor mine, G (6)	3.8	1035	5	37	207
H1	b.	Queensland, Dido, G (12)	1.25	45	812	277	0.06
H2	b.	Queensland, Dumbano, G (19)	2.78	127	374	220	0.34
H3	m.	Queensland, Elizabeth Creek, G (73)**	3.93	427	26	92	16
H4	m.	Queensland, Finlayson, G (6)**	3.98	388	43	102	9.1
H5	m.	Queensland, Mareeba, G (23)**	3.53	356	88	99	4.1
H6	m.	Queensland, Esmeralda, G (30)**	4.16	291	78	143	3.7
H7	m.	Queensland, Almaden, G (15)	2.47	146	196	165	0.74
H8	m.	Queensland, Herbert River, G (40)	3.47	233	101	149	2.3
J1	m.	St. Austell, G (35)**	3.84	886	51	43	17
J2	m.	Carmmenellis, A (1)**	4.3	400	80	10.8	5.0
K1	b.	Nova Scotia granites (5)	3.92	251	57	156	4.4
K2	(m.)	Nova Scotia granites (2)**	3.79	455	25	83	18.2
L1	b.	New Brunswick, St. George, G (1)	3.5	228	197	153	1.5
L2	(m.)	New Brunswick, Beech Hill, A (1)**	3.9	522	36	75	14.5
L3	m.	New Brunswick, Mt. Pleasant, G (1)**	5.9	742	13	80	57.1

* = geometric means; ** = tin-bearing granites.

b. = barren; (m.) = minor mineralization; m. = significant mineralization; G = granite s.s. and s.l.; QD = quartz diorite; GRD = granodiorite; QM = quartz monzonite; D = diorite; GA = granite-adamellite; A = adamellite. Sources of data: A1 = S.R. Taylor (1968); A2, A3 = Turekian and Wedepohl (1961); B = Armbrust et al. (1977); C = Olade and Fletcher (1975); D = Gunton and Nichol (1975); E = Lawrence (1975); F = Theodore et al. (1973); G = Groves (1972); H = Sheraton and Black (1973); J1 = Exley (1958); J2 = Butler (1953); K = Smith and Turek (1976); L = Dagger (1972).

TABLE 8-X

Comparison of K, Rb and Sr contents and K:Rb and Rb:Sr ratios in barren and mineralized andesitic rocks

Sample No.	Rock type and location (number of samples)	K (%)	Rb (ppm)	Sr (ppm)	K:Rb	Rb:Sr
A	average Circum-Pacific, Andesite	1.33	31	429	385	0.08
B1	b. Chile, average background andesite (12)	1.46	43	340	494	0.09
B2	m. El Teniente, propylitic (29)	1.53	85	180	331	0.26
B3	m. El Teniente, biotite, hypogene (54)	2.71	154	176	390	0.39
B4	m. El Teniente, biotite, supergene (22)	2.69	158	170	234	0.68
B5	m. El Teniente, leached capping (4)	3.26	154	212	29	5.31
B6	m. Rio Blanco, propylitic (2)	1.85	68	272	235	0.29
B7	m. Rio Blanco, ore zone (16)	4.80	140	343	71	1.97
C1	b. background, Copper Mountain (40)	1.33	8.3	1602	399	0.02
C2	m. Copper Mountain, south contact (51)	2.24	16.1	1392	649	0.03
C3	m. Copper Mountain, Mine Series (173)	1.74	18.4	947	919	0.02

b. = background; m. = mineralized. Sources of data: A = Turekian and Wedepohl (1961); B = Armbrust et al. (1977); C = Gunton and Nichol (1975).

in Sr (about three-quarters background in igenous rocks and down to about one-half background in andesite). In the andesites at El Teniente the Rb is about twice background in the propylitic zone, whereas K is only slightly above background, and therefore the K:Rb ratio is about one-half background. In the biotite zone Rb almost doubles but so does the K content, and therefore the K:Rb ratio is about the same as in the propylitic zone. The Sr content decreases by about one-third in the propylitic zone compared with background, apparently due to chloritization of pyroxene and a partial breakdown of plagioclase. In the biotite zone — where normally Ca and Sr may be expected to be depleted — the Sr content is in fact a little higher than in the propylitic zone because, although Ca silicate minerals have been replaced, Ca is retained as anhydrite.

In the supergene zone and the leached capping from above the ore (biotite zone) Rb shows no change from the biotite zone, and K increases in the leached capping. Strontium, however, decreases markedly in the supergene zone compared with the biotite zone, and shows a further almost 10-fold decrease in the weathered leached capping. These relations follow from the hydration of anhydrite to gypsum and its subsequent dissolution and hence loss of Ca and Sr, whereas K and Rb are retained in sericite and other clay-type minerals.

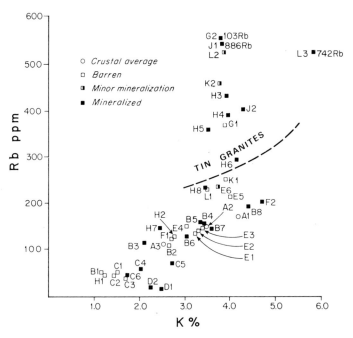

Fig. 8-27. Variation of K and Rb in barren and mineralized granites (see Table 8-IX for key to numbers).

222

Broad trends for K, Rb and Sr contents in intrusive rocks at El Teniente are similar to those described in andesites. The element distribution patterns in the volcanic and intrusive rocks at Rio Blanco and Los Bronces are also similar to those at El Teniente and differ only in a far greater depletion of Sr — which is attributable to the lack of anhydrite as an alteration mineral.

Potassium, Rb, and Sr contents of plutonic rocks associated with the Chilean porphyry deposits discussed above, the porphyry deposits of western Canada, various mineralized granites of Canada, Australia, and the British Isles (including the tin granites), and various background intrusions are included in Table 8-IX above. These data are used to show the variations in Rb and K in Fig. 8-27 and the variations of K:Rb and Rb:Sr ratios in Fig. 8-28. There is a good positive correlation between K and Rb if the tin granites are excluded (the Copper Mountain rocks fall below the main trend because of their very low Rb content, and the tin granites are distinguished from the others by marked increases in Rb relative to K; see also Chapter 5). In the porphyry granites there is a trend of increasing K and Rb content from non-mineralized to mineralized rocks *in any particular region*, but the absolute level of concentration is quite different and distinct in each region. There is, therefore, no universally applicable background level for these elements.

Consideration of the K:Rb and Rb:Sr ratios (Fig. 8-28) also shows that

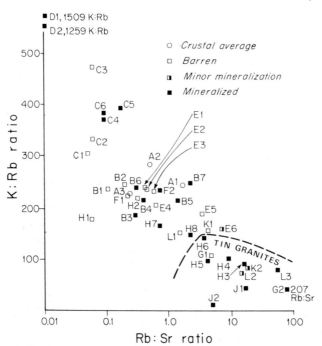

Fig. 8-28. Variation of K:Rb and Rb:Sr ratios in barren and mineralized granites (see Table 8-IX for key to numbers).

there are quite well-defined trends from non-mineralized to mineralized rocks *within* a particular region, but that no general background can be defined. The Rb:Sr ratio always increases from non-mineralized to mineralized rocks; the variation in the K:Rb ratios is less consistent, but generally the ratios decrease from non-mineralized to mineralized rocks. The tin granites are again distinctly different from other plutons by virtue of low K:Rb and high Rb:Sr ratios. The Copper Mountain rocks are off the main trend because of their uncharacteristic low Rb and high Sr contents.

Comparable K, Rb and Sr data for extrusive rocks associated with por-

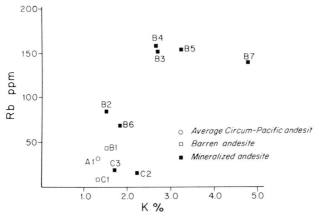

Fig. 8-29. Variation of K and Rb in barren and mineralized andesite (see Table 8-X for key to numbers).

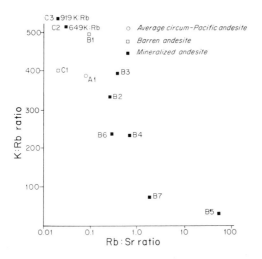

Fig. 8-30. Variation in K:Rb and Rb:Sr ratios in barren and mineralized andesite (see Table 8-X for key to numbers).

phyry deposits is limited to the Chilean examples and Copper Mountain. The relations of K and Rb contents (Fig. 8-29) and K:Rb and Rb:Sr ratios (Fig. 8-30) are similar to those described for igneous rocks — i.e., enrichment of Rb and K contents, decrease in K:Rb ratios, and increase in Rb:Sr ratios — in mineralized rocks compared to non-mineralized rocks. At Copper Mountain the K:Rb ratios are unusually high, and the Rb:Sr ratios are unusually low (again due to the very low Rb and high Sr contents).

In both igneous and volcanic rocks there appears to be adequate evidence to indicate a Rb enhancement and a Sr depletion associated with mineralization. The Rb enhancement and the Sr depletion appear to be related to the mineralizing process — and are not simply a reflection of increased fractionation. The enhancement and depletion are local in effect (except possibly in the case of tin), and a local background must be established. Moreover, there would be a distinct advantage in also determining K and Ca for control purposes.

CONCLUSIONS

Plutonic intrusions that host copper porphyry-type deposits generally appear to be enriched in Cu, although where detailed data are available (e.g., the Guichon Batholith in Canada) there is evidence that the deposits occur in Cu-deficient phases of the intrusion. Whereas the dominant control of element concentration is lithological (primarily variations in the content of ferromagnesian minerals), the superimposed hydrothermal alteration associated with porphyry deposits is generally clearly recognized. In the immediate vicinity of deposits the content of Cu and S is enriched for up to 900 m from the ore zone in all cases for which data are available. The evidence indicates that the enrichment of Cu is due largely to the presence of sulphide minerals and not to Cu substitution in silicate mineral lattices.

The large physical size of porphyry-type deposits and the characteristic mineralogical alteration associated with them has undoubtedly inhibited the use of rock geochemistry in exploration. There are, however, several neglected applications of rock geochemistry to the search for porphyry-type deposits. The alteration zones themselves are faithfully reflected by variations in the content of K, Ca, Rb, and Sr. Since quantitative measurement of either the major element pairs (K, Ca) or the trace element pairs (Rb, Sr) is more readily achieved than quantitative determination of the mineralogy in fine-grained alteration zones, measurement of either the major or the trace element pairs (preferably both) offers distinct advantages in the preliminary definition of alteration zones.

Mineralization is characterized by enrichment of K and Rb and depletion of Ca and Sr. In particular, a trend of decreasing K:Rb ratios and increasing Rb:Sr ratios from non-mineralized to mineralized intrusions appears to be a useful indicator. It is important to note that the absolute levels of element content and their ratios are quite different in different geographic areas; local background relations must be established. The very clear distinction between tin and porphyry plutons on the basis of absolute levels of K, Rb, and Sr, however, is apparently of universal application.

A second potentially extremely useful application of rock geochemistry is in the search for deeply-buried deposits based on the pronounced zoning exhibited by elements in a number of studies. In the Ely mining district in Nevada a regional-scale element zoning has been demonstrated from the centre of a 2-km-wide area of mineralization outwards to a distance of 7 km as follows: (Cu, S, Fe, Mo)—(Zn, Bi)—(Te, As, Sb)—Ag—Mn. On a mine scale there appears to be a distinct zoning from the centre of mineralization outwards as follows: (Cu, Rb)—(S, Mo)—Zn. The elements Cu and Rb (and generally also K) have peak values over the centre of the ore zone, whereas S reaches peak values at the periphery of the ore zones; all these elements, however, have positive anomalies over the entire ore zone and decline in concentration with increasing distance from mineralization. Molybdenum tends to have a simple positive anomaly at the margin of mineralization, and Zn and Sr have a pronounced negative anomaly over mineralization; Zn content increases to a moderate positive anomaly at the periphery, thereafter declining to background values with increasing distance from the deposit.

The zonal arrangement of the elements mentioned could obviously be used to determine the relation of a geochemical halo to a blind deposit and hence could be used to determine the probable position of mineralization. More work on the detailed distribution of trace elements around porphyry deposits could lead to the recognition of a specific zonal sequence that might be used very effectively to determine position, attitude, and possibly the depth of such deposits from surface data.

In terms of practical exploration, a sample density of 2—4 samples/km^2 should be adequate for preliminary investigations; a sample density of 20—30/km^2 is required for detailed follow-up exploration. Sample weights should be 1—2 kg. The most useful elements to determine are Cu, Zn, K, Ca, Rb, Sr (and S if facilities are readily available). A large range of other elements — e.g., As, Sb, Te, Ag, Au, Mn — may be useful in individual cases, but there are not adequate data available to make recommendations on their general applicability.

Chapter 9

LOCAL AND MINE SCALE EXPLORATION FOR VEIN AND REPLACEMENT DEPOSITS

INTRODUCTION

In this chapter geochemical responses to the smallest economic mineral exploration target — individual narrow veins and linear replacement deposits — are considered. As may be expected from the type of mineralization, deposits of precious metals are prominent among the investigations recorded; their geochemical response, however, does not appear to differ in character from that of base metals. Generally the maximum geochemical anomaly has a width of a few tens of metres and, where sufficiently detailed sampling has been done, elements display a characteristic logarithmic decay pattern away from mineralization.

Three main topics will be discussed: (1) logarithmic decay patterns (largely, but not entirely, based on underground fresh samples); (2) surface and weathered bedrock surveys; and (3) techniques of determining dimensions and distance of veins.

LOGARITHMIC DECAY PATTERNS

One of the earliest detailed investigations of the distribution of ore elements in wallrock was by J.S. Curtis (1884) who measured the amount of Ag (by Ag assay techniques) in limestone around the high-grade silver replacement bodies in the Eureka mining district of Nevada. The results of some of Curtis' data are shown in Fig. 9-1 where it is seen that anomalous contents of Ag extend for 20—30 m into the limestone of the wallrock. (An interesting feature is that the peak value was some distance from the contact.) Curtis noted that none of the samples contained any visible silver and were apparently indistinguishable from normal limestone in the region; he concluded that the Ag must have been introduced from solutions. Another early study (Finlayson, 1910b) around lead-zinc veins in limestone in England indicated a strong Pb and Zn anomaly that decays to background 5—10 m from the vein; the results of the study are shown in Fig. 9-2. The Zn maximum located

228

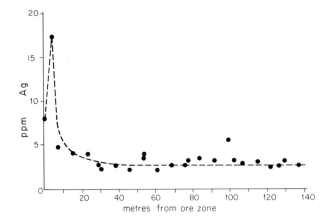

Fig. 9-1. Distribution of Ag in limestone, Eureka mining district, Nevada, U.S. (compiled from J.S. Curtis, 1884).

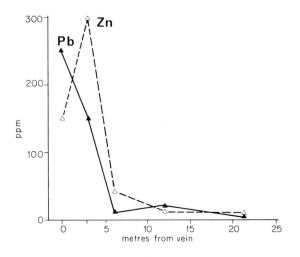

Fig. 9-2. Distribution of Pb and Zn in limestone adjacent to a lead-zinc sulphide vein, Nenthead, U.K. (compiled from Finlayson, 1910b).

some distance from the vein contact should be noted. Both of these early studies demonstrate the essential logarithmic decay pattern of elements away from mineralization.

One of the earliest detailed studies of wallrock dispersion *specifically* designed for exploration purposes is described by Morris and Lovering (1952). Surface outcrop and underground rock samples were collected around ore veins of various types in the Tintic district of Utah and analyzed for base metals. The authors concluded that the logarithmic decay pattern in the wallrock away from the veins strongly supported a diffusion process,

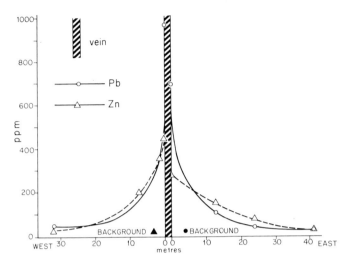

Fig. 9-3. Dispersion of Pb and Zn (HF-HClO$_4$ extraction) in quartz monzonite, Swansea vein, Tintic district, Utah, U.S. (compiled from Morris and Lovering, 1952).

and they noted that the ratio of metals in the wallrock reflected the composition of the ore bodies. The extent of the dispersion pattern appeared to be largely controlled by the composition of the wallrock; anomalous values were recorded up to about 25 m from a vein in quartz monzonite, but were noted up to only 3 m from a vein in dolomite.

The dispersion of Pb and Zn in quartz monzonite around the Swansea vein (in the southern part of the main Tintic district) is shown in Fig. 9-3. The ore in the Swansea vein is argentiferous cerussite, with some limonite, anglesite, and residual galena and pyrite in sugary quartz gangue. The vein walls are silicified, argillized, and pyritized; the vein selvage (less than 1 m thick) is fine grained quartz with sericite and pyrite. The vein was mined to a width of less than 2 m and to a depth of about 75 m. Alteration is visible up to 0.5—23 m on the east of the vein, but it is restricted to 1.5—3.0 m on the west of the vein; this asymmetry is reflected in the dispersion of Pb and Zn that appears to be slightly more extensive to the east of the vein. The dispersion pattern around the Swansea vein should be compared to that around the Carisa copper-gold ore body and the Eureka Hill copper-lead mine which are both in a dolomite host (Fig. 9-4). The Carisa deposit is siliceous ore of Cu carbonates and Cu arsenides, with appreciable quantities of Au and Ag and minor amounts of Pb; the unoxidized ore is argentiferous enargite and galena (no Zn minerals were reported). The ore at the Eureka Hill mine is galena, sphalerite, and enargite. The maximum dispersion in dolomite at Carisa is less than 1 m from the ore; and it is only about 3 m at Eureka (this is significantly less than in the igneous rock at the Swansea vein). Morris and Lovering (1952) attributed the restricted dispersion in

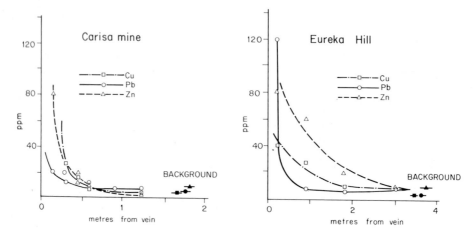

Fig. 9-4. Dispersion of Cu, Pb, and Zn (1N HCl extraction) in dolomite around the Carisa copper-gold and the Eureka Hill copper-lead deposits, Tintic district, Utah, U.S. (Redrawn with permission from Morris and Lovering, 1952, *Economic Geology*, vol. 47, 1952, fig. 13, p. 707 and fig. 15, p. 708.)

dolomite to the neutralizing and reactive effect of the dolomite on diffusing ore solutions, and to the greater porosity of the shattered and altered quartz monzonite at the Swansea vein.

The general logarithmic decay pattern for element dispersion in wallrock around vein deposits established by the early work described above is confirmed in some more recent studies by Bailey and McCormick (1974) around lead, zinc, silver, and copper lode deposits in the Park City district of Utah. The deposits occur in fault fissures of less than 1 m to about 25 m in thickness which range up to 760 m in length. The country rock is fine- to medium-grained fractured quartz sandstone (Pennsylvanian Weber Quartzite) with a high porosity and large amounts of intergranular carbonate (mostly calcite). Variations in the carbonate adjacent to the veins indicate that some of it is secondary; pyrolusite, associated with coarse grained calcite in many places, occurs in the wallrock up to about 1 m from the veins. The sedimentary sequence is intruded by Tertiary diorites and granodiorites.

The authors collected 411 chip samples underground adjacent to seven productive veins, including three of the richest in the district. Only 5 elements — Cu, Pb, Zn, Ag, and Mn analyzed by AAS — showed a systematic variation with distance from the veins. There were too few samples at individual veins for interpretation; the entire sample population is shown as a plot of element concentration against distance from veins in Fig. 9-5. The results are consistent with those of Morris and Lovering (1952) in the Tintic district: Cu has the narrowest halo (about 12 m), whereas Mn, Pb, and Zn halos extend for 15—18 m. The broad (24 m) halo for Ag is of doubtful validity because many of the samples were near the lower limit of analytical sensitivity (see, however, below).

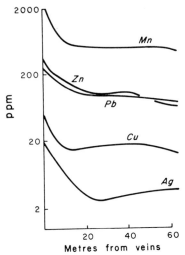

Fig. 9-5. Average smoothed curves for distribution of Mn, Pb, Zn, Cu, and Ag in country rock adjacent to mineralized veins, Park City district, Utah, U.S. (Redrawn with permission from Bailey and McCormick, 1974, *Economic Geology*, vol. 69, 1974, fig. 4, p. 380.)

Much more detailed sampling of wallrock around vein deposits in the Searchlight mining district of Nevada by Bolter and Al-Shaieb (1971) also substantiates the general logarithmic decay pattern and shows some additional features of interest. A total of 250 andesitic wallrock samples were collected from the vicinity of two gold-silver deposits. The Chief of the Hill vein has a vertical dip and is less than 1.5 m wide in andesite porphyry which shows no evidence of hydrothermal alteration; the grade of mineralization is 10—30 ppm Au and 1—37 ppm Ag. At the Duplex mine veins dip about 25°, and the wallrock shows evidence of strong hydrothermal alteration; the grade of mineralization is 10 ppm Au, 1—10 ppm Ag, 2—6% Cu, 0.2—1.0% Zn, and 1.0% Pb.

The distributions of Au and Ag (determined by neutron activation) in surface samples across the Chief of the Hill vein are shown in Fig. 9-6; although values are still above background at the end of the traverse (30 m from the vein), Bolter and Al-Shaieb stated that background is usually reached 30—45 m from the vein. The decay curve for Ag is clearly of the logarithmic type, but the detailed sampling indicates that the decay curve for Au is much less clearly so (because of the half-dozen samples close to the vein that are abnormally low in Au).

The distributions of Zn (determined by AAS after HF-HNO$_3$-HClO$_4$ digestion) in hanging-wall and footwall samples (collected underground) at the Duplex mine are shown in Fig. 9-7. The distribution of Zn in the footwall shows the characteristic logarithmic decay and reaches background 12—18 m from the vein; Bolter and Al-Shaieb stated that this is typical. An

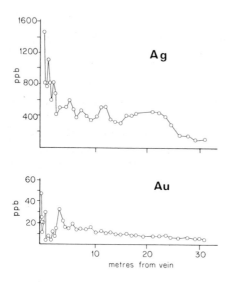

Fig. 9-6. Distribution of Ag and Au in surface andesite porphyry, Chief of the Hill Mine, Nevada, U.S. (Redrawn with permission from Bolter and Al-Shaieb, 1971, *Geochemical Exploration, Canadian Institute of Mining and Metallurgy, Special Volume,* 11, 1971, fig. 1, p. 289.)

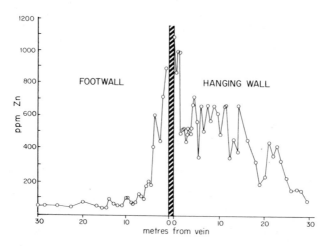

Fig. 9-7. Distribution of Zn in andesite-latite in underground cross-cut, Duplex Mine, Nevada, U.S. (Redrawn with permission from Bolter and Al-Shaieb, 1971, *Geochemical Exploration, Canadian Institute of Mining and Metallurgy, Special Volume,* 11, 1971, figs. 2 and 3, p. 290.)

exception to the pattern is shown by the distribution of Zn in the hanging wall which is clearly anomalous for more than 30 m from the vein; it also shows a marked relative depletion in about half a dozen samples near to the vein similar to that observed for Au at the Chief of the Hill vein. The authors concluded that this depletion could be caused by hydrothermal leaching, but they noted that in the same location not all elements show this depletion (e.g., compare Ag and Au in Fig. 9-6).

Bolter and Al-Shaieb concluded that in the area of their work Au and Ag generally are more widely dispersed in wallrock around veins than the base metals. Whether this conclusion has general applicability is not known. Certainly, however, the widths of precious metal dispersion at Searchlight is greater than for other metals at other localities described above, and the results of Bailey and McCormick (1974, see above) show that Ag may be more widely dispersed than base metals.

The rich silver veins (native silver, carbonates, Ni-Co arsenides) of the Cobalt district of Ontario in Canada occur as narrow (a few centimetres and less) veins over strike and dip lengths of less than 100 m. The veins are generally steep to vertical and cut Archaean Keewatin greenstones and associated rocks, the Proterozoic Cobalt Series conglomerate and quartzite, and the Nipissing diabase (which cuts both the Archaean and Proterozoic rocks). Dass et al. (1973) described the results of detailed mineralogical and geochemical investigations of the very narrow wallrock alteration zones around the veins; this alteration generally extends for 15—20 cm and rarely exceeds 30 cm. Of greater interest to the present discussion are the more widespread primary geochemical dispersions of a variety of elements into the host rocks. The distribution of Sb, As, Co, Ni, and Ag in greywacke of the Cobalt Series are illustrated in Fig. 9-8. Anomalous values extend 30—45 m

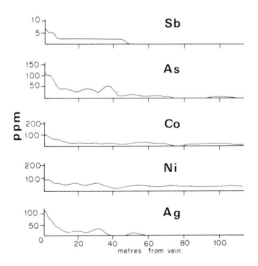

Fig. 9-8. Distribution of Sb, As, Co, Ni, and Ag in greywacke of the Cobalt Series, Little Silver Vein, Cobalt, Ontario, Canada. (Redrawn with permission from Dass et al., 1973, *Geochemical Exploration 1972*. Institution of Mining and Metallurgy, 1973, fig. 8, p. 31.)

234

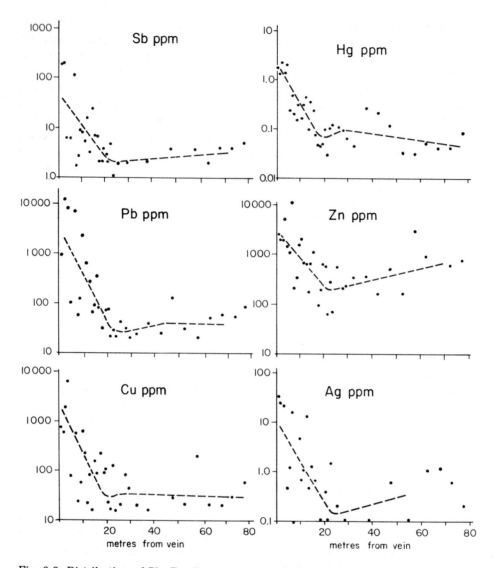

Fig. 9-9. Distribution of Pb, Zn, Sb, Cu, Ag, and Hg in rocks from cross-cut S2 at vein 2, Mavro Vouno, Mykonos, Greece (from Lahti and Govett, 1981).

from the veins — which is a considerable halo considering the very small width of the veins. The decay pattern is not clearly logarithmic and this, as well as the relatively broad dispersion, may be a function of the permeability of the greywacke. Similar patterns occur in the Keewatin rocks and the diabase, but the dispersion has a much more restricted extent; it is 15—16 m in the Keewatin rocks and less than 1 m in the diabase. Moreover, the decay pattern in Keewatin greenstones appears to more closely approximate a logarithmic pattern.

A final example of logarithmic decay patterns is provided by the dispersion of Cu, Pb, Zn, Ag, Sb, and Hg around the base metal-barite veins on Mykonos (see Chapter 6 for details of mineralization and geology). All elements decay to background levels at about 20 m from the vein into the hanging wall (Fig. 9-9). Some elements — especially Zn and Ag — then increase in concentration again (this may be due to secondary dispersion processes; see Lahti and Govett, 1981). Peak values of Cu, Pb, and Zn occur several metres from the vein; Zn shows the maximum displacement, with a peak value about 7 m from the vein.

SURFACE AND WEATHERED BEDROCK SURVEYS

A rock geochemical investigation has been described by Al-Atia and Barnes (1975) for gold and lead-zinc deposits that occur in rocks of Bala age (Upper Ordovician) about 83 km north of Swansea in South Wales (U.K.). The rock types are medium to dark shales with sandstones, grits, and conglomerates of both Bala and Llandovery (Lower Silurian) age; the boundary between the two age groups is difficult to distinguish due to the similarity of lithologies. There are no igneous rocks in the area, nor is there any obvious relation between mineralization and igneous activity. The Ogofau gold deposits were worked at least from Roman times until 1939. The gold occurs both free and in pyrite in quartz-pyrite veins along faults. A small amount of galena and sphalerite is present. The Nantymwyn lead-zinc deposits at Rhydymwyn occur as mineralized breccias along faults that cut grit banks in the Bala shales; mineralization becomes uneconomic as the faults pass into shales. The ore minerals are galena and sphalerite in a quartz gangue; there is no sign of replacement.

A total of 181 samples of shale were collected — 57 at the Ogofau deposits, 30 at the Nantymwyn mine, and 100 for background; they were analyzed for Cu, Pb, Zn, Rb, and K by XRF; Hg was determined by AAS. Frequency distributions for Pb, Zn, and Rb are shown in Fig. 9-10. At Ogofau 18% of the samples are anomalous in Zn and 30% are anomalous in Pb; there is no significant variation in the content of K and Hg compared to background rocks. The content of Rb is significantly enriched compared to background; 90% of the samples show Rb contents greater than the background mean plus two standard deviations, and 84% of the samples show Rb contents greater than the background mean plus three standard deviations. Both Rb and Zn reach their maximum values some distance from the veins, whereas Pb shows a simple symmetrical peak over the veins. At the Nantymwyn mine, despite the fact that samples could not be collected closer than 80 m to the veins, there are anomalous populations for Pb and Zn; Rb, however, is not anomalous (neither is Cu and Hg).

The genesis of the gold and the lead-zinc deposits is presumably different

236

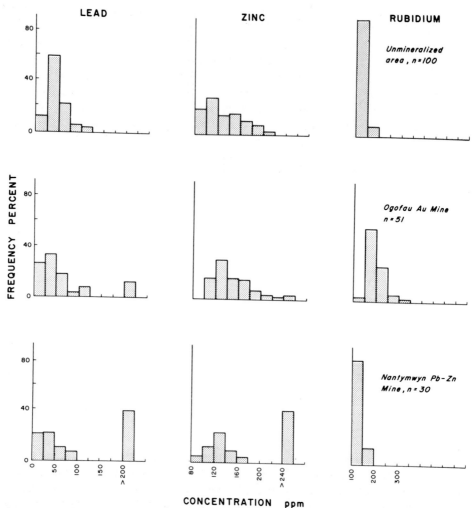

Fig. 9-10. Distribution of Pb, Zn, and Rb in shales around Ogofau gold mine, Nantymwyn lead-zinc mine, and an unmineralized area, Wales. (from Al-Atia and Barnes, 1975).

given the large differences in Rb content in rocks around the two deposits. Al-Atia and Barnes stated that the Rhydymwyn deposits are of very low temperature, rich in Pb and Zn, and poor in Fe, whereas the Ogofau deposit is a typically mesothermal epigenetic sulphide body, rich in Fe with some Pb, Zn, and Au. The intriguing aspect of the data from an exploration point of view is the pronounced enrichment of Rb at Ogofau.

The area was investigated more fully by Steed et al. (1976) who identified three distinct varieties of shale: normal (background) shales with only traces of pyrite; carbonaceous shales with abundant pyrite in the vicinity of the mine; and shales that are visibly altered and impregnated with pyrite within

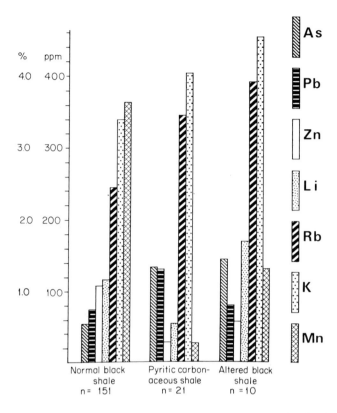

Fig. 9-11. Mean element contents of shales around the Ogofau gold mine, Wales (compiled from Steed et al., 1976). K in %; others in ppm.

the ore zone. In the veins and pyritic shales Au is present as inclusions in grains of pyrite and arsenopyrite; free Au occurs only near to the present surface. These varieties of shale are geochemically distinct, as shown in Fig. 9-11 which is based on 151 background shale samples, 21 pyritic carbonaceous shale samples, and 10 altered black shale samples (non-background samples were collected underground at the mine and from surface exposures within several hundred metres of the ore zone). The altered and associated pyritic shales are distinguished from background shales by enhanced values for Pb, As, and Rb, and by depletion of Zn and Mn. The altered shales have enhanced Li and are thus distinguished from the pyritic carbonaceous shales that have strongly depleted Li contents. The elements that give the most pronounced contrast are As, Rb, Mn, and Li. Contrary to the conclusion of Al-Atia and Barnes (1975), there is evidence of K enrichment in the pyritic carbonaceous shales and the altered shales, although the Rb:K ratio also increases from 72 in the background shales to 86 in pyritic carbonaceous shales and altered shales.

238

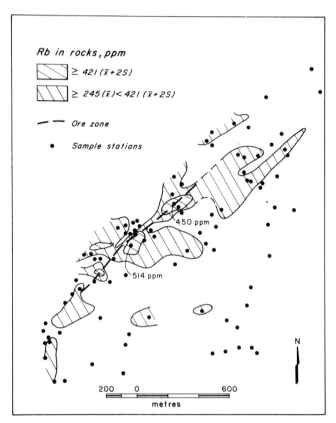

Fig. 9-12. Distribution of Rb in rocks, Ogofau gold mine, Wales. (Redrawn with permission from Steed et al., 1976, *Institution of Mining and Metallurgy, Transactions, Section B*, vol. 85, 1976, fig. 3, p. 114.)

The surface distribution of Rb in rocks is shown in Fig. 9-12. The ore deposit is clearly indicated by a zone of anomalous Rb content about 200—500 m wide. The distribution of As is not illustrated, but Steed and his co-workers demonstrated that As gives the broadest and best defined halo.

A local scale follow-up surface rock survey in the Mavro Vouno area on Mykonos (see above and Chapter 6) covered an area of about 4 km^2 (Fig. 9-13) where 1-kg chip samples were collected over an area of about 2 m^2 at intervals of 100 m along NW-SE lines 200 m apart; the samples were analyzed by AAS after concentrated HF-HNO$_3$-HClO$_4$ digestion (Lahti and Govett, 1981). The lowest contour of 70 ppm Zn represents the regional threshold. The distribution of Zn (as well as Pb and Ag) shows a well-developed E-W to NE-SW trend that is essentially parallel to the topographic contours; most, but not all, of the highest values occur in the volcanic-sedimentary sequence. The Zn anomaly at Mavro Vouno is divided into two, with relatively lower values on the higher ground. Other erratically distributed

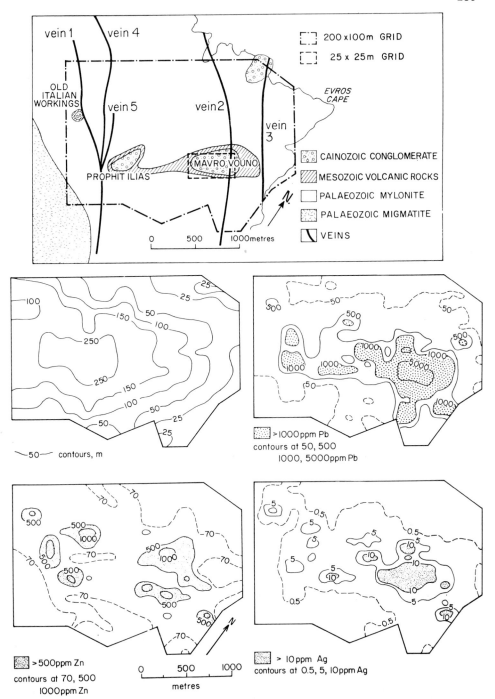

Fig. 9-13. Simplified geology, topography, and distribution of Pb, Zn, and Ag around Mavro Vouno, Mykonos, Greece (from Lahti and Govett, 1981).

high values west of Mavro Vouno appear to be related to veins 1, 4, and 5. Although the broad E-W trend of anomalous values parallel to the topography and the volcanic-sedimentary outlier is clearly dominant, there are more subtle NE-SW trends that are parallel to the vein system; these relations are more clearly evident in the distribution of Pb and Ag. The location of the mineralization at Mavro Vouno is most clearly indicated by the highest values for Pb and Ag (for both elements the lowest contour represents the regional threshold).

The broad and extensive anomalies (200–1000 m) are not believed to be due to primary dispersion related to emplacement of the veins senso stricto. When the veins pass from the granitic basement to the overlying volcanic sedimentary rocks they horsetail, and these latter rocks are impregnated with sulphides that are geochemically highly anomalous. In addition, oxidation and leaching of volcanic-sedimentary rocks — which occupy the highest ground — appear to have released metals that have become secondarily fixed in the upper levels of the surrounding (and topographically lower) granite. The geochemical anomalies are therefore partly primary (largely in the volcanic-sedimentary rocks) and partly secondary (largely in the granitic basement).

In many parts of the world weathering and oxidation over a long time period more severe than that on Mykonos results in fresh rock being inaccessable even by drilling to depths of more than 50 m. This situation occurs widely in Australia and a great deal of rock geochemistry is necessarily done on weathered bedrock. An example is the work at Costerfield, 130 km northeast of Melbourne where quartz-stibnite veins with gold and minor arsenopyrite and pyrite occur as fissure fillings along fault planes and in fractures in Lower Silurian mudstones (the example is from the compilation of case histories on geochemical exploration in Australia by Butt and Smith, 1980). The veins are 5–60 cm wide and average 10–15 m in length. Arseno-pyrite and stibnite are completely oxidized down to 30 m, and secondary products occur down to 70 m; hydromorphic dispersion of Sb and probable supergene enrichment of Au has been noted below 100 m. According to Hill (1980) concentrations of As and Sb in weathered bedrock (sampled by auger) greater than 20 ppm outline the fracture zones, with linear anomalies 20–200 m wide; values greater than 100 ppm outline zones of surface mineralization. As at Mykonos, the geochemical patterns are not due entirely to primary dispersion but are modified by secondary processes and, in this case, by contamination from mining.

Gold-bearing mafic volcanic rocks of the Vuda Valley, Fiji are also deeply weathered to depths of 5–10 m. Results of an exploration programme there, based on analysis of Au in weathered bedrock, are an instructive example of the dangers inherent in choosing an inadequate sampling interval because of incorrect geological interpretation; they also illustrate the use of geochemistry in re-directing geological thinking (Govett et al., 1980). The initial belief was that gold mineralization was of the low-grade, high-tonnage

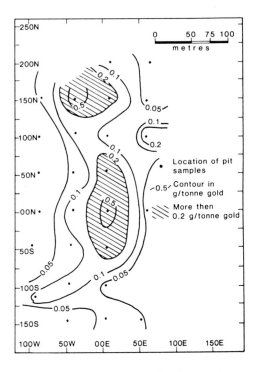

Fig. 9-14. Distribution of Au in weathered bedrock (pit sample), Delainasomo, Vuda, Fiji (based on data kindly supplied by Atherton Antimony NL).

disseminated type; accordingly, a sampling programme of weathered bedrock (collected from pits) on a 50-m grid was adopted at the Delainasomo prospect. As the results illustrated in Fig. 9-14 show, the Au values were disappointingly low considering that grab samples from that particular area, as well as random channel samples, had earlier shown values of up to 19 ppm Au. Despite the very low values from the sampling programme, contouring of the data revealed a pronounced north-south linear feature. This result prompted a re-interpretation of the geology, and it was demonstrated by close examination of an old adit at another area nearby that significant Au mineralization was confined to an intrusive dolerite dyke. A 0.1-ppm Au contour, based on weathered bedrock sampled at 10-m intervals, defined the trend of the dyke that contained an average of 7.3 ppm Au. The Delainasomo area was resampled to a depth of 2.5 m by auger. The distribution of Au (Fig. 9-15) now clearly defines four linear features 8—30 m wide trending approximately northeast; even more detailed sampling of 1800 m^2 at the centre of the area (at 5-m intervals) confirmed the validity of the assays, yielding 12 samples with more than 3 ppm Au. Subsequent detailed mapping revealed four intrusive dykes in a stream gully to the southwest of the sampled area that apparently trend directly towards the four linear

242

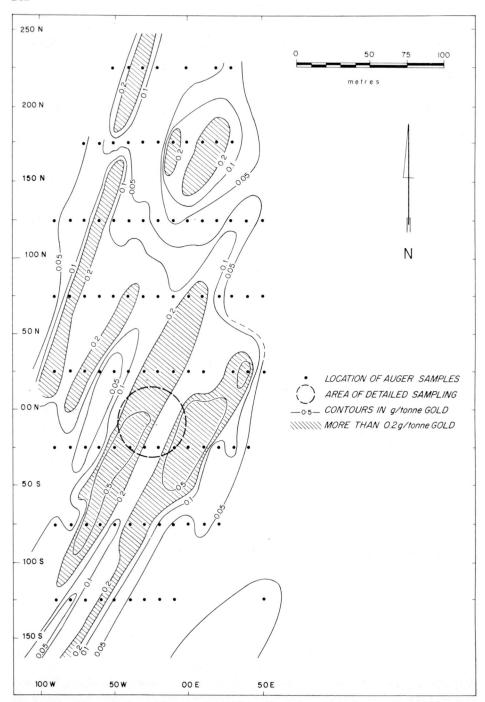

Fig. 9-15. Distribution of Au in weathered bedrock (auger samples), Delainasomo, Vuda, Fiji (from Govett et al., 1980).

geochemical zones defined by the 0.2-ppm Au contour in weathered bedrock. Diamond drilling (still in progress at the time of writing) confirmed the existence of dykes with Au mineralization beneath the geochemical anomalies.

TECHNIQUES OF DETERMINING DIMENSIONS AND DISTANCES OF VEINS

Work on Zn dispersion in carbonate host rocks around Mississippi Valley-type deposits in the Wisconsin zinc-lead district indicated the possibility of determining the size and distance of the nearest mineralization from the primary dispersion pattern (Lavery and Barnes, 1971; Barnes and Lavery, 1977). The basis of the authors' conclusions is that in background samples the Zn content is proportional to the amount of clay (determined as 12 M hydrochloric acid/insoluble residue) present according to the relation:

Zn (ppm) = 0.931 + 3.067 log % acid-insoluble residue

It is assumed that the difference between the inferred initial content of Zn (calculated from the amount of acid-insoluble material in a sample) and the analytically determined Zn represents the amount of Zn introduced by hydrothermal processes.

Core samples (2.2 cm diameter, 7.6 cm deep) were collected with a portable drill along two drifts between the Kitoe and Hayden ore bodies and the Sedgwick and North Gensler ore bodies. Two illustrative Zn dispersion patterns (corrected for background Zn content) are shown in Fig. 9-16. The dispersion from the North Gensler ore body extends about 53 m from the contact, whereas the dispersion from a 1.5-m vein between the Sedgwick

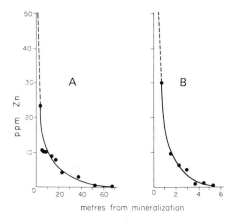

Fig. 9-16. Dispersion of Zn (corrected for background) in carbonate host rocks, Wisconsin zinc-lead district, U.S. (from Barnes and Lavery, 1977). A. North Gensler ore body. B. 1.5-m-wide vein between North Gensler and Sedgwick ore bodies.

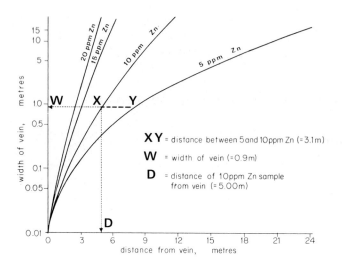

Fig. 9-17. Relation between width of vein, Zn dispersion, and distance of samples from vein (compiled from Barnes and Lavery, 1977).

and North Gensler ore bodies extends only 5.3 m from the contact. On the basis of these dispersion patterns and others from around veins of varying thicknesses, Barnes and Lavery (1977) constructed curves relating Zn content to distance from a vein and its thickness (Fig. 9-17). In the example shown in Fig. 9-17, if the distance between a sample containing 5 ppm Zn and 10 ppm Zn (both corrected for background) is 3.1 m on a smooth dispersion curve, the thickness of the adjacent vein (0.9 m) and the distance of the vein from the 10-ppm Zn sample (5.0 m) can be read directly from the graph. Silver veins greater than 1.0 m thick can be mined, and the authors calculated that at a sampling interval of 9 m and a threshold of 5 ppm Zn there is a 95% probability of the adjacent samples falling within the anomalous zone of veins 1.0 m or more thick; there was only a 33% probability of two adjacent samples falling in the anomalous zone of a 0.3-m (i.e., uneconomic) vein.

Successful application of the technique of Barnes and Lavery requires sensitive and reliable determination of Zn; on the basis of duplicate analyses by polarography of 20 samples, they stated that the average relative standard deviation was 1.0% (Lavery and Barnes, 1971). Moreover, the ability to normalize the Zn content by an independent estimate of background Zn *in each sample* obviously has considerable advantage in the recognition of small anomalies by removal of erratic values due to background fluctuations. Whether the technique is applicable in carbonate rocks elsewhere is not known, but it certainly holds promise for using rock geochemistry on both surface and drill core samples to detect narrow vein-type zinc-lead deposits in environments where dispersion is generally of restricted extent.

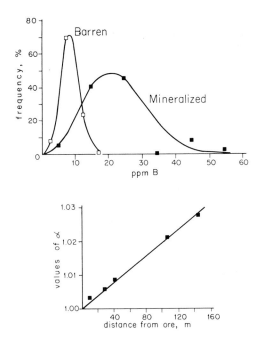

Fig. 9-18. Frequency distributions of B in barren and mineralized fault zones and variation in isotopic composition of B as a function of depth to ore. (Redrawn with permission from Shergina and Kaminskaya, 1965, *Geochemistry*, vol. 1, 1965, figs. 2 and 3, p. 40.) The term α is the value of $^{11}B:^{10}B$ at distance ΔD from ore divided by $^{11}B:^{10}B$ at the ore contact.

An intriguing application of the use of the element boron has been described by Shergina and Kaminskaya (1965). They showed that, compared to a background B content of 2—4 ppm, the B content of fault zones was 8 ppm and in ore-bearing fault zones the B content was 20—30 ppm in an undisclosed area in the U.S.S.R. (Fig. 9-18); the $^{11}B:^{10}B$ ratio changed as a function of distance from the ore. These relations are shown in Fig. 9-19 for B in fault material along which a "hydrothermal polymetallic" deposit is emplaced; the ratio of $^{11}B:^{10}B$ decreases as the ore zone is approached. In this particular example the variation in the ratio was a linear function of distance from the ore zone (Fig. 9-18) such that the distance from the ore, ΔD, is defined as:

$$\Delta D = 5156\,(\alpha - 1)$$

where:

$$\alpha = (^{11}B:^{10}B)_{\Delta D}\,/(^{11}B:^{10}B)_{\Delta D=0}$$

The value of $(^{11}B:^{10}B)_{\Delta D=0}$ is the value of $^{11}B:^{10}B$ at the contact of the ore; in this example it is 4.015. The authors point out that B probably occurs as

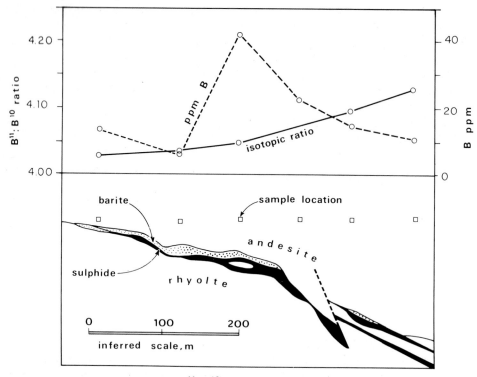

Fig. 9-19. Distribution of B and ^{11}B:^{10}B ratios in an ore-bearing fault. (Redrawn with permission from Shergina and Kaminskaya, 1965, *Geochemistry*, vol. 1, fig. 1, p. 39.) No scale given in original source; scale inferred from Fig. 9-18.

an isomorphous mixture in plagioclase, hornblende, and pyroxene and will be mobilized during hydrothermal alteration. The fact that ^{11}B is more mobile than ^{10}B — together with the possibility of isotope exchange between B in hydrothermal solutions and B in the host rocks — probably accounts for the observed variation in the ^{11}B:^{10}B ratios with proximity to ore.

CONCLUSIONS

Anomalous geochemical dispersion patterns *directly* related to vein-type deposits appear to be restricted to a few tens of metres in the wallrock. In all cases where an adequate number of samples have been collected the decay of the anomaly is logarithmic and is presumed to be controlled largely by diffusion processes. For this reason the intensity of the anomaly increases rapidly towards a vein. A deviation from strict adherence to this pattern is very commonly shown by Zn and, in some cases, by Cu, Pb, Rb, Ag, and Au where there is a zone of relatively low values adjacent to the vein and a second maxima is reached some distance from the vein. Since not all of the elements show this pattern in any particular case, the deviation is presumably due to the solubility of specific elements under the particular conditions of solution transport and host rocks.

Although veins vary in width from centimetres to metres, the width of primary halos around them are remarkably constant in the range of 20—40 m. On the basis of inadequate comparative data it appears that ore and associated elements give the same scale of responses; the results of a number of studies suggest, however, that the widest halo is given by Ag. The rarer elements have not been investigated, and it is not known whether they would give broader halos. As may be expected with diffusion-controlled processes, variation in the widths of dispersion have been observed in some situations that are attributable to the nature of the wallrock — reactive carbonate wallrock or impermeable wallrock may be expected to exhibit the narrowest anomalous element dispersion.

Anomalous patterns in surface rocks measure hundreds of metres in some instances. These broad halos apparently owe their origin to enhancement of the original primary halo by secondary processes operative in the zone of weathering.

Bedrock patterns are obviously of limited general surface exploration application for individual veins except in special circumstances when the anomalies are enhanced, but they could be extremely useful in drilling and underground control. The relations between the Zn decay pattern and the thickness of a vein and distance from the vein (described by Barnes and Lavery, 1977), and the relation between the isotopic composition of B and the distance from mineralization (described by Shergina and Kaminskaya, 1965) serve to indicate that a greater research effort may reveal systematic element distribution patterns that could be used to estimate both size and distance of a target from exploration samples.

The restricted width of geochemical anomalies in wallrock around veins dictates a close sample interval. Except in cases of anomaly enhancement by surface processes, samples should be collected at 5-m intervals (or at the most 10-m intervals) at right angles to the strike. In carbonate rocks sample intervals of 1—2 m are desirable.

Chapter 10

LOCAL AND MINE SCALE EXPLORATION FOR STRATIFORM
DEPOSITS OF VOLCANIC AND SEDIMENTARY ASSOCIATION —
PRECAMBRIAN, PROTEROZOIC, AND KUROKO DEPOSITS

INTRODUCTION

For many decades prior to the use of exploration rock geochemistry in
the search for massive sulphides geologists and geochemists had been study-
ing the form and nature of alteration pipes and wallrock alteration on footwall rocks. Alteration zones around Archaean massive sulphides on the
Canadian Shield and the Japanese Kuroko deposits are particularly well
developed and have been well documented. Since alteration zones are the
focus of extensive geochemical aureoles, the mineralogy and geochemistry
of alteration zones around Canadian Archaean deposits, as well as an example
of a Proterozoic deposit and the Cenozoic Kuroko deposits are reviewed
briefly in each of the sections of this chapter as a prelude to a description
of exploration rock geochemical results in rocks around deposits of these
ages; the reviews can also be used as background to the consideration of
exploration studies on massive sulphide deposits of other ages and types in
Chapters 11 and 12.

The general character of various categories of massive sulphide deposits
was outlined in Chapter 7; alteration zones were indicated in the host rocks
in the schematic diagram in Fig. 7-1 which illustrated the various types of
deposits. Extensive mineralogical alteration is normally confined to the footwall rocks (the Kuroko deposits are a notable exception; see below).

ARCHAEAN AND PROTEROZOIC DEPOSITS

Alteration zones

Canadian Archaean massive sulphides have been ably described by Sangster
(1972) and Sangster and Scott (1976) from whom the following description
is taken. The "standard" deposit has an upper, zoned massive sulphide mass
that passes down into stringer (vein, disseminated) ore of pyrite-chalcopyrite.
In the massive ore chalcopyrite increases towards the footwall and grades

into the stringer ore, whereas sphalerite and galena are more abundant in the upper part of the massive ore; galena increases towards the hanging wall. Pyrrhotite, where present, generally occurs in larger quantities in stringer ore than in the massive ore.

Not all sulphide zones, indeed not all of the sulphide minerals, are present in all deposits –– and not all deposits conform to the "standard" type — but the Cu-rich base and Zn-rich top is a fairly persistent feature of all massive sulphides of all ages. The alteration zone surrounds the stringer ore and increases in diameter upwards until it is coincident with that of the massive ore; in some cases the alteration pipe has been traced 1000 m stratigraphically below the massive ore. The main mineralogical feature of alteration is chloritization, followed in decreasing frequency by sericitization, silicification, and carbonation. Chlorite is commonly at the core of the pile and is succeeded outwards by sericite; disseminated sulphides are common in the alteration zone. The most common chemical changes are a relative increase of Mg, Fe, and S and a decrease of Na, K, and Si.

The Millenbach deposit, associated with a felsic volcanic centre in the Abitibi greenstone belt 8 km north of Noranda, Quebec, is an example of the typical Canadian Archaean model. It has about 3×10^6 tonnes of 3.5% Cu and 4.5% Zn. The deposit lies within quartz feldspar porphyry, with Amulet andesite as the immediate hanging wall; the quartz feldspar porphyry is succeeded downwards by the Millenbach andesite, which in turn is underlain by the Amulet rhyolite. The rocks have been metamorphosed to hornblende hornfels assemblages and then retrograded to greenschist facies; metamorphism has not affected the bulk composition of the rocks. According to

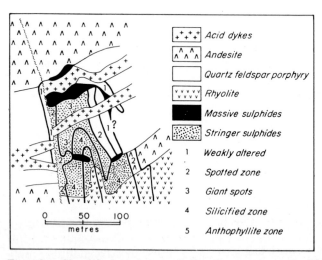

Fig. 10-1. Simplified geology and alteration zones at the Millenbach (main lens) deposit, Quebec, Canada (compiled from Riverin and Hodgson, 1980).

Riverin and Hodgson (1980) the alteration pipe transgresses lithological boundaries and has a core of massive chlorite (or anthophyllite) that grades outwards to a biotite-rich zone with a spotted texture; the latter zone was originally sericitic. These relations are shown in Fig. 10-1; geochemical trends and normative mineral zoning across one rock type (quartz feldspar porphyry) are shown in Fig. 10-2. The essential features of the chemical changes are strong depletion of Na and Ca in the outer and upper part of the alteration zone, but little change in the content of these elements in the inner zone; Mg and Fe content increase towards the core, the rate of increase being greatest in the inner core; K appears to have been added towards the margins and leached from the core of the alteration pipe.

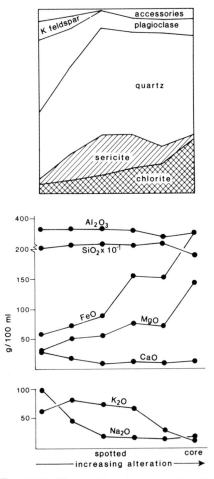

Fig. 10-2. Normative mineral zoning (top) and chemical variation (bottom) across alteration pipe in quartz feldspar porphyry in footwall of Millenbach deposit, Quebec, Canada (compiled from Riverin and Hodgson, 1980). Constant volume during metamorphism is assumed.

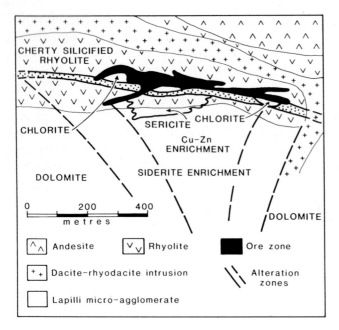

Fig. 10-3. Simplified geology and alteration zones in footwall rocks at the Mattabi mine, Ontario, Canada (compiled from Franklin et al., 1975).

The Mattabi deposit of the Sturgeon Lake area of Ontario is a typical zoned zinc-copper Archaean massive sulphide deposit with an atypical alteration pipe. The deposit has 12.9×10^6 tonnes of 7.6% Zn, 0.91% Cu, 0.84% Pb, 98 g/tonne Ag, and 0.22 g/tonne Au. The footwall of the various ore zones has a stringer zone overlain by massive sulphides of chalcopyrite-sphalerite; this is succeeded upwards by banded pyrite-sphalerite. The host rocks are slightly deformed and moderately metamorphosed felsic volcanic rocks. The mineralogy and geochemistry of the footwall rocks (rhyolite agglomerate and poorly sorted lapilli tuff) have been investigated by Franklin et al. (1975) who defined a zoned alteration pipe that transgresses lithology and extends at least 300 m below the deposit. This alteration pipe is illustrated schematically in Fig. 10-3. The typical chlorite-sericite-silicified zones are not well developed. Chlorite-enriched zones form only local pods in the immediate footwall of the ore; a silicified zone (15—23 m thick) underlies the ore body and extends along strike; a sericite-enriched zone has a restricted distribution beneath the silicified zone. The most notable feature of the alteration pipe is carbonation, with siderite forming the dominant carbonate mineral below the sulphides in a zone that narrows with depth; outside this zone dolomite is the chief carbonate mineral. Copper and Zn are enriched in a funnel-shaped zone within the siderite zone; 1—5% disseminated pyrite occurs in the footwall rhyolite for a distance of at least 900 m, and disseminated to massive (10—50%) sulphides (chiefly pyrite) occur over a

3-m zone in the rhyolite for a distance of at least 365 m. The siderite zone is characterized chemically by a marked depletion of Ca and Na, an enrichment of Fe and Mn, and a minor enrichment of Mg, CO_2, and S. The alteration zone differs from other Archaean massive sulphides (for example, those in the Noranda area) by abundance of siderite and paucity of chlorite. Franklin and his co-workers draw attention to the similarity of the chlorite and silicified alteration zones at Mattabi to the Japanese Kuroko deposits (see below).

A detailed study of the chemical and mineralogical changes in the wall-rocks at the Boliden Proterozoic deposit in Sweden illustrates the types of changes that appear to be typical of massive sulphide deposits (Nilsson, 1968). The sulphides occur within the Skellefte volcanic rocks (quartz porphyries, keratophyres, quartz keratophyres, and dacites); a systematic mineralogical zoning around the sulphides is transgressive to the original bedding. Three main alteration zones have been distinguished (Fig. 10-4).

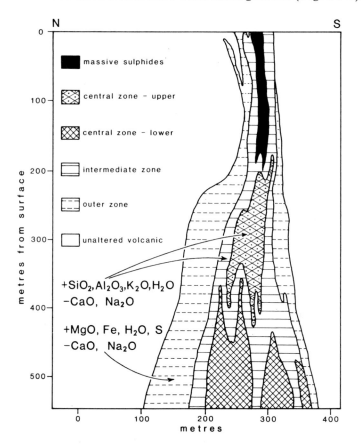

Fig. 10-4. Alteration zones and geochemical changes at the Boliden Mine, Sweden (compiled from Nilsson, 1968).

A central, most intensely altered, zone (which includes the sulphide body at the upper levels); andalusite-sericite rock passes down into andalusite-sericite quartz schist; at depths below about 400 m this zone passes through sericite quartz schist into silicic sericite quartz schist.

An intermediate zone of sericite quartz schist.

An outer zone of pyritic sericite quartz schist and chloritic sericite quartz schist which passes downwards to chloritic sericite quartz schist.

It was demonstrated that the alteration had occurred without a volume change; the altered envelope as a whole is depleted in Na, Ca, and Fe and is enriched in Mg, K, Al, Ti, S, F, and H_2O. In detail the central and intermediate zones are depleted in Na, Ca, Mg, and Fe and are enriched in Si, Al, K, and H_2O. The rocks of the outer zone are strongly depleted in Ca and Na and are enriched in Mg, Fe, S, and H_2O. In a vertical direction there is progressive enrichment of Al and depletion of Si towards the sulphide; at 500 m depth there is an actual addition of Si to the system.

The conclusion from the above brief review is that some or all of the elements Na, K, Ca, Mg, and Fe show a pronounced anomalous pattern around the deposits. Sodium and Ca are generally depleted, Mg and Fe are generally enriched, and K may be enriched or depleted depending on mineralogical alteration. Some exceptions have been noted (e.g., Mg is not noticeably enriched at Mattabi); whether an element is actually enriched or depleted depends upon the precise part of the alteration zone being considered (e.g., Mg at Boliden). Other variations will be noted in the next section.

It may be deduced, therefore, that any geochemical aureoles that persist beyond the alteration halos should also be reflected in variations in the same elements (i.e., Na, K, Ca, Mg, Fe). This has indeed been shown to be the case around massive sulphides in many parts of the world as will be shown in the remainder of this chapter and in Chapters 11 and 12.

Exploration studies

Apart from the work of Nichol and his co-workers (see below) there are few *specifically* exploration-oriented geochemical studies around Archaean massive sulphides; also few workers (again apart from Nichol) claim spectacular success. A case in point is the investigation by Turek et al. (1976) around the copper-zinc Fox ore body which is located about 43 km southwest of Lynn Lake, Manitoba. The remaining reserves in 1970 were 13.1 × 10^6 tonnes of 2.7% Zn and 1.84% Cu. Mineralization is within metasedimentary and metavolcanic rocks; the ore is massive (but locally disseminated), and the principal sulphide minerals are pyrite, pyrrhotite, chalcopyrite, sphalerite, arsenopyrite, and galena. There is an alteration envelope (which obliterates the primary rock character) in which the chief minerals are quartz, muscovite, sericite, biotite, amphibole, talc, and chlorite. The authors collected 109 samples from drill core and from underground workings.

Major elements and S were determined on powder pellets by XRF; AAS was used to determine Cu, Pb, Zn, Co, Ni, and Cr after an aqua regia digestion and for Hg after a cold HNO_3-H_2SO_4 digestion; Mo was measured colorimetrically after a H_2SO_4 digestion.

According to Turek and his co-workers there are no large-scale dispersion patterns, and only Cu and Co show an increase towards the ore zone; the alteration zone is characterized by higher Cu and Co and lower Hg than the other units. The authors also concluded that there is no lithological control of element concentration except for the amphibolite (andesite) which has low Si, Na, and K and high Ca and Cu. If this latter conclusion is accepted, the evidence of their element distribution along a cross-cut section contradicts their former conclusion that there are no extensive geochemical patterns.

Some of their results are given in Fig. 10-5. In the structural hanging wall (north) there is a 250-m halo defined by increasing contents of Cu, Co, and

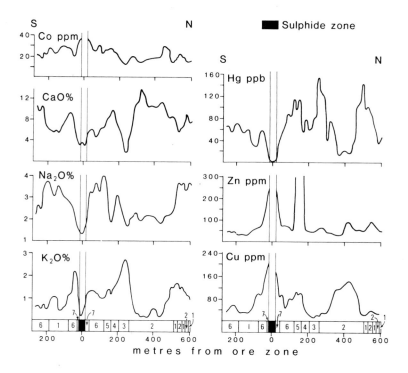

Fig. 10-5. Distribution along underground cross-cut, Fox deposit, Lynn Lake, Manitoba, Canada. (Redrawn with permission from Turek et al., 1976, *Canadian Institute of Mining and Metallurgy Bulletin*, vol. 69, 1976, fig. 2, p. 107.) *1* = quartz-hornblende gneiss (mafic intrusion); *2* = amphibolite (andesite); *3* = arkose with minor quartzite; *4* = quartz-hornblende-biotite, quartz-muscovite-biotite-sericite gneisses; *5* = argillite, quartz-hornblende-biotite gneiss; *6* = quartz-hornblende, quartz-hornblende-biotite gneisses; *7* = alteration zone-quartz-muscovite-biotite-sericite, biotite-amphibole gneisses.

Zn and decreasing contents of Hg, K (and less distinctly by Ni and Cr which are not illustrated in Fig. 10-5). Sodium increases over most of this distance and then decreases in the alteration zone (and is thus similar to the Brunswick No. 12 deposit; see Chapter 11). Calcium decreases strongly over 180 m. In the structural footwall (south) Na, Ca, and Hg decrease strongly over about 180 m, and K and Cu show a strong increase over this distance. The pattern, therefore, despite the stated conclusions of Turek et al. (1976), appears to conform well to the generalized pattern around massive sulphide deposits. The anomalous aureole is not large on the data given; moreover, since no measurements are given for the alteration envelope, it is not possible to state whether the geochemical aureole is more extensive than the mineralized alteration zone.

The importance of SiO_2 normalization to compensate for effects of lithological change and degree of fractionation on the distribution of other elements has been discussed in a regional context in Chapter 7. Whereas petrological variations may be expected to be less important, or at least more readily recognized and allowed for on a local scale, in fact large-scale and erratic fluctuations in element concentration occur; it is often difficult to distinguish effects due to alteration or dispersion related to mineralization from those due to lithological change. In these circumstances SiO_2 normalization has been shown to be extremely useful in the interpretation of local and mine scale geochemical data.

The first published example of this approach was by Descarreaux (1973) who showed that Na, after normalization against SiO_2 gave a strong negative anomaly extending more than 500 m from the Jay Copper Zone at Abitibi (Canada). At the Hanson Lake (Saskatchewan) copper-lead-zinc deposit Fox (1978) concluded that SiO_2 normalization was more effective than either factor analysis or trend analysis in extracting useful information from trace element data in drill core from volcanic rocks. A comparison of the raw Cu and Zn distribution with the SiO_2 normalized values is shown in Fig. 10-6. Fox points out that the SiO_2 normalization results provide the following advantages:

— a Cu anomaly in dacite and rhyolite between 75 m and 100 m (which may represent disseminated mineralization) not apparent in the raw data;

—– a possible Cu anomaly adjacent to mineralization not evident in the raw data;

— a clear Zn halo around the ore zone not apparent in the raw data;

— raw Zn peaks in basalt are reduced in significance.

Nichol and his co-workers have made rather more extensive use of SiO_2 normalization techniques in their studies of element distributions around Canadian Archaean massive sulphide deposits. They determined a wide range of elements on minus 200-mesh rock powders by XRF; for SiO_2, K_2O, Fe_2O_3, MgO, and TiO_2 there was a precision of better than ±5% at the 95% confidence level; for Al_2O_3 and Na_2O the precision was ±10%;

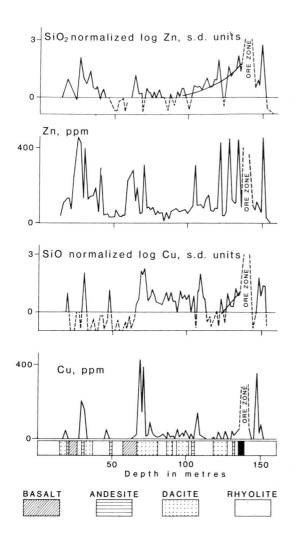

Fig. 10-6. Distribution of Cu and Zn, and SiO$_2$ normalized equivalents (in terms of standard deviations from mean of SiO$_2$ class) for drill core, Hanson Lake Mine, Saskatchewan, Canada. (Redrawn with permission from Fox, 1978, *Canadian Institute of Mining and Metallurgy Bulletin*, vol. 71, 1978, fig. 1, p. 112.)

for S and P$_2$O$_5$ it was ±20%. Cu, Pb, Zn, Co, and Ni were determined by AAS after an aqua regia digestion with a precision of ±15% at the 95% confidence level; the precision for Ag and Mn by the same technique was ±25%. At the East Waite deposit in the Noranda area Nichol et al. (1977) documented the usual Mg and Fe enrichment and Ca and Na depletion over an area twice the size of the deposit; they also showed that S, Cu, Zn, and Mn were enriched in the wallrock adjacent to the deposit. As is commonly

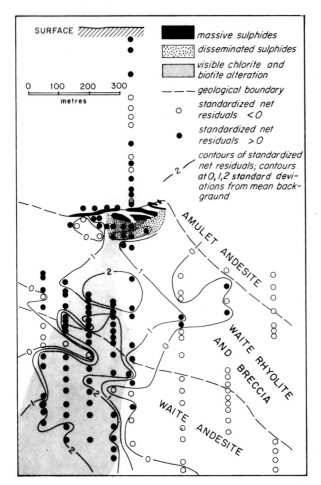

Fig. 10-7. Distribution of standardized net residuals, East Waite deposit, Quebec, Canada (compiled from Nichol et al., 1977).

observed, the patterns for individual elements tend to be erratic and discontinuous. To overcome these problems the residual Na_2O, CaO, Fe_2O_3, and MgO contents — after correction for SiO_2 content — were standardized and then summed to give a "standardized net residual". The distribution of this function (Fig. 10-7) outlines an alteration zone 600 m stratigraphically below the deposit. The distribution of a similar function at the South Bay Mine in the Uchi Lake Area showed an anomalous zone extending considerably beyond the limits of the mineralogical alteration zone to give a halo of about 100 m from the sulphides.

The Norbec deposit in the Noranda area (see Fig. 7-9 for location) is another typical Archaean massive sulphide investigated by Pirie and Nichol

(1980). The deposit lies at the contact between the footwall Waite Rhyolite and the hanging wall Amulet Andesite; both these rock units were sampled from drill core. Each lithological type was considered separately, and an average for each rock type for each drill hole was used to obtain a data point to plot the areal distribution of elements. Anomalous distribution patterns could be recognized only in the footwall rocks; generally Mg, Fe, and Mn are enriched and Na, Ca, and Si (in places) are depleted. The trace elements Cu, Zn, Ag, and S form a broad enrichment halo around the area of the deposit, but they do not provide a focus for mineralization.

The variation in MgO, CaO, and K_2O in background and in five different alteration types is shown in Fig. 10-8. According to Pirie and Nichol (1980) only two of the alteration types — amphibole and sericite — are associated with mineralization. It is clear from Fig. 10-8 that K_2O is of no value in the recognition of mineralization since the background range overlaps the range in all alteration types. Magnesia contents greater than the 60th cumulative percentile discriminate between alteration associated with mineralization and all other alteration (except for the sericite chlorite alteration). The distribution of CaO is interesting; the sericitic alteration associated with mineralization is indeed strongly depleted in CaO, but the amphibole alteration is *enriched* in CaO. Notwithstanding these observations, the main areas of mineralization clearly fall in zones of high MgO and low CaO (Fig. 10-9). The Norbec deposit is, however, almost completely encircled by high CaO.

Quite clearly, at Norbec whole rock geochemistry is providing better targets that petrographic data on alteration. The maximum diameter of a mineralogically identifiable alteration zone is 150 m; as shown in Table 10-I

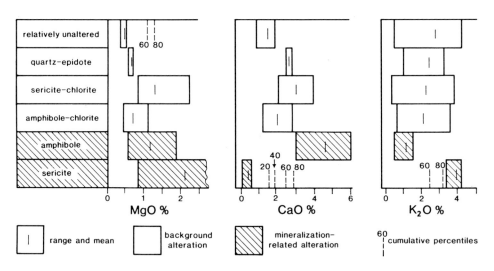

Fig. 10-8. Relation of MgO, CaO, and K_2O content to alteration type, Norbec deposit, Quebec, Canada (redrawn with permission from Pirie and Nichol, 1980).

260

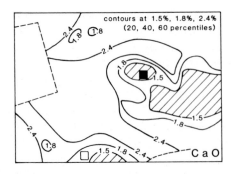

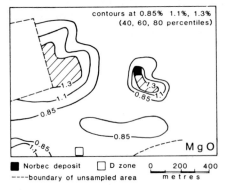

■ Norbec deposit ☐ D zone 0 200 400
----boundary of unsampled area metres

Fig. 10-9. Distribution of CaO and MgO in rhyolite breccia around the Norbec deposit, Quebec, Canada (redrawn with permission from Pirie and Nichol, 1980).

TABLE 10-I

Dispersion characteristics of certain major and trace elements in the texture "rhyolite breccia" around the Norbec deposit (from Pirie and Nichol, 1980)

	Width of dispersion related to mineralization (m)	Contrast of anomaly
SiO$_2$	375	−0.06
Al$_2$O$_3$	?	
Fe$_2$O$_3$	450	0.94
CaO	350	−0.98
MgO	150	2.15
K$_2$O	?	
Na$_2$O	300	−0.94
TiO$_2$	?	
Mn	375	2.04
S	300	4.16
Cu	200	7.62
Zn	200	2.19
Ag	300	3.82

? = uncertain relationship to mineralization

single-element geochemical halos range up to 450 m, and a multi-element halo (based on discriminant scores using S, MgO and Na_2O) has a width of 700 m around the Norbec deposit. As the data indicate, better contrast is obtained from trace elements, but major elements give more extensive halos.

The effect of textural variation in rocks on element contents was discussed in a regional context in Chapter 7. Based on comparisons of dispersion patterns in rhyolite breccias and more massive rhyolites, Pirie and Nichol (1980) concluded that in the Norbec area *generally* wider and better-defined halos occur in rocks that had an initially higher permeability (i.e., the rhyolite breccias). The relation, however, is not simple because their data apparently showed every variation — from a situation where no anomaly is apparent in the massive rhyolite (e.g., Mn) to the case of Na which gives a better defined halo in the rhyolite (Fig. 10-10). The authors do not offer an explanation for these differences, but it may be speculated that the Mn and Na distributions are caused by different processes during the deposition of the deposit; this is a topic requiring investigation that could yield significant results in terms of interpretation of exploration data.

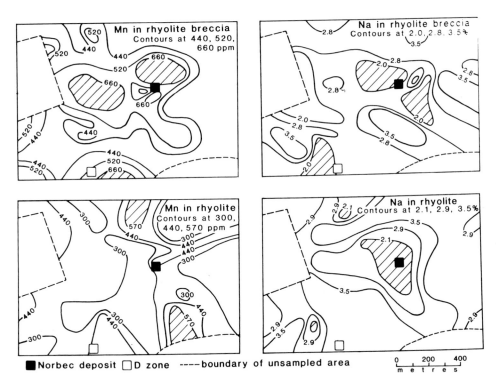

Fig. 10-10. Distribution of Mn and Na in rhyolite breccia and rhyolite around the Norbec deposit, Quebec, Canada (redrawn with permission from Pirie and Nichol, 1980). Contours for Mn at 40, 60, 80 percentiles; contours for Na at 20, 40, 60 percentiles.

The conclusion from the Norbec study is that, with proper precautions concerning lithological and textural variations, large target zones are clearly defined by whole rock geochemistry. Pirie and Nichol (1980) concluded that samples on a 200-m grid would be adequate to define the areas of known mineralization.

An investigation around the small Louvem deposit (21 km east of Val d'Or in the Abitibi belt) showed that similar dispersion patterns occur around copper-pyrite deposits. Spitz and Darling (1978) described an anomalous halo 210 m wide and at least 420 m long based on surface samples (with some additional samples from drill core). The halo was enriched in Fe and S and depleted in Na, Ca, and CO_2; within the halo, and corresponding with an altered zone around the deposit, there is a 100 m by 50 m halo with anomalously high Mg and H_2O values. I.G.L. Sinclair (1977), however, published data on the Detour deposit in the Abitibi belt (at Lac Brouillan, 100 km west of the town of Matagami) which showed a different pattern. The deposit lies within felsic volcanic rocks and has about 32×10^6 tonnes of 2.30% Zn, 0.39% Cu, and 36 g/tonne Ag. The ore zone appears to be enveloped in rocks *deficient* in Mg and K over several hundred metres, although there is evidence of an enrichment in both elements flanking the zone of depletion. Sinclair suggested that the volcanism may have been sub-aerial, and that the mineralization is of a replacement rather than a syngenetic stratiform type (i.e., that the deposit may not be a typical massive sulphide). On the other hand, the pattern shown at the Detour deposit may simply appear different because of the *scale*; thus, only a few samples are more than 200 m from massive sulphides. Goodfellow (1975a) noted K depletion close to mineralization at the Brunswick No. 12 deposit (see Chapter 11); Riverin and Hodgson (1980) described K depletion at the core and K enrichment at the margin of the Millenbach alteration zone (see above), and Nilsson (1968) described MgO depletion at the core of the alteration zone below the Boliden deposit (also discussed above).

The enormously rich Broken Hill lead-zinc deposits in Australia have been the subject of geological controversy since their discovery; their structure and stratigraphy is still debated, but there is now sufficient evidence to believe that they had a volcanic-sedimentary origin. These deposits (originally more than 200×10^6 tonnes of about 25% combined metal) occur in the Proterozic high-grade metamorphosed granulite facies Willyama Complex; they are interpreted as occurring in a hinge of a downward-facing F_2 structure. Metasediments are the most abundant rock type; interlayered pelitic, psammitic, and quartz-feldspathic rocks occur stratigraphically below the sulphides, and pelitic metasediments overlie the sulphides. Iron formations occur at horizons stratigraphically equivalent to the sulphides. Plimer (1979) has published a major study that convincingly demonstrates geochemical variations in the host rocks cannot be related to metamorphic or structural features; he concluded that they are attributable to original alteration

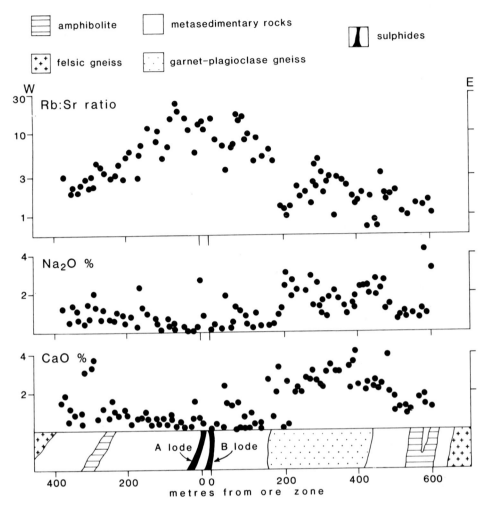

Fig. 10-11. Distribution of Rb:Sr ratios, Na_2O, and CaO across the strike from sulphide ore bodies, Section 84, NBHC mine, Broken Hill, New South Wales, Australia (compiled from Plimer, 1979).

associated with deposition of the sulphide bodies. The author recorded widespread slight depletion of Na and enrichment of total Fe, Mn, and Ti, and showed that the degree of depletion and enrichment increases towards the sulphide bodies. The geochemical alteration is most intense in the footwall within 500 m of the orebodies, and the rocks are depleted in Na, Ca, Sr, and Mg and are enriched in SiO_2, K, Rb, Mn, Pb, S, and possibly total Fe and Ti.

The trends for CaO, Na_2O, and the Rb:Sr ratios are illustrated in Fig. 10-11. The pattern for Rb:Sr is particularly prominent. Plimer and Elliott

(1979) demonstrated for several deposits at Broken Hill that the Rb:Sr ratio increases towards sulphide mineralization, and that the increase is independent of lithology. The geochemical halos at Broken Hill, despite the structural complications and high-grade metamorphism, conform to the general pattern for massive sulphides; the only exception is the lack of a positive Mg halo, which indicates that magnesium metasomatism was, for some reason, not a feature of ore deposition in this area.

KUROKO DEPOSITS

Alteration zones

The Kuroko massive sulphides of Japan, because of their occurrence in young (Miocene) and relatively undisturbed and unmetamorphosed strata, are extremely well described and are commonly used as a standard of comparison for older deposits. The actual sulphide zones are in many ways broadly similar to the Canadian Archaean deposits inasmuch as they are characteristically zoned and pass down into a stringer and stockwork zone; the host rocks are also felsic lavas and tuffs.

The generalized zoning sequence, from top to bottom (Lambert and Sato, 1974), is:
– ferruginous chert
– barite ore, sometimes with minor amounts of calcite, dolomite, and siderite; stratiform
– black ore (kuroko); sphalerite-galena-chalcopyrite-pyrite; stratiform
– yellow ore (oko); pyrite-chalcopyrite (spalerite-barite-quartz); stratiform
– pyrite ore (ryukako); pyrite-(chalcopyrite); usually stratiform; also occurs as veins and disseminations
– gypsum ore (sekkoko); gypsum-anhydrite-(pyrite-chalcopyrite-sphalerite-galena-quartz-clays); stratabound and less commonly as veins
– siliceous ore (keiko); pyrite-chalcopyrite-quartz; stockwork that grades laterally into mineralized lavas and tuffs.

The average composition of these deposits in the Hokuroku district (the most important area of Kuroko deposits) is 20% Cu, 5% Zn, 1.5% Pb, 1.5 g/tonne Au, and 95 g/tonne Ag. The deposits have distinct alteration halos that persist for several hundred metres into the hanging wall. Four alteration zones are generally recognized (Iijima, 1974; Shirozu, 1974; Lambert and Sato, 1974) forming a mushroom-shaped aureole about the funnel-shaped ore deposits (Fig. 10-12). The zones are as follows:

Zone I (the outer alteration) grades into the regional zeolite facies alteration and is characterized by montmorillonite. It is enriched in Ca, Mg, and slightly in Fe; Na is depleted.

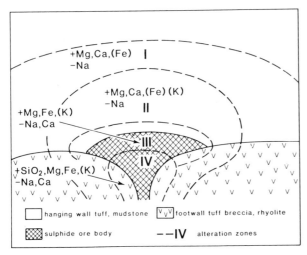

Fig. 10-12. Schematic illustration of alteration zones around Kuroko massive sulphide deposits (compiled from Iijima 1972, Shirozu, 1974, and Lambert and Sato, 1974). Enrichment (+) and depletion (−) of major elements in the zones is indicated; see text for description of alteration zones.

Zone II is up to 200 m thick in hanging wall rocks; the alteration minerals are sericite, montmorillonite, chlorite, albite, K-feldspar, and quartz. Chemically it is enriched in Mg and Ca, slightly enriched in K and Fe; Na is depleted.

Zone III surrounds the ore zone and is 1—20 m thick in the hanging wall; the characteristic alteration minerals are sericite, interstratified sericite-montmorillonite, and Mg-chlorite. The zone is enriched in Mg, strongly enriched in Fe, slightly enriched in K; Ca and Na are depleted.

Zone IV is associated with the stockwork and is characterized by quartz and sericite with minor Mg-chlorite; SiO_2, Mg, K, and Fe are enriched, and Na and Ca are depleted.

Exploration studies

The conclusions from the above discussion of the alteration zones around Kuroko massive sulphides is that geochemical characteristics are remarkably similar to those around Archaean deposits. Some or all of Na, K, Ca, Mg, and Fe show anomalous concentrations; Na is generally depleted, Mg and Fe are generally enriched, and K and Ca may be either enriched or depleted depending upon mineralogical alteration. There are, however, few exploration-oriented geochemical studies of the Kuroko deposits published in English.

The results of most of the studies that have been published show that the $K_2O : Na_2O$ ratio progressively increases towards the orebodies over several hundred metres. In the footwall rhyolite and dacite rocks around the deposits of the Akita district, Tono (1974) has shown a very pronounced difference in

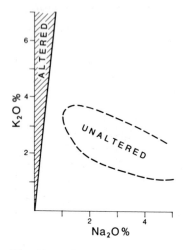

Fig. 10-13. Variation of K_2O and Na_2O content in footwall dacite and rhyolite around Kuroko deposits, Akita district, Japan (modified from Tono, 1974).

the $K_2O:Na_2O$ ratios between altered and unaltered rocks (Fig. 10-13). He indicated that the depletion of Na and Mn, and the enrichment of K, Cu, Pb, Zn, As, Tl, Mo, and Hg are characteristic geochemical patterns associated with mineralization. Izawa et al. (1978) recorded increases in S and As over several hundred metres.

Convincing rock geochemical halos around the Fukasawa deposit, obtained by sampling footwall rhyolite at 5-m intervals in drill core (averaging 15 such samples) are shown in Fig. 10-14. Anomalous zones, up to 2 km long and 1 km wide, are very distinctly defined by depleted Na_2O and CaO contents and are rather less distinctly defined by enrichment of MgO and K_2O contents.

An early paper that is concerned specifically with exploration aspects reports that the contents of Ag, Cu, Ni, Ba, Cr, V, Co, and Pb increase regularly towards the ore zones (H. Ishikawa et al., 1962). The data shows a strong logarithmic decay to background values over distances of 10—40 m and is, therefore, illustrative of probable diffusion-controlled wallrock dispersion of the type discussed for vein deposits in Chapter 9.

CONCLUSIONS

Massive sulphides of the two contrasting types chosen for discussion — Archaean and Proterozoic, and Kuroko — characteristically have pipes or funnel-like alteration zones in the footwall beneath the sulphides. The Kuroko deposits also generally have an alteration aureole in the hanging wall. The alteration is zoned most commonly with chlorite at the core in

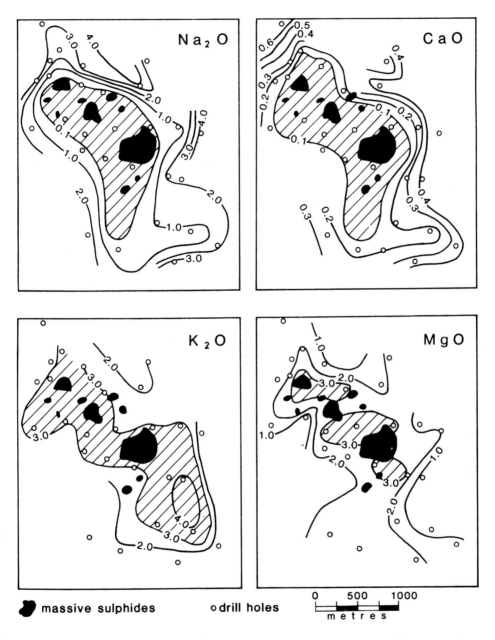

Fig. 10-14. Distribution of Na_2O, K_2O, CaO, and MgO in footwall at Fukasawa Mine, Japan. (Redrawn with permission from Y. Ishikawa et al., 1976, *Mining Geology*, vol. 26, 1976, fig. 4, p. 112.) Contours in %.

268

Archaean and Proterozoic deposits and with sericite at the core of the Kuroko deposits. The most common alteration types are chloritization, sericitization, silicification, and carbonation. The mineralogical alteration is associated with distinct geochemical patterns which, like the alteration, transgress lithological boundaries. The main features are:

— depletion of Na and enrichment of Fe, which appears to be a universal characteristic;

— commonly depletion of Ca, but Ca can also be enriched, especially towards the outer edge of the anomalous halos;

— characteristically an enrichment of Mg, but the degree of enrichment depends upon the amount of chlorite-forming Mg metasomatism (in a number of cases Mg is actually depleted towards the centre of mineralization);

— variable and generally unreliable behaviour of K (it may be enriched, depleted, or show no trend);

— strong enrichment of the ore elements — Cu, Pb, and Zn — in a restricted zone in the area of most intense alteration;

— enrichment of a variety of other minor elements (S and Mn are most commonly cited) in individual cases.

Where adequate data have been published it is clear that major element halos, particularly multi-element halos, have dimensions that significantly exceed the size of mineralogical alteration. Moreover, the main geochemical trends appear to be more regular in many cases than petrographic mineralogic work would indicate.

Geochemical halos in hanging wall rocks of Archaean and Proterozoic deposits have not been specifically recorded; in Kuroko deposits halos in the hanging wall exceed 200 m. In footwall rocks of Archaean and Proterozoic deposits recorded geochemical halos persist stratigraphically for up to 700 m below the ore, and for at least twice the lateral width of the alteration zone. The more usual dimensions of geochemical halos for single elements are several hundred metres, but multi-element halos (particularly after correction for petrographic variation, e.g., by normalization for SiO_2 content) these dimensions are doubled. Exploration data for Kuroko deposits are sparse, but it appears that geochemical halos may be far more extensive and are up to several kilometres in size.

Sampling intervals of 150—200 m for surface samples should be adequate to define mineralization. A sample weight of about 1 kg made up of chip samples is desirable. Obviously particular care must be taken to sample comparable rock types. Also, given the variation in response attributed to differences in textures, care must be taken to sample comparable textural varieties; limited data suggest that rocks with an initial high permeability (e.g., pyroclastics rather than massive layers) have the best defined geochemical halo.

Chapter 11

LOCAL AND MINE SCALE EXPLORATION FOR STRATIFORM DEPOSITS OF VOLCANIC AND SEDIMENTARY ASSOCIATION — NEW BRUNSWICK DEPOSITS

INTRODUCTION

Detailed rock geochemical studies designed specifically to develop exploration techniques were undertaken on five of the Palaeozoic zinc-lead (copper) massive sulphide deposits in northern New Brunswick (Canada) by the author and his co-workers during the period 1969—1977. These studies represent one of the most intensive investigations in one district (115,000 analyses on 13,000 samples) and are therefore described in detail. Moreover, since the investigations were all conducted on similar sampling and analytical bases, the geochemical responses can be readily compared to provide a useful study of the variations likely to be encountered around essentially similar deposits in a single geological province. The account is derived from published work as cited in this chapter and from unpublished Ph.D. theses (University of New Brunswick) on the Heath Steele B zone (Whitehead, 1973b); the Heath Steele ACD and B zones, and Key Anacon (Wahl, 1978); Brunswick Mining and Smelting No. 12 (Goodfellow, 1975b); and Caribou (Gandhi, 1978).

Most of the data were derived from drill core. A total of approximately 15 cm of core was taken over 0.5—1.5 m (depending on lithological homogeneity) at intervals of 30 m to within about 60 m of sulphide horizons; samples were then collected at every 10—20 m. Surface samples of 1 kg (chips) were collected below the weathering crust. All samples were scrubbed in deionized water prior to crushing and grinding. Most of the analyses were performed on AAS (unless otherwise stated below) for total element content by HF-based digestion, and for partial element content by hot concentrated HNO_3 digestion. Bulk samples of each lithology were prepared as internal standards and standardized against a variety of international standards; these internal standards were used to maintain accuracy, and precision of results (better than ±20%).

DESCRIPTION OF THE DEPOSITS

The regional geology of the northern New Brunswick deposits is given in

270

TABLE 11-I

Comparative stratigraphic sequence at Brunswick No. 12, Heath Steele ACD and B zones, Caribou, and Key Anacon deposits*

Brunswick No. 12 (Goodfellow, 1975b)	Heath Steele B zone (Whitehead, 1973b)	Heath Steele ACD zone (Wahl, 1978)	Caribou (Gandhi, 1978)	Key Anacon (Saif, 1977)
Upper metasedimentary	mafic volcanic	rhyolite and QFP	mafic volcanic	upper metasedimentary
			rhyolite porphyry, QFP	
Mafic volcanic	rhyolite and QF crystal tuff	rhyolite and QFP	tuffaceous rhyolite, QF sericite schist	mafic volcanic
Iron formation	iron formation	iron formation	chlorite schist	iron formation
Massive sulphides QFP and quartz sericite schist	*massive sulphides* QFP and rhyolite tuff, minor metasedimentary	*massive sulphides* metasedimentary and QFP	*massive sulphides* metasedimentary, minor acid tuffs, with minor mafic volcanic units	*massive sulphides* acid metatuffs
Metasedimentary	metasedimentary	QF sericite augen schist		lower metasedimentary

* QF = quartz feldspar; QFP = quartz-feldspar porphyry.

Chapter 7. The stratigraphic sequences at each of the five massive sulphide deposits considered here are compared in Table 11-I

At Heath Steele two sulphide zones were considered: the ACD zone, and the B zone. The ore reserves at the B zone are 23×10^6 tonnes of 5.0% Zn, 2.0% Pb, and less than 1.0% Cu. The ore zones are enclosed within a sedimentary and quartz feldspar porphyry sequence that has been folded against a rigid rhyolite dome south of the mine area. Quartz feldspar porphyry occurs in both the hanging wall and the footwall; unaltered samples from both stratigraphic positions are indistinguishable megascopically and microscopically. The quartz feldspar porphyry is similar texturally, mineralogically, and chemically to the quartz feldspar augen schist at the Brunswick No. 12 deposit (see below). It is intensely altered (feldspar phenocrysts are replaced by sericite, and amounts of chlorite and sericite increase in the matrix) in both the hanging wall and footwall at the ACD zone; at the B zone the footwall is weakly altered, and the hanging wall is not altered at all. Sedimentary rocks (siltstone, greywacke, and pebble conglomerates) are mostly confined to the footwall, with only minor amounts in the hanging wall. The general fine-grained texture and high content of primary chlorite and sericite (up to 35%) masks hydrothermal alteration which is only reflected in increasing quantities of disseminated sulphides (up to 15% of the rock).

The Brunswick No. 12 deposit has at least 100×10^6 tonnes of 9.3% Zn, 3.8% Pb, 0.3% Cu, and 79 g/tonne Ag. The main sulphide minerals are pyrite, sphalerite, galena, pyrrhotite, chalcopyrite, tetrahedrite, and bornite. The sulphide zones lie essentially within a siliceous sericitic chlorite schist, although in places the immediate footwall is the felsic volcanic unit. The latter is an interbedded sequence of quartz-feldspar augen schist, quartz augen schist, and crystal tuff that occurs at the stratigraphic top of this sequence. There is a stockwork zone of pyrite and pyrrhotite stringers and intense metasomatic alteration below the sulphide zone. A silicified zone occurs at the contact between the sulphide and the footwall sedimentary rocks and tuffs; it extends 30 m into these rocks. Chlorite and sericite alteration in the footwall rocks is more extensive, persisting to 60 m into the footwall within the lateral extent of the sulphide zone.

The Key Anacon deposit has about 1×10^6 tonnes of 5% Zn, 2% Pb, and less than 1% Cu. The main sulphide minerals are pyrite, pyrrhotite, sphalerite, galena, and chalcopyrite. Although the Key Anacon deposit is generally similar to the other deposits in the Bathurst district, Saif et al. (1978) have pointed out that it lies well to the east of the main ore-bearing volcanic pile (see Fig. 7-23). The volcanic rocks are very fine-grained and occur within a narrow band, in contrast to the thick volcanic sequence at Brunswick No. 12. The rock units in the area of the mine (in descending sequence) are: upper metasedimentary rocks, andesite metatuffs, iron formations and associated metasedimentary rocks, sulphide bodies, acid metatuffs, and lower metasedimentary rocks. The acid metatuffs contain up to 20% pyrite where they underlie the ore zone — which Wahl (1978) considered may represent a stockwork.

The Caribou deposit has 25×10^6 tonnes of 5.0% Zn, 2.05% Pb, and 31 g/ tonne Ag. The ore zone is about 1120 m long, 6—20 m thick, and has been traced to a depth of 400 m. The deposit occurs as three tabular lenses en echelon and steeply dipping around the limbs of a local, northerly plunging synform (Fig. 11-6 below). Two of these lenses occur on the west limb of the fold, and the third lens occurs in the east limb. The sulphides are extremely fine-grained. The main minerals in the ore zones are pyrite, magnetite, sphalerite, galena, and chalcopyrite, with minor amounts of arsenopyrite, marcasite, pyrrhotite, bornite, tetrahedrite, and hematite. The ore bodies are zoned vertically, with Zn-Pb in the hanging wall separated from Cu in the footwall by a low-grade pyritic zone (this zone is best displayed in the northern sulphide lens on the west limb). There is also a lateral zoning, with a transition to massive pyrite towards the end of the sulphide bodies. The stratigraphic sequence (in downward succession) is rhyolite and quartz feldspar porphyry, augen schist, sulphides, chloritic and sericitic phyllite, graphitic schist, spilitic and greenstone mafic metavolcanic rocks, and interbedded chloritic phyllite. A thin chloritic schist (which Gandhi, 1978, equates with iron formations at other deposits) occurs between the sulphides and the handing wall silicic metavolcanic rocks. Chloritization and sericitization are evident in both the footwall and hanging wall rocks; there is, however, no discernible pattern that can be related to the ore zones, and it is not possible to distinguish between possible hydrothermal alteration and that due to greenschist facies metamorphism.

TRACE ELEMENT GEOCHEMICAL RESPONSE

A large number of different trace elements were measured in the rocks around the deposits; generally it was found that *individual* elements do not give well-defined halos of any significant extent. The ore elements — Cu, Pb, and Zn — are generally higher close to the sulphide horizons (Fig. 11-1), but the patterns are poorly defined. Copper is generally enriched in footwall rocks and depleted in hanging wall rocks, and the behaviour of Pb and Zn is variable compared with regional values (see detailed discussion below). The deposits have associated Hg anomalies, but their interpretation is difficult. Goodfellow (1975b) showed that cold-extractable Hg (and As) is highly concentrated below the Brunswick No. 12 deposit, and that there is a large Hg anomaly (which seems to be independent of lithological variations) located vertically above the deposit (Fig. 11-2). The Hg content also increases towards mineralization at the Caribou deposit (Fig. 11-1), but erratic high values of up to 400 ppb occur distant from mineralization — apparently due to dissipation along fractures and faults.

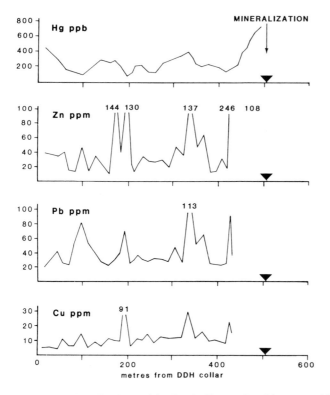

Fig. 11-1. Distribution of hxCu, hxZn, and cold-extractable Hg in hanging wall silicic volcanic rock, Caribou, New Brunswick, Canada (DDH 72 compiled from Gandhi, 1978).

The most successful use of trace elements around the deposits has been through a multi-element approach, either through simple element ratios or by additive and multiplicative data processing, or through the use of multi-element statistical data processing (chiefly by linear discriminant functions — LDF). Some examples are described below.

Additive halos at Heath Steele deposits

Wahl (1978) determined the total element content in drill core around the Heath Steele deposits by inductively coupled plasma emission spectometry of Cu, Pb, Zn, Co, Cr, P, Sr, Ni, V, and Ag. In both quartz-feldspar porphyry and sedimentary rocks Cu, Pb, Zn, Co, Cr, P, and Ag generally increase in concentration towards the ore zone, whereas Ni, V, and Sr generally decrease in concentration towards mineralization. There are, however, marked differences in absolute concentrations between hanging wall and footwall and between the ACD zone and the B zone — this is shown by the geometric means (*gm*) listed in Table 11-II.

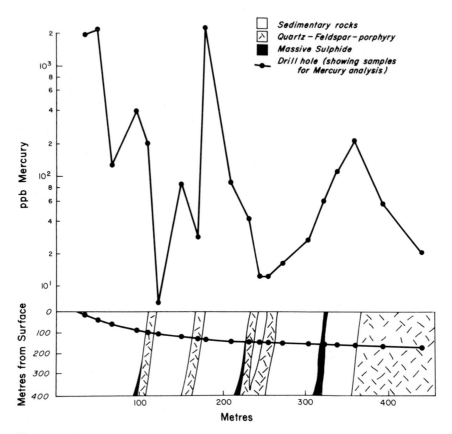

Fig. 11-2. Distribution of Hg in rocks around the Brunswick No. 12 deposit, New Brunswick, Canada (compiled from Goodfellow, 1975b).

To maximize the halo all element concentrations were first normalized to footwall B zone concentrations. For example, the normalized value for Cu for any particular sedimentary sample is given by:

$$\frac{(Cu \text{ in sample}) - (gm \text{ of Cu in all B zone footwall sedimentary rocks})}{(gm \text{ of Cu in all B zone footwall sedimentary rocks})}$$

A standardized factor was then calculated by summing the normalized values of all elements that increase in concentration towards mineralization and subtracting the sum of normalized values of all elements that decrease in concentration towards mineralization, i.e.,

standardized factor $= \Sigma$ normalized $(Cu, Pb, Zn, Co, Cr, P, Ag) - \Sigma$ normalized (Ni, V, Sr)

The more positive the standardized factor, the greater is the anomaly. Histograms of the standardized factor for quartz feldspar porphyry at the ACD zone and the B zone are shown in Fig. 11-3. The greatest anomalies occur in

TABLE 11-II

Geometric mean values for trace elements in rocks at Heath Steele ACD and B zones after total digestion (from Wahl, 1978)

	Footwall sedimentary rocks		Footwall quartz-feldspar porphyry		Hanging-wall quartz-feldspar porphyry	
	ACD $n = 135$	B $n = 65$	ACD $n = 69$	B $n = 89$	ACD $n = 40$	B $n = 87$
Cu	155	65	49	15	69	15
Zn	532	470	382	281	340	251
Co	49	32	12	13	19	14
Cr	140	98	58	28	60	30
P	767	659	760	562	782	732
Pb	37	23	89	7	39	36
Ag	4.1	4.5	4.4	3.3	4.5	3.9
Ni	43	45	25	25	27	29
V	89	99	32	33	33	31
Sr	14	31	28	44	24	43

the hanging wall and footwall at the ACD zone; the smallest anomaly is in the footwall of the B zone. The anomaly in sedimentary rocks is similarly much greater at the ACD zone than at the B zone as shown in Fig. 11-4. The extent of the halos is summarized in Table 11-III.

Trace element halos at Brunswick No. 12

The footwall acid volcanic rocks at Brunswick No. 12 are depleted in Ni and have anomalously high contents (compared with background) of Cu, Pb, Zn, and Co. Footwall sedimentary rocks have anomalously high Cu, Pb, Zn, and Co. In the hanging-wall sedimentary rocks Pb and Zn are enriched (although to a lesser extent than in the footwall), and Cu is at or below background levels.

An interesting example of a multi-element halo (characterized by a depletion of Cu and Ni and an enrichment of Zn and Co) was detected by Goodfellow (1975b) in an alkali basalt lying stratigraphically above the Brunswick No. 12 deposit, but separated from the ore zone by at least 100 m of sedimentary and acid volcanic rocks. There is a hemispherical alteration zone of 400 m radius in the basalt; the alteration is characterized by chlorite, microcrystalline quartz, and calcite, with argillaceous alteration in the rocks closest to the deposit. LDFs were calculated for 24 samples of altered basalt and 24 samples of unaltered basalt well away from mineralization using hot-extractable Cu (hxCu), hxZn, hxCo, and hxNi as variables. The LDF was used to classify all 129 drill core and surface samples and to calculate the probability of each sample belonging to the altered basalt group (i.e., anomalous group). The probabilities were contoured; they showed an intensive (probability of being anomalous greater than 70%) hemispherical halo with a

276

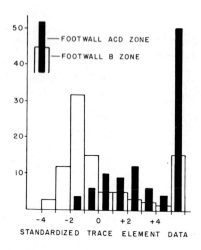

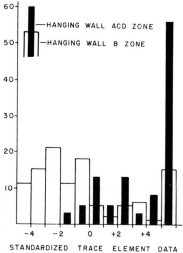

Fig. 11-3. Frequency distributions of standardized factor for Σ normalized (Cu + Pb + Zn + Co + Cr + P + Ag) − (Ni + V + Sr) in acid volcanic rocks at Heath Steele ACD and B zones, New Brunswick, Canada (redrawn with permission from Wahl, 1978).

radius of about 800 m directly stratigraphically above the deposit (Fig. 11-5). The magnitude of the geochemical contrast is indicated by the fact that background samples have a probability of less than 20% of being anomalous.

Use of cluster analysis and LDF at Caribou

Gandhi (1978) attempted to eliminate subjective judgement in the selection of "anomalous" and "background" population groups by classifying the hanging wall acid volcanic rocks at the Caribou deposit by Q-mode cluster

A and C ZONES

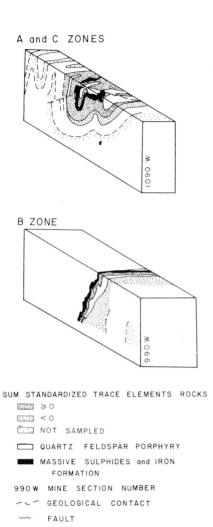

B ZONE

SUM STANDARDIZED TRACE ELEMENTS ROCKS

⬚ ⩾ 0

⬚ < 0

⬚ NOT SAMPLED

▭ QUARTZ FELDSPAR PORPHYRY

■ MASSIVE SULPHIDES and IRON
FORMATION

990 W MINE SECTION NUMBER

⌒⌣⌒ GEOLOGICAL CONTACT

⌇⌇ FAULT

Fig. 11-4. Spatial distribution of standardized factor for Σ normalized (Cu + Pb + Zn + Co + Cr + P + Ag) − (Ni + V + Sr) in footwall sedimentary rocks at Heath Steele ACD and B zones, New Brunswick, Canada (compiled from Wahl, 1978).

analysis on the basis of hxCu, hxPb, hxZn, hxCo, and hxNi. The samples cluster into seven main groups (Fig. 11-6). The groups 1A, 1B, and 1C are closely related and cluster together within an amalgamated distance of 0.665; these are all distant from mineralization and are clearly background samples (Fig. 11-7). The groups 2A, 2B, and 2C are quite distinct from the background samples and, in fact, lie close to mineralization. The samples in group 2B are the most closely related to each other and link with the hierarchial dendogram only at an amalgamated distance of 1.627; this group lies closest to the main mineralization. Group 1D occupies an intermediate

TABLE 11-III

Dimensions (in metres) of geochemical halos around massive sulphide deposits, New Brunswick, Canada. It is not possible to calculate the dimensions of all halos known to be present because of lack of drill data in parts of the various deposits (from Wahl, 1978)

Deposit	Horizon and rock*	Elements				
		Ca, Mg, K, Na, Fe	Cu, Pb, Zn, Ni, Co, Ag, Cr, V, Sr, P	Cu, Pb, Zn, Co, Ni	Cu, Zn, Co, Ni	Pb, Zn
BMS 12	FW acid volcanic rocks	450 × 915				450 × 915
	HW basic volcanic rocks				800 × 1600	
Heath Steele ACD	FW acid volcanic rocks + sedimentary rocks	>100 × >1140	>100 × >1140			
	HW acid volcanic rocks	>100 × >915	>90 × >915			>90 × >915
Heath Steele B	FW acid volcanic rocks	450 × 900	275 × 900			
	HW acid volcanic rocks	120 × 455	30 × 455		360 × 1200	
Caribou	HW acid volcanic rocks	200 × 800		200 × 800		
Key Anacon	FW acid volcanic rocks	>205 × 1800		205 × > 210?		>90 × >915

* FW = footwall; HW = hanging wall.

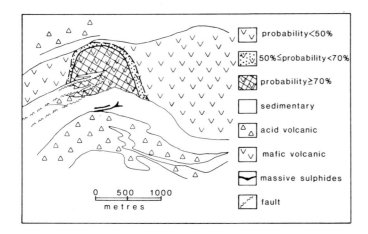

Fig. 11-5. Surface contour of probability of mafic volcanic rock being anomalous (calculated from LDF with hxCu, hxZn, hxCo, and hxNi as variables) above the Brunswick No. 12 deposit, New Brunswick, Canada (compiled from Goodfellow, 1975b).

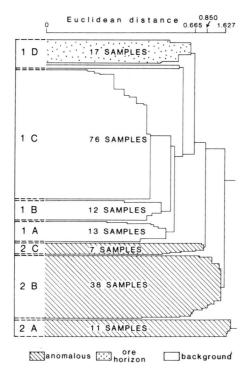

Fig. 11-6. Dendogram of Q-mode cluster analysis of silicic volcanic rocks based on hxCu, hxPb, hxZn, hxNi, and hxCo, Caribou, New Brunswick, Canada (redrawn with permission from Gandhi, 1978).

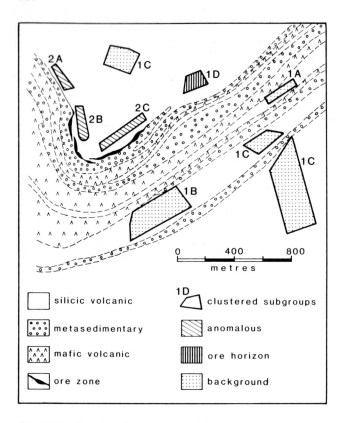

Fig. 11-7. Simplified geology and location of clustered subgroups (see Fig. 11-6) of silicic volcanic rocks at Caribou, New Brunswick, Canada (compiled from Gandhi, 1978).

position between the anomalous and background samples on the dendogram; it is significant that this group, although it is about 400 m from mineralization, contains samples from the same stratigraphic level as the ore horizon.

Cluster analysis can obviously be a useful technique for preliminary sorting of data. It is an entirely objective technique that does not involve any a priori knowledge of sample locations with respect to mineralization; as such it can be used to sort samples for processing by LDF.

From the anomalous and background groups defined by cluster analysis, Gandhi selected 48 samples from the anomalous groups and 130 samples from the background group to constitute a "training set" for LDF using hxCu, hxPb, hxZn, hxCo, and hxNi as variables. The degree of separation of the two groups is shown graphically in Fig. 11-8 where the first canonical variable (which measures the difference between groups) is plotted against the second canonical variable (which measures the variations within groups that are independent of group differences). The remaining samples were used as a 'testing set" to judge the effectiveness of the calculated LDF. The

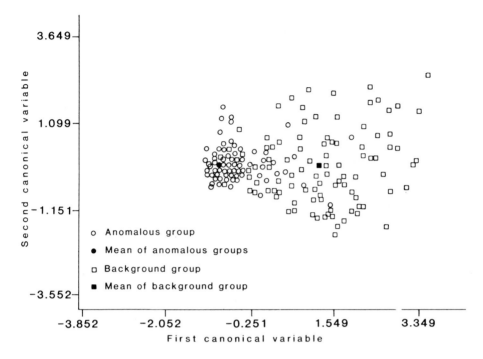

Fig. 11-8. Canonical variables for anomalous and background silicic volcanic rocks calculated for hxCu, hxPb, hxZn, hxCo, and hxNi, Caribou, New Brunswick, Canada (redrawn with permission from Gandhi, 1978).

TABLE 11-IV

Confusion matrix for LDF classification of hanging-wall silicic volcanic rocks at Caribou, based on hxCu, hxZn, hxPb, hxCo, and hxNi (from Gandhi, 1978)

	Training set		Testing set	
	anomalous	background	anomalous	background
Anomalous	44	4	35	5
Background	2	128	4	86
Correctly classified (%)	94.6		93.1	

"confusion matrix" of both data sets is shown in Table 11-IV. The degree of correct classification is very good, and the similarity of the percentage of correctly classified samples in the training and testing sets indicates that the calculated LDF is not biased.

The LDF defined a halo in the acidic volcanic rocks for 200 m stratigraphically above the ore zone and about 800 m laterally from it. The relatively small halo (compared with halos around other deposits in the Bathurst district, see Table 11-III above) is presumably due to the narrow width of mineralization and the fact that samples were available only from the upper stratigraphic level.

Lead-zinc LDF halos at Heath Steele, Brunswick No. 12, and Caribou

At the Heath Steele B zone a hemispherical halo zone extending 360 m stratigraphically above the ore zone and 1200 m laterally from the ore was defined in the hanging-wall quartz-feldspar porphyry by LDF using the ore elements Pb and Zn as variables. The statistical probability that a particular sample belongs to the anomalous population was calculated and plotted (Whitehead and Govett, 1974). An example of this approach is shown in Fig. 11-9, where a gradual increase in the probability of the quartz feldspar porphyry being anomalous occurs as the Heath Steele B zone ore horizon is approached.

The relation between Pb and Zn and the 50% probability of a sample being anomalous is shown in Fig. 11-10 for the Heath Steele B zone. Clearly an increase in Pb relative to Zn is the characteristic of anomalous samples. At Brunswick No. 12 acidic volcanic rocks occur only in the footwall, and attempts to apply the Heath Steele B zone discriminant functions to the rocks at Brunswick No. 12 failed to define an anomaly. This result was expected since the footwall quartz-feldspar porphyry at the Heath Steele B zone deposit was defined as essentially a background group in the derivation

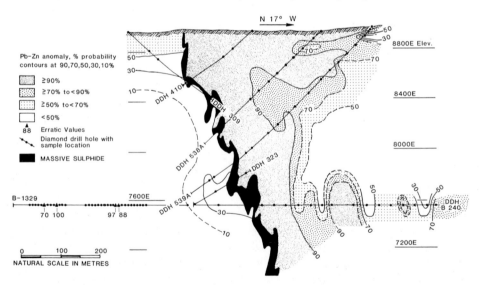

Fig. 11-9. Contours of probability of acid volcanic rocks containing anomalous contents of Pb and Zn around the Heath Steele B zone, New Brunswick, Canada (compiled from Whitehead and Govett, 1974).

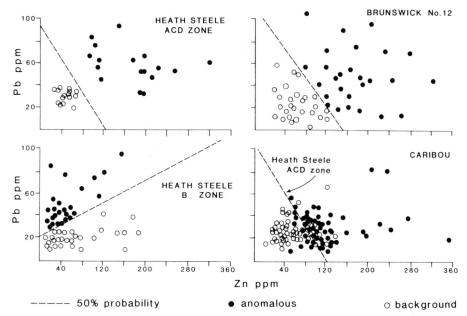

Fig. 11-10. Variation in Pb and Zn contents in acidic volcanic rocks in relation to 50% probability line between anomalous and background conditions (calculated by LDF with Pb and Zn as variables) for New Brunswick, Canada massive sulphide deposits (compiled from Whitehead and Govett, 1974; Goodfellow, 1975b; Wahl, 1978; and Gandhi, 1978).

of the functions (Whitehead and Govett, 1974). Accordingly, Goodfellow (1975b) determined another Pb-Zn function for the Brunswick No. 12 footwall quartz-feldspar porphyry rocks that defined a halo that extends 450 m below the ore zone and 950 m laterally from it. The 50% probability line dividing anomalous from background samples is at right angles to the corresponding line for the Heath Steele B zone (Fig. 11-10) and shows that in this case anomalous samples are characterized by both high Zn and high Pb.

At the Heath Steele ACD zone there are quartz-feldspar rocks in both the hanging wall and the footwall; nevertheless, the Heath Steele B zone function failed to define a halo zone even in the hanging wall. Therefore, a new function was derived for the ACD zone data; it was almost identical to the Brunswick No. 12 function — as may be seen by a comparison with the 50% probability dividing line between anomalous and background samples (Fig. 11-10). The hanging wall halo is greater than 90 m × 915 m (the precise extent cannot be determined because of the present erosion level). A footwall halo is not well defined, possibly because 100—150 m of sedimentary rocks lie between the footwall quartz feldspar porphyry and the ore zone.

The contents of Pb and Zn in background and anomalous hanging wall acidic volcanic rocks at the Caribou deposit are also shown in Fig. 11-10 in relation to the Heath Steele ACD zone 50% probability line (the samples were classified as either anomalous or background on the basis of other elements; see above). Although the Heath Steele ACD zone 50% discrimination does not give a perfect classification of the Caribou samples, they clearly are more closely allied to the Heath Steele ACD zone—Brunswick No. 12 population than to the Heath Steele B zone population; the Pb-Zn geochemical response at the latter deposit is obviously exceptional.

Distribution of water-soluble elements, conductance, and pH

An interesting study of water-soluble trace elements, specific conductance, and pH of aqueous rock powder slurries was made by Goodfellow and Wahl (1976) using the techniques described by Govett (1973, 1975, 1976a). They determined Na, K, Ca, and Mg by AAS on the filtered solution obtained by stirring 10 ml of deionized H_2O with 1 g minus 200-mesh rock powder for 1 minute. Water-extractable F and Cl were determined by specific ion electrode on the filtered leachate obtained by stirring 5 ml deionized H_2O with 1 g minus 200-mesh rock powder for 1 minute. Specific conductance and pH were determined on the slurry obtained by stirring 1 g minus 200-mesh rock powder with deionized H_2O for 1 minute.

The average contents of water-soluble Na, K, Ca, Mg, Cl, F, and H ion and specific conductance for footwall acid volcanic rocks at Brunswick No. 12 and the footwall and hanging-wall acid volcanic rocks at Heath Steele are given in Table 11-V. Compared to background values the altered acid volcanic rocks are enriched in K, Ca, and Mg, are depleted in Cl, and have a higher specific conductance at both deposits. Sodium shows minor depletion at Brunswick No. 12 and minor enrichment at Heath Steele; H ion is enriched at Brunswick No. 12 and depleted at Heath Steele (background H ion in the Heath Steele area is an order of magnitude higher than in the Brunswick No. 12 area). The spatial trends at Brunswick No. 12 towards the ore zone (illustrated along a drill hole in Fig. 11-11 and Fig. 11-12) show an increase in K, Mg, H ion, and specific conductance, and a decrease in Na, Cl, and F. Calcium shows an initial large increase in concentration but declines close to mineralization (although values remain anomalously high).

Over the first 200 m from the ore in the Heath Steele footwall (i.e., a comparable distance to the whole of the Brunswick No. 12 traverse) K and H ion increase, and Na, Cl, and F decrease in concentration; Mg increases slightly, and specific conductance and Ca show no variation. Over the whole sampled distance of 400 m in the Heath Steele footwall K shows a continuous increase and Na declines erratically (but all values are anomalously high except for the sample closest to the sulphides); Cl and F decline slowly and erratically, and H ion shows no trend (i.e., the increase in H ion seen over the 200 m closest to ore is probably not significant); specific conductance shows a very small increase. The spatial distribution in hanging-wall rocks at Heath Steele (Figs. 11-11 and 11-12) shows an increase in Na, F, and H

TABLE 11-V

Average concentration of water-extractable elements and conductance in volcanic rocks at Brunswick No. 12 and Heath Steele deposits* (from Goodfellow and Wahl, 1976)

	Na (ppm)	K (ppm)	Ca (ppm)	Mg (ppm)	Cl (ppm)	F (ppm)	H^+ (ppb)	Conductance (micromohs/cm)
Brunswick No. 12 deposit								
Unaltered quartz-feldspar augen schist	36	49	4	5	74	1.9	0.0126	68
Altered quartz-feldspar augen schist	31	141	58	25	37	1.2	0.1000	327
Heath Steele deposit								
Unaltered quartz-feldspar porphyry	27	29	9	1	—	—	0.2520	38
Altered footwall quartz-feldspar porphyry	49	108	15	7	35	1.5	0.0020	186
Altered hanging-wall quartz-feldspar porphyry	34	151	73	8	29	1.5	0.0080	415

* H^+ and conductance measured on 1 g/100 ml slurry.

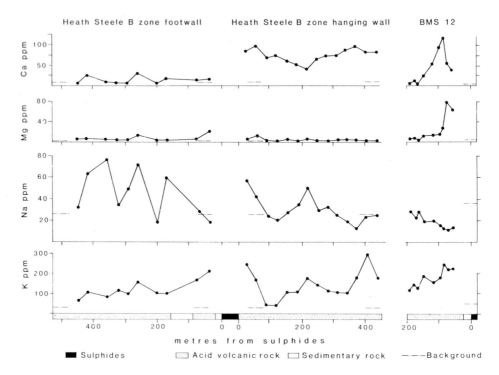

Fig. 11-11. Distribution of H_2O-soluble K, Na, Mg, and Ca in rocks at Heath Steele B zone and Brunswick No. 12 deposits, New Brunswick, Canada (compiled from Goodfellow and Wahl, 1976).

ion towards the ore zone; K steadily declines towards the ore zone and then increases again about 80 m away; specific conductance also steadily declines towards the ore zone to a minimum at about 200 m and then increases; and Ca, Mg, and Cl show no significant variation.

At Brunswick No. 12 the extensive halo in the footwall defined by total Pb, Zn, and major elements (see below) is equally well defined by anomalously high content of water-soluble Mg, Ca, and K, low contents of water-soluble Na and Cl, and high H ion concentration and specific conductance. The 800-m-radius hxCu, hxZn, hxCo, and hxNi anomaly defined in the hanging-wall alkali basalt is similarly precisely defined by high water-soluble Cl and low water-soluble K in the basalt.

At the Heath Steele B zone a hemispherical halo of 800 m radius in the hanging wall (which is larger than the 360 m × 1200 m halo defined by Pb-Zn described above) is defined by anomalously high contents of water-soluble Mg, K, and Ca, high specific conductance, and low H ion. The footwall halo defined by major elements (see below) is also equally well defined by anomalously high water-soluble Mg, Ca, K, and specific conductance.

As pointed out by Goodfellow and Wahl (1976) the water-extractable

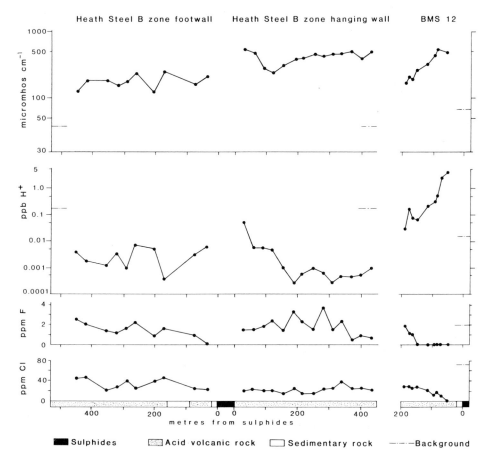

Fig. 11-12. Distribution of H_2O-soluble Cl and F, and H ion and specific conductance of aqueous slurries of rock powders at Heath Steele B zone and Brunswick No. 12 deposits, New Brunswick, Canada (compiled from Goodfellow and Wahl, 1976).

elements are presumably derived from chlorides, carbonates, sulphates, and sulphides rather than from silicate minerals. Sodium and Cl are presumed to be present as NaCl (since $[Na] \simeq [Cl]$) in fluid inclusions; the deficiency of these elements near to mineralization may be due to destruction of inclusions by the alteration processes. Calcium and Mg are probably derived from secondary carbonates in the alteration zones. There is no obvious explanation for the water-soluble K — although it could be due to secondary K_2SO_4.

These water-soluble anomalies offer an intriguing prospect for a very simple, rapid and cheap exploration technique, although confirmatory evidence of their general applicability is needed from other deposits. Gandhi (1978) undertook a limited study of water-soluble F and Cl at the Caribou deposit, and found that in the hanging-wall rocks the F content generally declined towards the sulphide bodies, whereas the Cl content either increased or decreased. Despite the inconsistent behaviour of Cl, the mean value of 31 ppm indicates a deficiency of Cl compared with background data from Heath Steele and Brunswick No. 12. Fedikow (1978) reported that the specific conductance in the footwall and hanging-wall rocks of the Sullivan deposit in British Columbia is an order of magnitude (80—150 micromohs/cm) greater than in the background rocks (background has a limited range of 8-10 micromohs/cm).

MAJOR ELEMENT HALOS

Brunswick No. 12 deposit

At the Brunswick No. 12 deposit mineralogical alteration (chloritization, sericitization, and silicification) is limited to about 60 m below the deposit and laterally for the extent of the sulphides. An anomalous major element halo, however, was defined by Goodfellow (1975a) extending about 450 m into the footwall acid volcanic rocks below the deposit and for a similar distance north and south of the deposit. He defined two main zones passing outwards from the sulphides: an inner strongly altered zone (corresponding to the mineralogical alteration) which includes the stockwork zone where Fe exceeds 7%; and an outer geochemical halo zone where mineralogical alteration is weak or absent. The geochemical trends in these zones are:

— Mg increases and Ca decreases in content towards the deposit (Fig. 11-13); Mg generally has a lower mean concentration in the strongly altered zone than in the geochemical halo zone, but increases in concentration towards the ore deposit; the Mg:Ca ratios increase fairly consistently towards the deposit.

— Mn decreases slightly (relative to background) in the geochemical halo zone, but it shows higher concentrations in the strongly altered zone; concentrations in the latter zone increase towards the deposit.

— Fe generally increases towards the deposit, especially in the strongly altered zone.

— K and Na have a complicated behaviour, although the Na:K ratio consistently decreases towards the deposit (Fig. 11-14). The Na content increases slightly in the geochemical halo zone, but it shows a marked decrease in the strongly altered zone; the content of K is generally lower than regional background in both the geochemical halo and strongly altered zone. The mean content of K in the strongly altered zone is lower than that in the geochemical halo zone, but the concentration may increase or decrease towards the deposit.

The footwall sedimentary rocks show a similar response to the acid volcanic rocks; there is a depletion halo of Na, K, Ca, and Mn; Mg is enriched.

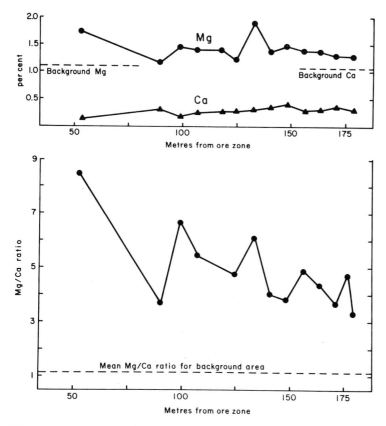

Fig. 11-13. Distribution of Mg, Ca, and Mg:Ca ratio in footwall acid volcanic rocks as a function of distance from the Brunswick No. 12 deposit, New Brunswick, Canada. (Reproduced with permission from Govett and Goodfellow, 1975, *Institution of Mining and Metallurgy, Transactions, Section B*, vol. 84, 1975, fig. 4, p. 137.)

There are no acid volcanic rocks in the hanging wall, but the sedimentary rocks are depleted in Na, K, and Ca (although less so than in the footwall) and are enriched in Mg and Mn; the content of Mn increases laterally from the deposit, reaching a concentration of 4% compared with 0.5% in the ore hanging wall. There is a general depletion halo of hxNa around the deposit that is apparently little affected by variation in lithology (Fig. 11-15). An alkali basalt located 100 m stratigraphically above the ore zone is depleted in Mg and Ca and is enriched in Fe (the halo in the basalt is discussed more fully above).

Heath Steele deposits

At the Heath Steele ACD zone Wahl (1978) recognized a strongly altered zone in both hanging-wall and footwall rocks similar to that in the footwall at Brunswick No. 12; there is no well-defined stockwork zone, although

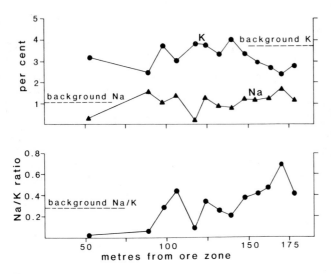

Fig. 11-14. Distribution of Na, K, and Na:K ratio in footwall acid volcanic rocks as a function of distance from the Brunswick No. 12 deposit, New Brunswick, Canada, (from Goodfellow, 1975a).

pyritization occurs below the C-1 zone. At the B zone deposit there is no mineralogical alteration in the hanging wall, and alteration is weak and patchy in the footwall. The major elements trends at both the ACD and the B zones are similar to those at the Brunswick No. 12 deposit and differ only in the behaviour of K — which shows significant increases in concentration towards the sulphide zones. The effect of the high K at Heath Steele on the relation between K, Ca + Na, and Fe + Mg can be seen on the ternary diagram in Fig. 11-16 (it may be noted that highly altered sericite-chlorite acid volcanic rocks from the footwall zone of disseminated sulphides at Key Anacon fall in the strongly altered zone on this diagram). The major element trends in sedimentary rocks are similar to those in the volcanic rocks.

Whereas the major element trends in the acid volcanic rocks are similar at the Heath Steele ACD and B zones, there are differences between the deposits that are most clearly shown by the Mg:Ca ratios (as illustrated in Fig. 11-17). In the hanging wall at the B zone about 70% of the samples have Mg:Ca ratios of less than 1.0 (regional background is 1.2), compared to only 18% of corresponding samples at the ACD zone which have Mg:Ca ratios of less than 1.0; this difference is reflected by the spatial extent of the halo which extends only 50 m above the ore horizon at the B zone but extends more than 100 m at the ACD zone. The difference in Mg:Ca ratios between the B and the ACD zones is not so marked in the footwall, although the ratios are also higher in the ACD zone than in the B zone; the mean ratios for both zones are higher than the corresponding footwall rocks. Differences in the geochemical response between the two zones are also indicated by variations in other major elements (and, as noted above, in trace elements).

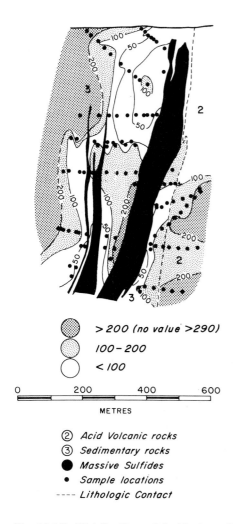

> 200 (no value >290)

100 - 200

< 100

```
0        200       400       600
════════════════════════════════
            METRES
```

② Acid Volcanic rocks
③ Sedimentary rocks
● Massive Sulfides
• Sample locations
---- Lithologic Contact

Fig. 11-15. Distribution of hxNa in sedimentary and acid volcanic rocks around the Brunswick No. 12 deposit, New Brunswick, Canada (compiled from Goodfellow, 1975b).

It is suggested by Wahl that the geochemical differences are a reflection of the fact that the ACD zone is a proximal deposit and that the B zone is a distal deposit (this is discussed at greater length below). Although the present erosion surface and insufficient drilling data at the ACD zone precludes determination of the full extent of the anomalous zone, it has a greater dimension that that at the B zone (Table 11-III), as would be expected.

Caribou deposit

Gandhi's (1978) work at the Caribou deposit showed that Na and K trends in hanging wall silicic metavolcanic rocks are different to those

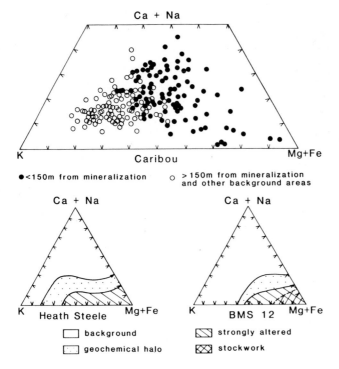

Fig. 11-16. Ternary diagram of the variation in K, Ca + Na, and Mg + Fe in acid volcanic rocks with proximity to sulphide mineralization at Heath Steele, Brunswick No. 12, and Caribou deposits, New Brunswick, Canada (compiled from Gandhi, 1978; Goodfellow, 1975b; and Wahl, 1978).

formed around other deposits in the Bathurst district. Sodium *increases* and K *decreases* with proximity to mineralization, thus causing an increase in the Na:K ratios (Fig. 11-18). He stated that there is no discernable mineralogical change accompanying the variation in K and Na, but the K content and the K:Na ratio is higher in rhyolites at Caribou than in the majority of similar rocks in the Bathurst district. The behaviour of Ca and Mg is normal for the area, and the Mg:Ca ratio increases towards mineralization. The concentrations of Fe and Mn increase towards mineralization, and there is no variation in the SiO_2 content. Although the Caribou rocks are clearly divided into an anomalous and a background group on a ternary plot of K, Ca + Na, and Mg + Fe, they do not follow the same trend shown at the Brunswick No. 12 and the Heath Steele deposits (Fig. 11-16). In Fig. 11-19, which is a ternary diagram of Ca, K + Na, and Mg + Fe, however, there is a clear trend from the Na + K corner to the Fe + Mg corner with proximity to mineralization; this is consistent with the trend around volcanic-sedimentary deposits of the Canadian Shield reported by Roscoe (1965), Descarreaux (1973), and Nichol et al. (1977).

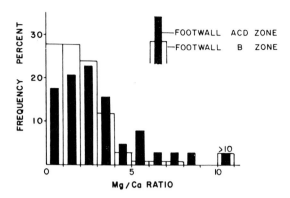

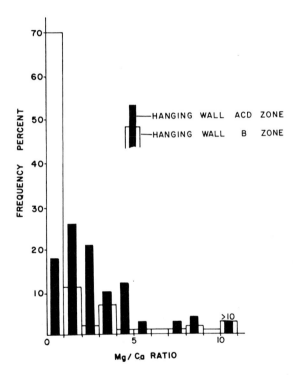

Fig. 11-17. Frequency distributions of Mg:Ca ratios in silicic volcanic rocks at Heath Steele ACD and B zones, New Brunswick, Canada (Redrawn with permission from Wahl, 1978).

The best combination of major elements to distinguish anomalous samples at Caribou was found to be K, Na, and Mg. The use of these variables in discriminant analysis (and the plotting of the percent probability of a sample

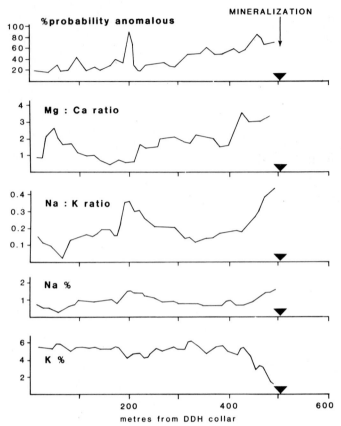

Fig. 11-18. Distribution of K, Na, Na:K ratios, Mg:Ca ratios, and probability of a sample being anomalous (based on LDF with K, Na, and Mg as variables) in hanging wall silicic volcanic rocks, Caribou, New Brunswick, Canada (compiled from Gandhi, 1978).

being anomalous) gave a fairly regular trend of increasing anomalous values towards mineralization (see Fig. 11-18 above). The halos defined by major elements extend for about 200 m stratigraphically above the deposit and about 800 m laterally. There are inadequate drill data to determine the extent of the anomalies in footwall rocks.

ANALYTICAL CONSIDERATIONS

An important operational conclusion is the demonstration that hot concentrated HNO_3 sample digestion gives element distribution patterns comparable to those obtained by the much more costly and time-consuming total sample digestion (Wahl, 1978). A comparison of results from the two digestion techniques for sedimentary rock samples from drill core 1500 m

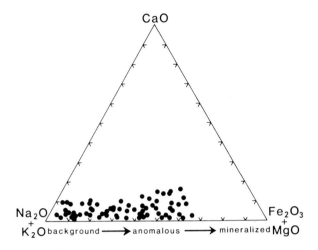

Fig. 11-19. Ternary diagram of CaO, Fe_2O_3, MgO, and $Na_2O + K_2O$ in silicic volcanic rocks, Caribou, New Brunswick, Canada (redrawn with permission from Gandhi, 1978).

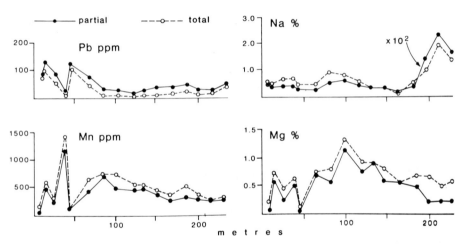

Fig. 11-20. Comparison of total (HF-HNO_3) and partial (HNO_3) analyses for Mn, Pb, Mg, and Na in sedimentary rocks at Heath Steele, New Brunswick, Canada (redrawn with permission from Wahl, 1978). Note that partial Pb is higher than total Pb; this is due to determinations being made by AAS without background correction.

west of the westernmost limit of massive sulphides at the Heath Steele ACD zone is shown in Fig. 11-20; this drill hole was selected to minimize the effects of sulphide minerals. There is a very close correspondence in patterns given by the total and partial digestion techniques, and extraction by the latter technique is high for all elements except Na and K. Similar results were obtained in acid volcanic rocks in the hanging wall and footwall of the Heath Steele B zone deposit.

TABLE 11-VI

Percentage of total element content of rocks at Heath Steele extracted by partial (i.e., hot, concentrated HNO_3) digestion (from Wahl, 1978)

Element	Percent extraction		
	sedimentary rocks, W. Grid	acid volcanic rocks, B zone	
		hanging wall	footwall
Cu	99	50	94
Zn	40	19	38
Mn	72	77	55
Fe	76	50	69
Na	7	2	3
K	4	5	13
Ca	87	76	79
Mg	70	36	43

The percentage extractions for sedimentary and acid volcanic rocks are shown in Table 11-VI. Of particular interest is the higher extraction of Cu, Zn, Fe, Na, K, and Mg in the footwall compared to the hanging-wall rocks. The higher extraction of Cu, Zn, and Fe is probably due to the greater proportion of sulphide minerals in the footwall; the higher extraction of Na, K, and Mg is probably a reflection of the greater alteration in the footwall.

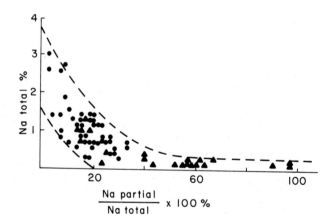

▲ Samples taken near MINE

● Samples taken further from MINE

Fig. 11-21. Variation in proportion of Na extracted by hot nitric acid (Na partial) as function of total Na and proximity to ore zone in footwall acid volcanic rocks at Brunswick No. 12, New Brunswick, Canada. (Reproduced with permission from Govett and Goodfellow, 1975, *Institution of Mining and Metallurgy, Transactions, Section B*, vol. 84, 1975, fig. 5, p. 138.)

Some elements also show a variation in percent extraction with distance from sulphides. In the hanging wall there is an increase in percent extraction of Fe, Mn, and Mg towards the sulphides; in the footwall the percent extraction of Na increases towards the sulphides but the extractability of Mg and K decreases towards the sulphides. The increase in extractability of Na towards the sulphides in footwall acid rocks seems to be a general feature, since it was reported at Brunswick No. 12 by Govett and Goodfellow (1975). This is illustrated in Fig. 11-21.

Whereas Wahl (1978) demonstrated that the general disposition of major

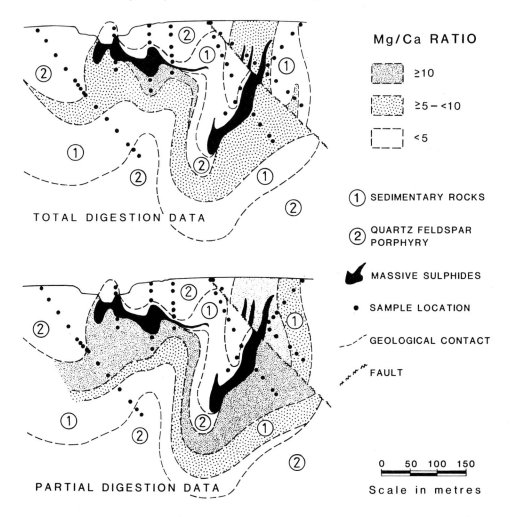

Fig. 11-22. Comparison of Mg:Ca halo obtained by total and partial digestions of footwall sedimentary rocks at the Heath Steele ACD zone, New Brunswick, Canada (redrawn with permission from Wahl, 1978).

element halos was similar for partial and total sample digestion techniques, the dimensions of the halos varied. In the footwall sedimentary rocks at the Heath Steele ACD zone halos defined by Mn:Fe, Na:K, and Mg:Ca ratios are greater for partial digestion than for total digestion techniques (this is illustrated for Mg:Ca ratios in Fig. 11-22). In acid volcanic rocks in both the footwall and the hanging wall at the Heath Steele B zone the extent of the halos for Mn:Fe, Na:K, and Mg:Ca ratios are greater for total sample digestion than for partial digestion. The difference in extent and intensity of halos obtained by the two sample digestion techniques is small, however, and halos are quite adequately defined by partial digestion techniques.

GENERALIZED MODEL AND INTERPRETATION OF GEOCHEMICAL DISPERSION

The type and extent of element dispersion in rocks around massive sulphide deposits will depend upon the precise mechanism of formation of the sulphide deposit, the physico-chemical nature of the host rocks (especially the degree of permeability), the physico-chemical environment of deposition if in a sedimentary basin, and the mechanism of element dispersion. Alteration due to hydrothermal activity from fumarolic fluids has been recognized by many workers in various places (Tatsumi and Clark, 1972; Descarreaux, 1973; Simmons, 1973) and has the same general geochemical characteristics as around the New Brunswick deposits — broadly an enrichment of Fe, Mn, Mg and, in some cases, K, and a depletion of Ca and ususally Na. The extent and disposition of alteration zones depends upon the primary permeability as well as the chemical reactivity of the rocks. At Brunswick No. 12 the pyroclastic rocks below the sulphide zone presumably had a high primary permeability, and the alteration zone there is extensive; the alteration zones in the more massive and less permeable rhyolite underlying the deposits of the Abitibi Volcanic Belt in the Canadian Shield are much less extensive (Descarreaux, 1973).

Factors that will influence disposition of elements in a sedimentary basin are Eh and pH, temperature gradient due to hydrothermal brines, composition and solubility of the brines, the topography of the basin, and the confining pressures. In the immediate area of sulphide deposition pH is likely to be low and Eh negative, conditions which favour Mn remaining in solution. Further away from the area of sulphide deposition conditions become less acidic and more oxidizing, causing Mn to become less soluble and to precipitate in larger amounts relative to Fe than in the immediate vicinity of the sulphides. This effect is illustrated by the study by Whitehead (1973a), who showed that Mn is concentrated lateral to the Heath Steele B zone deposit and is also enriched in the hanging wall relative to the footwall.

Towards the end of sulphide deposition a decrease in sulphur fugacity is to be expected that would lead to a corresponding increase in Eh and pH. These conditions would cause an increase in the solubility of Cu and a decrease in the solubility of Pb; thus, it would be expected that Cu should be depleted and Pb enriched in hanging-wall rocks relative to footwall rocks. This is generally true of the New Brunswick deposits. Dispersion into post-ore rocks may occur after deposition of the sulphides; the nature of such dispersion will depend on whether it occurs before or after consolidation of the overlying rocks. If dispersion occurs after consolidation, primary permeability is reduced and dispersion may be largely confined to favourable fractures, shears, and contacts. The geometry of the resultant halos will depend upon the disposition of channel ways, chemical reactivity of the rock, the chemical character of the medium of transport, and probably the composition of the sulphide deposit. The *mechanism* of dispersion into post-ore rocks is conjectural. Whereas late-stage hydrothermal fluids can always be invoked, it is probable that upward moving (and probably saline) waters of compaction played a significant role.

It was Wahl (1978) who first drew attention to the importance of distinguishing between proximal and distal deposits in interpreting rock geochemical data around massive sulphides in northern New Brunswick. He recognized the probable significance of the difference in the Pb-Zn halo in the hanging wall at the Heath Steele B zone and at other deposits. This, and other geochemical differences, lead to the conclusion that the Heath Steele B zone is a distal deposit. The case for this conclusion (Wahl and Govett, 1978) is summarized below.

— The footwall sedimentary halo defined by total trace element variation is more intensive and extensive at the ACD zone than at the B zone (Figs. 11-3 and 11-4 above); it has its maximum intensity below the C-1 zone. The same pattern occurs in the footwall quartz-feldspar porphyry at the ACD zone, and there is none in the hanging wall at the B zone (apart from the Pb-Zn halo defined by discriminant analysis). The differences are not related to the amount of massive sulphides present — which is large at the B zone but is relatively minor at the C-1 zone. The differences do suggest, however, that the C-1 zone was a focus of hydrothermal activity that continued after the deposition of sulphides, whereas there was little hydrothermal activity around the B zone. The obvious conclusion is that the ACD zone deposits lie close to a fumarolic vent (near to the C-1 zone that also has the largest quantity of disseminated pyrite in the footwall), whereas the B zone was formed at a greater distance from the fumarolic source by downslope migration of heavy brines.

— The ratio of Mn:Fe at the Heath Steele B zone and at Brunswick No. 12 was shown to decrease towards the sulphide zones (Whitehead, 1973a; Goodfellow, 1975b). The Mn:Fe ratio is relatively larger in the sedimentary rocks around the Heath Steele B zone than around the ACD zone (Fig. 11-23). This suggests that the source is closer to the ACD zone where relatively more Fe than Mn is quickly precipitated, whereas relatively more Mn is able to travel further than Fe.

— Major element alteration in the footwall sedimentary rocks that lie

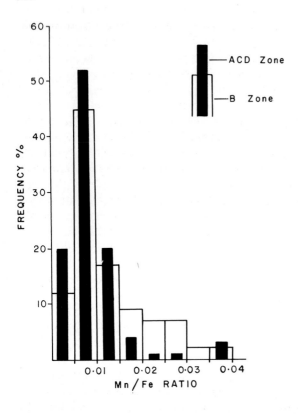

Fig. 11-23. Frequency distributions of Mn:Fe ratios in footwall sedimentary rocks at Heath Steele ACD and B zones, New Brunswick, Canada (redrawn with permission from Wahl, 1978).

immediately below the sulphides is similar at both the Heath Steele ACD and B zones. There are differences between the two zones, however, in the quartz-feldspar porphyry. The footwall and the hanging wall at the ACD zone is intensely altered.

— The trace elements Sr, Ni, and V are higher in the footwall sedimentary rocks at the Heath Steele B zone than at the ACD zone. There is evidence to suggest that low values of these elements indicate proximity to hydrothermal activity. Boström and Peterson (1966) found Sr to be low in areas of high heat flows at the East Pacific Rise; Kuroda (1961) reported that Sr (as well as V) decreased in concentration with increasing hydrothermal alteration around the cupriferous pyrite deposits at the Hitachi mine in Japan (relative loss of Sr associated with hydrothermal alteration has been noted by workers in many situations); Govett (1972) noted loss of Ni as a characteristic of alteration zones around the massive sulphide deposits of Cyprus.

— The geometric mean content of Cu in the footwall sedimentary rocks at the Heath Steele ACD zone is 155 ppm, compared to 65 ppm in the corresponding rocks at the B zone (Table 11-II above). At Brunswick No. 12 data indicate a geometric mean Cu content of 210 ppm in the footwall sedimentary rocks and a regional geometric mean of about 50 ppm in sedimentary rocks. Thus, the Cu content in the footwall at the Heath Steele ACD zone is similar to that at Brunswick No. 12 — where Goodfellow (1975a) established the presence of a stockwork zone — and in the footwall of the B zone the Cu content is only a little above background. The footwall rocks around stockwork zones beneath the Cyprus massive sulphide deposits are similarly characterized by abnormally high contents of Cu (Constantinou and Govett, 1973).

The trace and major element halos — and the mineralized alteration — at the Heath Steele ACD zone are similar to those at Brunswick No. 12; the alteration is less intense at the ACD zone, but it is concluded that the source vent is nearby. The ACD zone is, therefore, a proximal deposit. Alteration at the ACD zone is hydrothermal, modified by secondary processes (downward seepage of brines and upward movement of pore water due to compaction) in comparison to the B zone where it is essentially secondary in nature.

On present evidence the Brunswick No. 12 deposit appears to be a true proximal deposit with a well-developed stockwork zone in the footwall. The Key Anacon deposit may also be a proximal type, but there are inadequate data to confirm this (Wahl, 1978). There is no evidence of a stockwork at the Caribou deposit, but there are few drill data in the footwall; the geochemical characteristics of the deposit are intermediate between those of Heath Steele ACD and B zones, but possibly more closely similar to the ACD zone (e.g., compare the Pb-Zn hanging-wall halos in Fig. 11-12 above). Although Jambor (1979) has classified all of the New Brunswick deposits discussed here as proximal types on the basis of their sulphide mineralogy, the total evidence reviewed above seems to clearly contradict his conclusion for the Heath Steele B zone deposit.

CONCLUSIONS

The results of the investigations around the massive sulphide deposits in New Brunswick allow a generalized model of geochemical response to be formulated; this is illustrated schematically in Fig. 11-24. Although there are some differences in element behaviour between deposits, these can plausibly be related to variations in the environment of deposition of the sulphide deposits attributable largely to their location relative to primary brine sources (i.e., the degree of proximal or distal characteristics). The main characteristics of the model are as follows:

— Mg enrichment and Ca depletion are characteristic of footwall and hanging-wall rocks. In proximal deposits the anomalies are similar in extent

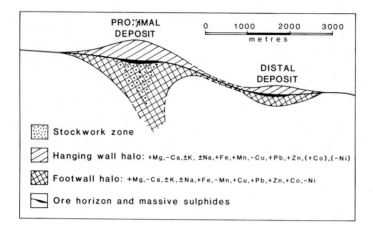

Fig. 11-24. Schematic diagram illustrating form of geochemical anomalies in rocks around massive sulphide deposits in New Brunswick, Canada.

and intensity in both footwall and hanging-wall sequences; in distal deposits the halo in the hanging wall is restricted to the immediate vicinity of the deposit.

— K and Na generally exhibit anomalous concentrations compared with background conditions, but they show considerable variation between deposits. In most cases the Na:K ratio decreases towards mineralization regardless of the absolute levels of the elements relative to background. The exception to this is at Caribou where the Na:K ratio increases and where there is an enrichment of Na and a depletion of K in hanging-wall rocks.

— The content of Fe increases towards sulphide zones, and it is especially high in footwall rocks of proximal-type deposits. Manganese increases in concentration in hanging-wall rocks towards sulphide deposits; maximum concentrations characteristically occur lateral to deposits (especially in sedimentary rocks), and the immediate hanging wall may be relatively deficient in Mn. The content of Mn in the footwall rocks is generally lower than in hanging-wall rocks and may be deficient relative to background, although the concentration still increases towards the sulphide zone. In sedimentary rocks the Mn:Fe ratio decreases towards the sulphides; this ratio reaches its lowest levels in the vicinity of proximal-type deposits.

— Cu is relatively depleted in hanging-wall rocks compared with footwall rocks; it may be also absolutely depleted in the hanging wall relative to background levels. In the footwall Cu is anomalously high, especially in proximal type deposits.

— Pb and Zn are both anomalously high in footwall and hanging-wall rocks; the content of Pb is generally higher in hanging-wall rocks than in lithologically comparable footwall rocks.

— Co is enriched and Ni is depleted in footwall rocks. On the basis of limited data it appears that similar relations are also true, although less marked, in hanging-wall rocks.

— Water-soluble K, Ca, and Mg give extensive positive anomalies in both hanging-wall and footwall rocks; conductance of water slurries of rock powders also gives an intensive and extensive anomaly in hanging-wall and footwall rocks.

There are obviously differences in relative response between proximal and distal deposits as well as between hanging-wall and footwall rocks around the New Brunswick massive sulphide deposits. It is doubtful, however, that there are adequate data even on the New Brunswick deposits to define *absolute* parameters to discriminate between the various categories of response. Nevertheless, from the point of view of general exploration, the extent of the halos exhibited by a variety of major and trace elements is such that deposits buried to depths of 500—600 m should have a surface expression. The general hemispherical shape of halos means, however, that a flat-lying deposit buried at 300 m could be detected by sampling on a 150-m grid, whereas a flat-lying deposit buried to a depth of 600 m would not be detected.

The massive sulphide deposits of the Bathurst district are, in fact, steeply dipping, and, since the lateral halo is generally more extensive than the vertical halo, a 200-m sampling grid should be adequate to detect a proximal deposit up to 500 m below the surface. Simple halos (i.e., single elements or element ratios) are only clearly defined for major elements. Thus, the extensive (205 m x 1800 m) halo listed in Table 11-III for Key Anacon is based on Mg:Ca ratios, Mn:Fe ratios, and Na in surface rocks; single trace elements are not anomalous at the surface. Simple trace element halos are confined to the immediate proximity of the sulphides and are generally recognizable for only tens of metres; subtle trace element halos identifiable by multivariate statistical techniques can normally be recognized and are of similar extent to the simple major element halos.

The results of the detailed studies of the New Brunswick deposits are obviously of considerable local value in exploration for other deposits and additional ore zones around existing mines. The data also have more general application in other areas. The results demonstrate the kinds of variations in response to expect between distal and proximal deposits; the detailed examination of responses for individual elements show that some of them (e.g., Mn and K) may be enriched in one part of the aureole and depleted in another part (depending upon stratigraphic position and lateral distance from sulphides). It was the convincing data from New Brunswick that gave the writer the confidence to design the studies in Turkey and Australia discussed in Chapter 12 in the belief that extensive geochemical responses would be found.

LOCAL AND MINE SCALE EXPLORATION FOR STRATIFORM DEPOSITS OF VOLCANIC AND SEDIMENTARY ASSOCIATION — CYPRUS, TURKEY, AND OCEANIA

CYPRUS DEPOSITS

One of the earliest uses of rock geochemistry for exploration for massive sulphides was based on the work by the writer in Cyprus in 1967—1968; this work formed the basis for the later studies of the New Brunswick deposits described in Chapter 11. Since many of the techniques used were common to both the New Brunswick and the Cyprus deposits and have already been described in Chapter 11, the account of the Cyprus work here is brief; it is included because the geological setting of the Cyprus deposits is quite different from that of other massive sulphides described in this book (the difference in geological setting of Turkish, Fijian, and Australian deposits also accounts for their inclusion in this chapter).

The cupriferous pyrite deposits of Cyprus lie near the top of the Troodos volcanic series which consists of basaltic pillow lavas and intrusive dykes; the latter increase in intensity downwards. The sulphide deposits are characteristically overlain by the Ochre Group, a Mn-poor, Fe-rich sediment which commonly contains sulphides as bands and fragments. The Ochre Group does not normally extend beyond the boundaries of the sulphide deposits (Constantinou and Govett, 1972). Where the succession is complete the Ochre Group is overlain by very basic pillow lavas (which in places rest directly on sulphides) which are defined as Upper Pillow Lavas (UPL); the pre-ochre and pre-mineralization lavas are defined as the Lower Pillow Lavas (LPL). The stratigraphic relations are illustrated in Fig. 12-1 and are discussed more fully by Govett and Pantazis (1971), Constantinou and Govett (1972), and Searle (1972).

The sulphide deposits (after correction for local tilting and faulting) all have a saucer-shaped disposition and are characterized by a distinct vertical zoning from top to bottom (see Constantinou and Govett, 1973 for a detailed discussion):

— *Zone A* is massive ore, with more than 40% sulphur. The upper part of the zone is referred to as "conglomerate ore" and is fragmental (with sulphide blocks in a matrix of friable iron disulphides); it contains interstitial copper sulphides, sphalerite, quartz, and gypsum and other sulphates;

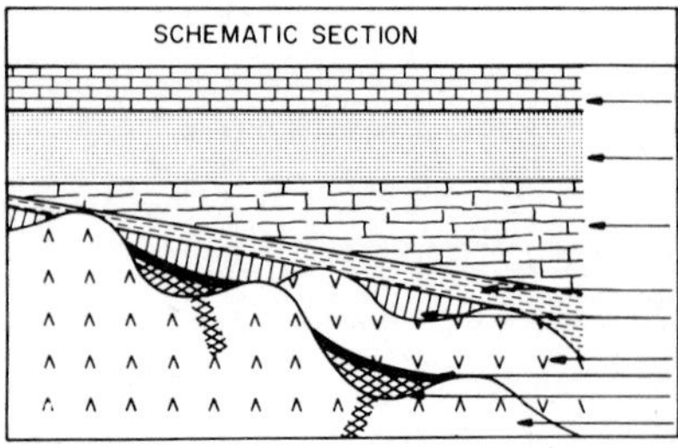

306

SCHEMATIC SECTION	LITHOLOGY	FORMATION	AGE
	REEF LIMESTONE	KORONIA	MIDDLE MIOCENE –UPPER MIOCENE
	MARLS– MARLY CHALKS	PAKHNA	MIDDLE MIOCENE
	CHALKS–CHALKS{ CHERTS– MARLS	LEFKARA	LOWER MIOCENE –UPPER CRETACEOUS (MAESTRICHTIAN)
	ARGILLITES,GRITS,CHERT UMBER	PERAPEDHI	UPPER CRETACEOUS (CAMPANIAN)
	PILLOW LAVA(post-mineralization) OCHRE SULPHIDE PILLOW LAVA(pre-mineralization)	TROODOS VOLCANIC SERIES	UPPER CRETACEOUS ?

Fig. 12-1. Schematic stratigraphic relations of Cyprus cupriferous pyrite deposits. (Reproduced with permission from Constantinou and Govett, 1972, *Institution of Mining and Metallurgy, Transactions, Section B*, 1972, vol. 81, fig. 2, p. 36.)

collomorphic pyrite is common. The lower part of the zone is the "compact ore" which is composed mainly of large blocks of pyrite coated with chalcopyrite and covellite along fractures.

— *Zone B* is a pyrite-quartz and red jasper zone with 1—2% Cu, containing 40% S at the top grading down to 30% S at the bottom. Chalcopyrite and sphalerite occur interstitially in the pyrite and, less commonly, as minute inclusions in the pyrite. Zone B is not present in all deposits.

— *Zone C* is a stockwork zone with less than 30% S (which decreases downwards) with disseminated pyrite in lavas and pyrite within fractures. The amount of Fe in disseminated pyrite within the lavas is about the same as the total Fe in unaltered lavas, and there is no chalcopyrite or sphalerite in the disseminated pyrite. Although the main zone of alteration is normally below the centre of the ore bodies, in many places sulphide disseminations extend for several kilometres stratigraphically below the ore bodies.

The geochemical responses in both pre-ore (LPL) and in post-ore (UPL) lavas around the deposit are broadly similar to the regional scale anomalies described in Chapter 7 — i.e., depletion of Cu and enrichment of Zn. In addition there is a significant enrichment of Co and usually a depletion of Ni.

There are some variations in response between deposits, as shown in Fig. 12-2 for Skouriotissa (2.5×10^6 tonnes of 2.35% Cu), Mathiati (3×10^6 tonnes of 0.24% Cu), Agrokypia A (65,000 tonnes of 1.5% Cu grading down to 0.4% Cu), and Agrokypia B (4.5×10^6 tonnes, of which 0.5×10^6 tonnes is high-grade Cu and Zn massive ore). The Cu, Zn, and Co contents of the sulphides are shown in Table 12-I. The samples at Mathiati and Skouriotissa are from the surface within about 1 km of the open pits and stratigraphically above the ore; samples at Agrokypia A are from the west face of the open pit above the ore zone; samples at Agrokypia B are from a surface traverse about 140 m above the ore zone.

In post-ore rocks Cu is depleted at Skouriotissa, Mathiati, and Agrokypia A, but it is considerably enriched at Agrokypia B. The latter feature is probably a reflection of the very high Cu content in the ore zone at Agrokypia B,

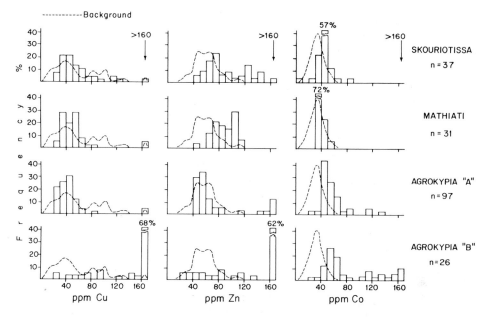

Fig. 12-2. Frequency distributions of Cu, Zn, and Co in pillow lavas around the Skourio-tissa, Mathiati, Agrokypia A, and Agrokypia B deposits, Cyprus.

TABLE 12-I

Composition of ore from Skouriotissa, Mathiati, Agrokypia A, and Agrokypia B ore bodies, Cyprus (from Govett, 1972)

	Cu (%)	Zn (ppm)	Ni (ppm)	Co (ppm)
Skouriotissa	2.15	747	24	330
Mathiati, A zone	0.28	1050	44	53
B zone	0.09	10,450	36	33
Stockwork	0.15	770	27	62
Agrokypia A, stockwork	0.16	83	61	133
Agrokypia B, A zone	0.03—12.2	520—78,000	15—45	20—50
B, stockwork	0.036	520	36	81

and the fact that all of the samples are much closer to ore than the samples from the other relatively Cu-rich deposit (Skouriotissa). The lavas around all of the deposits show an enrichment in Zn. The greatest Zn enrichment is found at Agrokypia B (which has the highest sphalerite content in the sulphides); the lowest Zn enrichment occurs at Agrokypia A (which has the lowest Zn in the ore). Cobalt is enriched at all deposits; the lowest Co enrich-ment is at Mathiati, which also shows the lowest enrichment of Co in the ore.

TABLE 12-II

Comparison of Cu content (ppm) in the pillow centre and adjacent interstitial material between pillows, Cyprus (halo samples are from within 1 km of mineralization, wallrock samples are from within 10 m of ore)

	Number of samples	Centre of pillow	Interstitial material
Background samples	8	89	87
Halo samples	13	39	163
Wallrock samples	23	51	291

All of the element trends described are based on metal content in the centre of the pillows; limited work on the silty interstitial material shows some different trends, particularly for Cu. The contents of Cu in pillow centres and interstitial material for background samples, samples within 1 km of the deposits (halo samples), and samples from wallrock within 10 m of ore are shown in Table 12-II. In the centres of pillows the contents of Cu show an overall decline in concentration with proximity to mineralization, but reach a minimum in the halo samples. In the interstitial material, however, Cu content shows a marked *increase* with proximity to mineralization; this is due to the interstitial material being enriched relative to pillow centres by a factor of 4 in the halo samples and by a factor of 6 in the wallrock samples. There is no enrichment of Cu in interstitial material of background samples.

The interstitial material presumably provided the only ready channels for rising mineralizing solutions, and it is tentatively suggested that the observed enrichment of Cu in interstitial material reflects the passage of Cu-rich solutions. Whether the depletion of Cu in the pillows represents a primary magmatic feature, or is the result of leaching by the mineralizing solutions, cannot be determined on present data. Notwithstanding doubts concerning genetic explanations of the distribution of Cu, it is probable that analysis of interstitial material between pillows (i.e., the probable "plumbing" system) would be an effective means of defining mineralization.

The distribution of both Cu and Zn within lavas becomes increasingly erratic as mineralization is approached, as illustrated in Table 12-III. The distribution of Cu is erratic even in background areas where it has a variation of up to 50%; this increases to almost 130% at the 95% confidence level near ore. In background areas Zn is much less variable than Cu — up to 20% — but becomes much more variable with increasing alteration and is similar to Cu near ore deposits. This relationship has been found to be a general feature of the behaviour of ore elements in rocks around massive sulphides.

The differences in element concentration in post-mineralization lavas around the cupriferous sulphide deposits compared to background levels are generally small and there is considerable overlap between the background and the anomalous populations. In this type of situation a multi-element

TABLE 12-III

Mean ($\bar{x}$, ppm) and coefficient of variation (cv, %) of Cu and Zn in centres of eight adjacent pillows at various localities, Cyprus (from Govett and Pantazis, 1971)

	Cu		Zn	
	$\bar{x}$	cv ·	$\bar{x}$	cv
Background	110	25	54	10
	49	9	73	7
Altered	42	19	62	9
	109	14	55	22
	25	28	62	16
Near gossan	67	28	75	25
Near ore	67	64	180	62

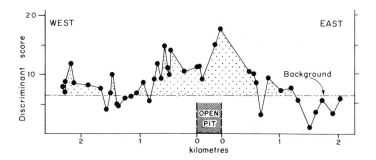

Fig. 12-3. Discriminant scores for anomalous function (Cu, Zn, Co as variables) for pillow lavas around the Mathiati open pit cupriferous pyrite deposit, Cyprus. (Redrawn with permission from Govett and Goodfellow, 1975, *Institution of Mining and Metallurgy, Transactions, Section B*, vol. 84, 1975, fig. 2, p. 135.)

approach is obviously appropriate. Govett (1972) showed that extensive kilometre-scale halos could be defined by discriminant analysis (and also by multiplicative graphical methods). An example is shown in Fig. 12-3 by the plot of discriminant scores for anomalous function based on Cu, Zn, and Co across the Mathiati deposit; a well-defined anomaly with a radius of more than 1.5 km clearly defines the location of the ore zone.

TURKISH DEPOSITS

Cupriferous pyrite deposits occur in the mafic rocks of the southeastern Anatolian Ophiolite belt of the Ergani-Maden region in Turkey. The succession is divided into two groups: the lower Guleman Group of Jurassic/

Cretaceous age which comprises peridotite, banded gabbro, and basalt (in ascending order) metamorphosed to the greenschist facies; and an upper Maden Group of an interdigitating mixture of mafic volcanic rocks, volcano-clastics, mudstones, and limestones of Maestrichtian/Upper Eocene age metamorphosed to prehnite-pumpellyite facies. As the basis of a Ph.D. thesis supervised by the writer at the University of New Brunswick, Erdogan (1977) studied the geochemical dispersion around four small deposits: (1) the Helezür deposit in the basalt unit of the Guleman Group; (2) the Main Open-pit and its displaced extension; (3) the Mihrap Dägi deposits at the base of the Maden Group; and (4) the Hacan deposit towards the top of the Maden Group. The Main Open-pit is a typical Cyprus-type deposit, but since it is surrounded by much older rocks due to structural complexities, there are few rock outcrops suitable for an investigation of geochemical dispersion; therefore the following discussion is limited to the small uneconomic Helezür and Hacan deposits.

The Helezür pyritic deposit is a thin concordant lens (50—60 cm thick) extending about 100 m within basaltic pillow lavas, flows, and tuffs. There is a massive sulphide zone at the top of the deposit which grades down into a mineralogically altered zone of chlorite, vein and disseminated sulphides, silica, and sericite. The sulphide mineral is pyrite; chalcopyrite occurs as veinlets within and between pyrite crystals, replaced in places by sphalerite. In the zone of massive sulphides the Cu content is 0.6—0.8%, and the Zn content is 500—1950 ppm. Three traverses across the deposit (Fig. 12-4) were sampled by taking about 20 chip samples (total weight 1 kg) of un-weathered surface rock; the samples were analyzed by AAS for Cu, Pb, Zn,

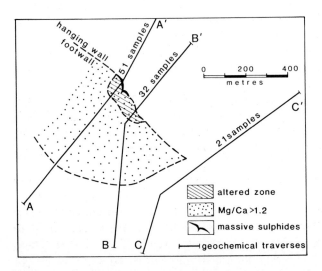

Fig. 12-4. Distribution of Mg:Ca halo around the Helezür sulphide deposit, Turkey (redrawn with permission from Erdogan, 1977).

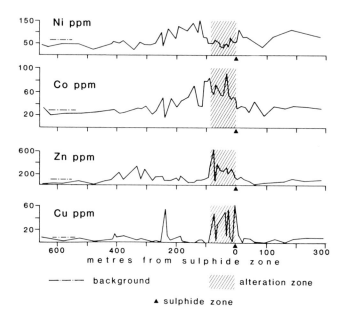

Fig. 12-5. Distribution of Cu, Zn, Co, and Ni in rocks of traverse A-A' (Fig. 12-4) around the Helezür sulphide deposit, Turkey (compiled from Erdogan, 1977).

Co, and Ni after hot concentrated HNO_3 digestion (Fig. 12-5) and for Na, Ca, Mg, Fe, and Mn after HF-$HClO_4$ digestion (Fig. 12-6).

A geochemical halo zone of about 400 m into the footwall is defined by high Mg:Ca ratios, as shown in Fig. 12-7 (the mineralogically altered zone extends only 80 m into the footwall); it is estimated that the halo zone also extends about 300 m laterally from the deposit. Other elements also give anomalous halos in the footwall: Na (Fig. 12-6) and Ni (Fig. 12-5) are depleted; Co forms a broad zone of anomalously high values (Fig. 12-5); Cu and Zn (and also Pb) are erratically high in the area of mineralogical alteration (Fig. 12-5); and Fe is strongly enriched in the altered zone (Fig. 12-6). The hanging wall is unaltered, and there is no apparent geochemical response. Away from the geochemical halo the footwall and hanging-wall basalts are chemically and petrologically indistinguishable.

At the Hacan deposit there are only disseminated and vein sulphides, and the zone of massive sulphides appears to be missing. The deposit occurs within a basalt pillow lava and tuff sequence, with considerable brecciation of the lavas. The footwall basalt is chloritized to a depth of about 50 m below the sulphide zone, and alteration extends laterally 200—250 m. There is quite an extensive halo (at least 1000 m long) defined by Mg and Fe enrichment and by Ca and Na depletion; trace element halos are not well defined, although Ni and Cu are depleted throughout the sampled extent of

312

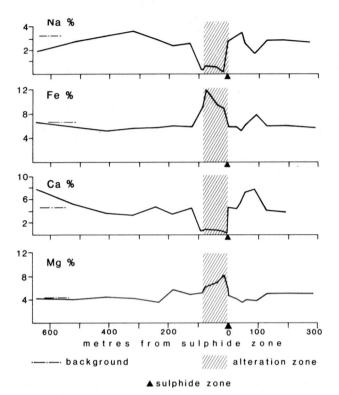

Fig. 12-6. Distribution of Mg, Ca, Fe, and Na in rocks of traverse A-A' (Fig. 12-4) around the Helezür deposit, Turkey (compiled from Erdogan, 1977).

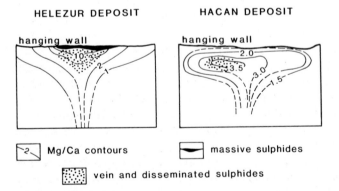

Fig. 12-7. Interpretation of Mg:Ca contours around the Hacan and Helezür deposits, Turkey, prior to structural deformation (redrawn with permission from Erdogan, 1977).

the footwall basalts compared to the hanging-wall basalts (there are no petrological differences between footwall and hanging wall).

The maximum alteration at the Hacan deposit occurs 30—60 m below the top of the footwall basalts. A reconstruction of the Mg:Ca ratio contours at the Helezür and the Hacan deposits prior to tilting (see Fig. 12-7) illustrates the essential difference in the alteration patterns. Erdogan (1977) speculated that the disposition of Mg:Ca contours mimic the pattern of isotherms at the time of sulphide deposition; the closed "mushroom" pattern at Hacan would favour boiling of the hydrothermal fluids at the sea floor, thus preventing deposition of massive sulphides, whereas the upwards open pattern at Helezür would favour sulphide deposition at the sea floor. This is quite a plausible explanation since the comparable size of the geochemical halos at both deposits suggests that the hydrothermal system was comparable at both locations.

AUSTRALIAN DEPOSITS

Mount Morgan and Mount Lyell

The Mount Morgan and the Mount Lyell sulphide deposits have received detailed investigations by two of the writer's research students at the University of New South Wales (Fedikow, 1982; Sheppard, 1981). The sampling, analytical, and interpretative techniques used were broadly similar to those used for the New Brunswick deposits; the results are, therefore, comparable with the results described in Chapter 11.

The Mount Morgan pyritic copper-gold deposit in south-central Queensland is now almost completely worked out; originally it had about 68×10^6 tonnes of 0.75% Cu and 3.4 g/tonne Au. Although minor remnants of stratiform sulphides have been found, essentially all that remains of the deposit is a discordant pipe (with more than 50% sulphides) that cuts across the stratigraphy beneath an extensive gold-rich gossan. The ore body was 300 m thick, 600 m long, and 200 m wide. Pyrite was the dominant sulphide, with lesser amounts of chalcopyrite, sphalerite, pyrrhotite, and magnetite; gold was disseminated throughout the ore as ultrafine to sub-microscopic inclusions and localized telluride shoots. The deposit occurs within a belt of Middle Devonian volcanic quartz-feldspar porphyry and rhyolitic and dacitic tuffs, with minor carbonate and chert rocks. Most of the mineralization was within the Lower Mine Porphyries (LMP), which are mostly massive quartz-feldspar porphyry and quartz porphyry. The quartz porphyry occurs in a pipe structure below the sulphides and extends to a known depth of 750 m; it is considered to be an alteration product of the quartz-feldspar porphyry. The LMP is conformably overlain by the Banded Mine Sequence (BMS), which is a laminated tuffaceous sequence interbedded with 5- 100-cm bands of hematite-magnetite jasper in the upper part of the 250-m sequence. The BMS is host to the upper part of the mineralization. The upper stratigraphic unit is the Upper Mine Porphyries (UMP), which are poorly-sorted fragmental and massive quartz-feldspar porphyry with fragments of chert, jasper, and quartz-feldspar porphyry.

Approximately 1200 samples of LMP and BMS rocks were taken from drill holes extending to 1230 m south of the ore zone; 300 surface rock samples of UMP were collected to a distance of about 1400 m south of the ore zone. Samples of jasper were digested with HNO_3-HF-$HClO_4$ for determination of Cu, Pb, and Zn by AAS; major elements were determined by XRF. All other samples were digested with HNO_3 for measurement of Cu, Pb, Zn, Ni, Co, Fe, Mn, Na, K, Ca, and Mg by AAS. Gold was determined by fire assay, H_2O by gravimetry, and Ba by XRF. Precision was better than 20% for all elements.

The contents of Mg, Zn, Au, and H_2O increase and the contents of Na, K, and Ca decrease towards the ore zone in drill core samples of BMS and LMP (Fig. 12-8). The trends are generally better defined in BMS than in LMP, and the most regular patterns are given by Na, K, and Au. Barium and Pb (not illustrated) also increase in concentration towards mineralization. The surface samples of UMP show an increase in Fe, Mn, and Mg contents and a decrease in Na and K contents towards mineralization; these trends are best illustrated as element ratios (Fig. 12-9). The jaspers show a rather less extensive halo of increasing Mn, Zn, Cu, Fe, and Mg contents, and a decrease of Na, Ca, and K contents with proximity to the ore.

It should be noted from Figs. 12-8 and 12-9 that (except for Zn, Cu, and Mn in the jaspers) there is little indication that elements have reached background at the southern limit of the drilling (1230 m from the deposit) or at the southern limit of the surface sampling (1400 m from the deposit). This is best illustrated by Au, which has a mean concentration in LMP and BMS of about 80 ppb 1230 m south of the deposit compared with an anticipated background of 1—4 ppb in unaltered rhyolite and dacite.

An X-ray diffraction study of samples from the LMP showed that in the immediate vicinity of the mine (within about 170 m) there is a siliceous aureole around the deposit which is succeeded outwards by a sericite and then a chlorite aureole. There is a progressive increase in plagioclase feldspar with increasing distance from the mine (to about 860 m from the mine). The depletion of Na and Ca towards mineralization is presumably the result of the progressive destruction of feldspar. Electron microprobe studies showed a progressive increase in the Mg content of chlorite towards the deposit; this feature, as well as an increase in the abundance of chlorite towards the deposit, accounts for the Mg pattern. The increase in H_2O towards mineralization probably reflects the increase in hydrous minerals — chiefly chlorite and sericite. The increasing concentrations of Cu, Pb, Zn, and Au with proximity to mineralization is probably related to a pyritic halo (this could be substantiated by estimating sulphide-held Fe by a sulphide-selective leach).

The comprehensive study at Mount Morgan has demonstrated extensive geochemical halos characteristic of massive sulphides elsewhere, as well as

315

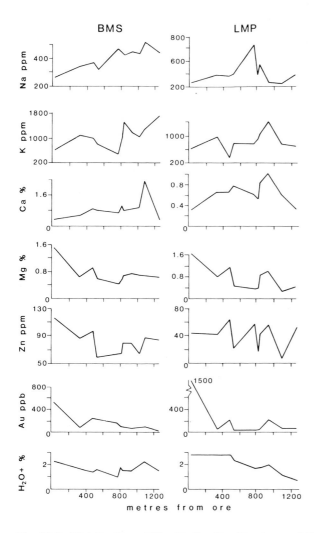

Fig. 12-8. Distribution of Na, K, Ca, Mg, Zn, Au, and H_2O in drill core of BMS (Banded Mine Sequence) and LMP (Lower Mine Porphyries) around the Mount Morgan deposit, Queensland, Australia (compiled from Fedikow, 1982).

producing some mineralogical evidence to explain the halos. An important result in the context of Australian conditions is the demonstration that halos are detectable even in weathered surface rocks.

Present production from the Mt. Lyell deposits in Tasmania is from stringer and disseminated pyrite-chalcopyrite ore; the massive sulphide part of the deposits was small and has been mined out. The major source of present ore is the Prince Lyell ore body from which 68×10^6 tonnes of 0.79% Cu has been mined; an estimated 22×10^6 tonnes of 1.44% Cu remains. The

316

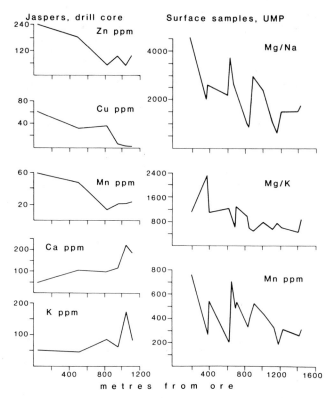

Fig. 12-9. Distribution of Zn, Cu, Mn, Ca, and K in jaspers (drill core), and distribution of Mn and Mg:K and Mg:Na ratios in surface samples of UMP (Upper Mine Porphyries) around the Mount Morgan deposit, Queensland, Australia (compiled from Fedikow, 1982).

ore body (as defined by a 0.8% Cu cut-off) is 360 m long, 10—95 m wide, and open at depth (greater than 480 m). The main sulphide minerals are pyrite and chalcopyrite, with quartz, sericite, chlorite, and iron oxides as alteration minerals.

Sheppard (1981) made a detailed study of the geochemical dispersion patterns around the Mount Lyell deposits based on more than 20,000 determinations on 1200 underground and drill core samples. Copper, Pb, Zn, Mn, Fe, Na, K, Ca, and Mg were determined by AAS following hot concentrated HNO_3 digestion; Ba, Sr, Rb, Sn, and Mo were determined by XRF; CO_2 was determined by Leco furnace.

The Mount Lyell deposits lie within the Middle to Upper Cambrian Mount Read volcanic sequence. In the vicinity of the Prince Lyell deposit the sequence consists of nearly vertical intermediate tuffs and lavas (possibly also intrusive bodies) intercalated with acid pyroclastic rocks and lavas. A 1-km-long section of the footwall of the Prince Lyell deposit was sampled

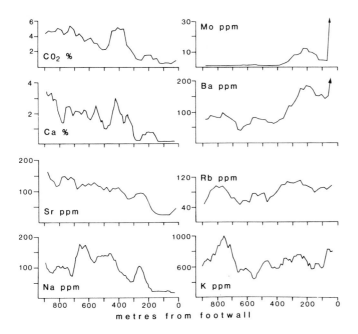

Fig. 12-10. Distribution of Na, K, Rb, Sr, Ca, Ba, Mo, and CO_2 (5-point moving average, 91 sample stations) in footwall volcanic rocks of the Prince Lyell deposit, Tasmania, Australia (compiled from Sheppard, 1981).

underground in the West Lyell haulage tunnel. The distribution of Na, K, Rb, Sr, Ca, Ba, Mo, and CO_2 are shown as 5-point moving averages in Fig. 12-10 (the moving average technique is used to reduce noise due to rapid changes in lithology).

About 700 m beneath the ore body in the footwall there is a major structural break (clearly indicated by an abrupt change in the trends of Na, K, Rb, and Ba) that terminates the mine volcanic sequence. From this point to the ore body there is a fairly steady decline in the contents of Na, Ca, Sr, and CO_2; there is extreme depletion of these elements and CO_2 for 250–300 m immediately below the deposit. The contents of Ba and Rb increase towards the ore body over 700 m, K increases over about 550 m, and Mo increases over about 300 m. Plagioclase, calcite, and dolomite decline in amounts towards the sulphides; this would account for the distribution patterns of Na, Ca, and CO_2. Sericite increases in amount towards the ore body, explaining the increase in K and Rb contents. It is not known whether Ba is present as barite or possibly in a mica lattice. The Mo halo is only slightly more extensive than a recognizable pyrite envelope that extends about 200 m from the deposit (Mo probably occurs as molybdenite).

The distributions of Cu, Zn, Mn, and Mg are shown in Fig. 12-11. The concentration of Cu increases erratically towards the deposit over about

318

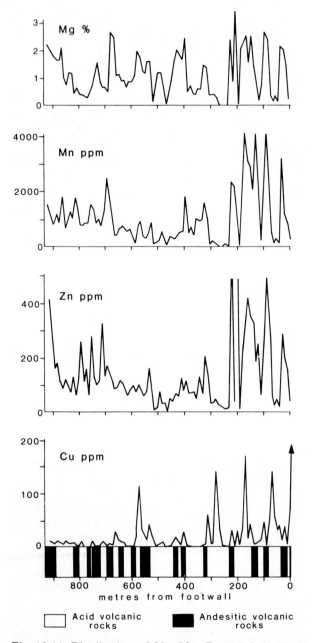

Fig. 12-11. Distribution of Mg, Mn, Zn and Cu in footwall volcanic rocks of the Prince Lyell deposit, Tasmania, Australia (compiled from Sheppard, 1981).

700 m; the peak values are about equally divided between the acidic and andesitic lithologies (Cu is probably present as chalcopyrite). Peak values for Zn, Mn, and Mg, however, correlate strongly with the andesitic lithologies; this suggests that these elements are present in silicate lattices and that their distribution is largely controlled by the relative greater abundance of ferromagnesian minerals (dominantly chlorite) in andesite. Both Zn and Mn are strongly anomalous in the andesites for about 200 m below the deposit; the distribution of Mg shows no discernible trend (except possibly a decrease in concentration in the acidic rocks) with proximity to mineralization.

The lack of suitable sections (and the presence of other ore bodies nearby) restricted the determination of the lateral extent of the halos and detailed investigation of the geochemical dispersion in the hanging wall. A drill hole 300 m south of the Prince Lyell deposit does show, however, a trend of decreasing Na, Sr, Ca, and CO_2 contents and an increase in K and Rb contents for 200—300 m of the footwall and the along-strike equivalent of the ore horizon.

Sheppard (1981) has also investigated responses around a number of other Mt. Lyell deposits. The general response is similar to that around the Prince Lyell deposit, and differences that do occur are attributable to differences in ore and alteration minerals. For example, at the 12 West deposit the main alteration mineral in the footwall is pyrophyllite rather than sericite, and K and Rb *decrease* in concentration towards the deposit. Interestingly, Sr *increases* in concentration in the ore horizon; Sheppard attributes this to substitution of Sr for the extremely abundant Ba in the virtual absence of Ca.

Woodlawn

The Woodlawn massive sulphide deposit in New South Wales was discovered during a drainage and soil geochemistry survey. The mineralization, ore mineralogy and chemistry, and mineralogical and geochemical alteration of the host rocks have been described by Malone (1979), Ayres (1979), and Petersen and Lambert (1979), respectively. The deposit is a typically zoned (from Pb-rich at the top, down through a Zn-rich zone, to a Cu-rich zone at the base) banded massive sulphide deposit, with stringer Cu-pyrite ore beneath and lateral to the deposit. There are 1×10^6 tonnes of ore grading 9% Zn, 3.4% Pb, and 1.76% Cu. The sulphides occur within a Middle to Upper Silurian sequence of felsic volcanic rocks and sedimentary rocks that are metamorphosed to the lower greenschist facies; the felsic volcanic rocks in the vicinity of the deposit are mainly pyroclastics with bedded chert. Silicification, sericitization, and chloritization are widespread; there is a virtual absence of feldspar in felsic volcanic and sedimentary rocks for up to 200 m above the ore, at least 100 m below the ore, and 500 m laterally from the ore. Chlorite schist is common below the ore, and in places chlorite is the main component, commonly associated with pyrite; sericitic muscovite is common close to mineralization.

The main geochemical characteristics are enrichment of Fe, Mg, S, Si, and H_2O and depletion of Na, Ca, and Sr. Hydrothermally altered rocks close to mineralization are sporadically enriched in Ag, Ba, Bi, Cd, Cu, Mn, Pb, Sn, and Zn. Petersen and Lambert (1979) defined a series of alteration zones around the deposit; these are summarized below and shown in Fig. 12-12.

— *Zone I* is characterized by chlorite schists, chert, and chloritized volcanic and sedimentary rocks. It is developed within the main mineralized zone and around other massive and stringer mineralized zones. It is analogous to the Kuroko siliceous stockwork, argillaceous lenses, and bedded cherts. Geochemically it is enriched in Mg, Fe, and SiO_2 and is depleted in Na, Ca, and Sr; K is slightly depleted.

— *Zone II* envelopes zone I and is characterized by abundant quartz, sericite, and chlorite; moderately abundant pyrite, minor chalcopyrite, sphalerite, and galena; and essentially no feldspar or primary ferromagnesian minerals. The zone is comparable to the Kuroko sericite—Mg-chlorite zone. Iron and Mg are enriched (but less than in zone I), SiO_2 is enriched, and Na, Ca, and Sr are depleted.

— *Zone III* occurs in the central volcanic pile to the south of the deposit and is characterized by quartz, sericitic muscovite, albitic plagioclase, chlorite, some pyrite and other sulphides as veins and disseminations, and minor K-feldspar. The zone is not pervasively altered and is geochemically variable.

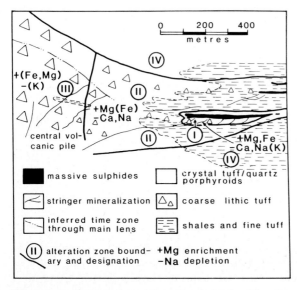

Fig. 12-12. Schematic section through the Woodlawn deposit (pre-deformation), New South Wales, Australia, showing main rock types and alteration zones (modified after Petersen and Lambert, 1979).

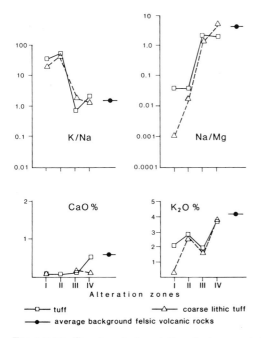

Fig. 12-13. Geochemical variation in two rock types around the Woodlawn deposit, New South Wales, Australia (compiled from Petersen and Lambert, 1979). Background is more than 2 km from the mine.

Magnesium and Fe are sporadically enriched, Na is markedly higher than in zone II, Ca is slightly higher than in zone II, and K is slightly depleted.

— *Zone IV* occurs stratigraphically above and below zones II and III and is characterized by quartz, albitic feldspar, K-feldspar, sericitic muscovite, and chlorite. The zone may be compared to the Kuroko outer feldspar-bearing clay zones. There is minor silicification and sporadic Mg enrichment.

The main mineralogical trends with proximity to mineralization are an increase in chlorite (and the Mg:Fe ratio in dolomite) and sulphide minerals. As in many other cases, geochemical trends are enhanced by element ratios. As shown in Fig. 12-13, Ca is depleted in zones I, II, and III (and slightly in zone IV for the rock types shown), but it shows little trend; K is similarly depleted in zones I, II, and III for the rock types shown. The variations are much more clearly defined for Na:Mg ratios which decrease by orders of magnitude towards the mineralization; the K:Na ratios show almost two orders of magnitude increase towards mineralization in zones I and II. The trends in sedimenatry rocks are similar to those in the felsic volcanic rocks, although they are rather less clearly defined.

Limerick

A small occurrence (less than 7% combined Pb and Zn over a width of

322

less than 2 m) of massive sulphide mineralization within middle to late
Silurian rhyolitic and dacitic pyroclastic rocks with interbedded sandstones,
siltstones, and shales at Limerick in New South Wales has been interpreted
as a distal-type deposit by Dunlop et al. (1979). Chip samples of the least
weathered rock (weathering extends to a depth of 30 m) were collected at
5- to 30-m intervals along four 500-m traverses normal to the strike over a
strike length of 1 km; drill core was also sampled. The total content of Mg,
Ca, Na, and K and HNO_3-$HClO_3$-soluble Cu, Pb, Zn, Fe, and Mn were determined by AAS.

Results for one of the traverses across the mineralized zone are shown in
Fig. 12-14 (there is considerable gradational change between totally sedimentary and totally volcanic rocks, and no attempt has been made to distinguish

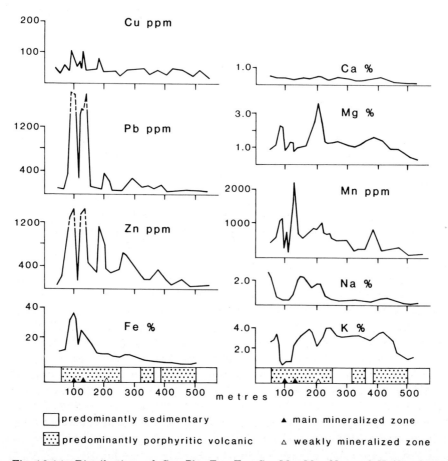

Fig. 12-14. Distribution of Cu, Pb, Zn, Fe, Ca, Mg, Mn, Na, and K (3-point moving
average) in surface rock chip samples across the Limerick deposit, New South Wales,
Australia (from Dunlop et al., 1979).

these variations in the data presented here). In the vicinity of the ore zone Mn, K, and Na are depleted; Fe, Pb, and Zn show strong peaks. In the footwall rocks Zn, Fe, and Mn show an increase in concentration towards mineralization over about 400 m; Na tends to increase in concentration towards mineralization, showing a strong peripheral peak within 100 m of mineralization. The variation in Mg content is correlated with chlorite content and does not seem to have been substantially redistributed during the weathering processes. Calcium contents, however, are uniformly low and are at least an order of magnitude lower than in drill core — this is attributed to weathering.

The higher contents of K are associated wtih sericite, occurring predominantly in the fine grained rocks of the hanging wall. The low contents of Na and K (and, in some cases of Ca) associated with high Fe contents are believed to be due to acidic leaching consequent upon oxidative weathering of Fe sulphides. An interesting point raised by Dunlop and his co-workers is the strong Pb-Zn anomaly in surface rocks that is not evident in drill core; as they have pointed out, this must be a secondary phenomenon (secondary anomalies in surface rocks on the island of Mykonos in Greece are also described in Chapter 9).

An important conclusion of this study is that geochemical patterns can be recognized in highly weathered surface rocks even around a distal type deposit without pervasive alteration. Patterns are not obscured by an heterogeneous sample population due to rapid variations in lithology.

Ajax prospect

In Chapter 9 it was remarked that the depth of weathering in many parts of Australia makes the use of conventional rock geochemistry impractical, but that weathered bedrock is proving to be a useful substitute; indeed, one of the recently discovered major massive sulphide deposits in Australia was found and defined by this technique (unfortunately data are not yet available for publication). However, a number of examples of the use of weathered bedrock in exploration for massive sulphides in Australia were given by Butt and Smith (1980). Results from one of them — the Ajax prospect 22 km southeast of Mount Morgan in Queensland — are described below on the basis of a description by Large (1980).

The mineralization occurs within a south-plunging anticline of intensely altered rhyolitic lithic tuffs and ash tuffs within the Middle Devonian Moongan Rhyolite (Fig. 12-15). In the mineralized tuffs there is 5—40% disseminated pyrite; the primary mineralization is pyrite, chalcopyrite, sphalerite, and minor native gold. The grade is 0.5% Cu, 2.5% Zn, 0.03% Pb, 10 g/tonne Ag, and 0.5 g/tonne Au. Weathered bedrock is 0.5—7.0 m below the surface. Samples were taken on a 25 m × 50 m grid (reduced to 10 m × 10 m over the anomalous zone).

324

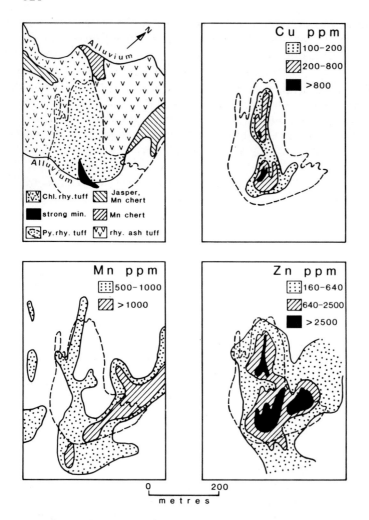

Fig. 12-15. Geology and distribution of Cu, Zn, and Mn in weathered bedrock at the Ajax copper-zinc-silver prospect, Queensland, Australia (from Large, 1980).

The distribution of Cu, Zn, and Mn in weathered bedrock is shown in Fig. 12-15. The distribution of Mn defines the fold structure of the mineralized horizon and the overlying Mn-rich exhalite (cherts and jaspers). Copper is anomalous directly over the mineralized zone; the distribution of Pb (not illustrated) is essentially the same as Cu. The distribution of anomalous Zn is far more widespread, and, as suggested by Large (1980), presumably reflects secondary dispersion processes.

THE WAINALEKA DEPOSIT, FIJI

The Tertiary age Wainaleka massive sulphide deposit in the south of Vitu Levu, Fiji, has been investigated by Rugless (1981). Mineralization has a strike length of 100 m with an average width of 5 m; it persists to a depth of 120 m. The average grade is 5% Zn, 1.5% Cu, and 10 g/tonne Ag. The Zn-rich top overlies massive pyrite that grades into disseminated sphalerite and chalcopyrite. A low-grade stockwork zone of stringers and disseminations of chalcopyrite and pyrite occurs as a 20-m zone below the mineralized lens. Gypsum veining occurs around the mineralization.

The deposit occurs in a volcanic fragmental unit which mantles a sodic rhyolite domal complex within predominantly basic to intermediate lavas and volcanoclastic rocks. The volcanic sequence has low-potassium island arc tholeiitic affinities. The geological relations are shown in Fig. 12-16. There is strong phyllic (quartz-sericite) alteration associated with mineralization at the top of the domal complex; this phyllic zone is 200—250 m wide and has a strike length of 2000 m. The phyllic alteration zone passes into an argillic (clay-sericite) zone that has a strike length of 4000 m and extends

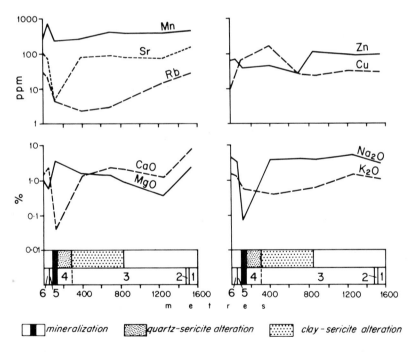

Fig. 12-16. Distribution of Mn, Sr, Rb, Cu, Zn, CaO, MgO, Na_2O, and K_2O in surface rock, Wainaleka deposit, Fiji (compiled from Rugless, 1981). Footwall: 1 = sodic rhyolite; 2 = basalt; 3 = rhyolite; 4 = fragmental unit. Hanging wall: 5 = andesite; 6 = sodic rhyolite.

800 m into the footwall from mineralization. The hanging wall intermediate to acid volcanic rocks have propylitic (chlorite—calcite ± epidote ± albite ± zeolite) alteration for up to 100 m from the mineralization.

Relatively unweathered surface samples were collected across the strike from creek beds. They were analyzed for SiO_2, Al_2O_3, K_2O, Na_2O, CaO, and MgO by XRF and for Cu, Pb, Zn, Ag, Co, Mn, Rb, and Sr by AAS after a total HF-$HClO_4$-HNO_3 digestion. Drill core samples of the hanging-wall and the footwall fragmental units were analyzed for Cu, Pb, Zn, Ag, Co, Mn, Rb, Sr, Na, K, Al, Mg, Fe, and Ca by AAS after a total HF-$HClO_4$-HNO_3 digestion.

There are strong wallrock halos of the logarithmic decay type over about 30 m in the footwall for Cu, Zn, Pb, Ag, Co, Fe, K, and Rb; Mn, Al, and Mg (and, to a lesser extent, Na and Ca) are depleted. In the hanging wall Mn and Sr are enriched for about 30 m.

Large surface halos (Fig. 12-16) that extend for about 1200 m into the footwall are shown by depletion of K_2O and Rb (and, less distinctly, by Mn) and by enrichment of MgO; K_2O and Rb show a slight increase in the phyllic zone, but they are still below regional concentration. The phyllic alteration zone is strongly depleted in CaO, Na_2O, and Sr and is enriched in Cu. Zinc (and Co) are depleted for 500—600 m below the deposit. A similar traverse about 1500 m west of the deposit shows pronounced halos of decreasing Na_2O and Co contents and increasing Mg contents towards the stratigraphic horizon of the mineralization over a distance of about 600 m.

The local scale anomalies in the footwall at the Wainaleka deposit are similar to those around most massive sulphide deposits except for the strong depletion of Zn and Co. Extensive halos for MgO, K_2O, and Rb that persist well beyond the mineralogical alteration zones are clearly recognizable, and the focus of mineralization is defined by CaO, Na_2O, and Sr.

CONCLUSIONS

There is a broad similarity of geochemical responses in the wide variety of massive sulphides considered in this chapter to the Archaean deposits of the Canadian Shield, the Kuroko deposits, and the New Brunswick deposits. Although major elements were not determined in the Cyprus study, the geologically similar cupriferous pyrite deposits of Turkey showed strong footwall halos of increasing contents of Fe and Mg and decreasing contents of Na and Ca with proximity to mineralization. This indicates that the ophiolite-associated Cyprus type deposits may be expected to fit the general patterns of geochemical response for massive sulphides described in Chapters 10 and 11.

With the exception of the Turkish deposits, all deposits for which data are available have hanging wall halos. The fact that the Turkish deposits have no discernible hanging-wall halos may be due to the extremely small size of the occurrences investigated.

The data from Cyprus — which shows that Cu *decreases* in pillow lavas and *increases* in interstitial material with proximity to mineralization — suggests the possibility that the former is a regional effect and the latter a direct result of the mineralizing process. It will be recalled that different responses were obtained in rhyolite breccia and massive rhyolite around the Norbec deposit on the Canadian Shield (Chapter 10), and that there have been references in several chapters to variations in response around massive sulphides as a function of textural variations. The matter demands investigation; it is quite probable that geochemical variations due to different phases of mineralizing processes can be identified and isolated by more careful selection of sample type. This approach holds the promise of more clearly discriminating between regional and local scale anomalies and hence enhancing anomalies and improving interpretation.

Both the Mount Morgan and the Wainaleka deposits have clearly defined halos of decreasing K. The mineralogical or geochemical explanation of this feature has not emerged from the work done on the deposits, but a similar decline in K around the 12 West deposit at Mount Lyell was attributed to the development of pyrophyllite rather than sericite as an alteration mineral. In the Mount Morgan and Mount Lyell studies specific mineralogical investigations have been made in an attempt to provide an explanation for the observed geochemical trends. At Mount Morgan an X-ray diffraction and electron microprobe study established mineralogical zones of siliceous alteration passing successively outwards into sericite and chlorite alteration to a maximum distance of 860 m from the ore zone; in this alteration halo plagioclase steadily decreases in amount towards the deposit, chlorite increases, and the content of Mg in the chlorite increases. These observations provide a rational explanation for the increase in Mg and the decrease in Na and Ca contents with proximity to mineralization. The important practical conclusion from these data is, however, that geochemical halos extend for a minimum of 600 m further than mineralogical alteration can be detected by X-ray techniques.

In the three cases where H_2O was measured (Mount Morgan, Prince Lyell, and Woodlawn) it showed an increase in content towards the mineralization and was a useful indicator of the development of hydrous minerals (sericite, chlorite). The intensity and extent of the Au halo at Mount Morgan is also encouraging — and a little surprising, considering the relatively small size of the samples for the measurement of Au; the results certainly suggest that determination of Au may be useful in exploration for volcanic-hosted gold deposits.

The surface rocks of all of the areas described in this chapter are considerably weathered, especially around the Australian and Fijian deposits. Nevertheless, extensive geochemical halos are discernible, albeit with secondary modifications (especially close to mineralization, as at Limerick). Where weathering is extreme, as at the Ajax prospect (and samples of partly decomposed weathered bedrock must be obtained by rotary or percussion drilling) the responses are clearly influenced by secondary processes. In these circumstances, the rock has a response rather similar to that of residual soil — the less mobile elements form discrete anomalies more or less coincident with mineralization, and the more mobile elements give broad anomalies.

The focus of mineralization in the deposits considered in this chapter is characteristically indicated by extreme depletion or enrichment of major elements or an increase in content and variance of the ore elements. All of the deposits (except the Cyprus deposits and Ajax prospect for which there were no major element data) could have been detected by enrichment or depletion halos of Na, K, Ca, Mg, and Fe at a sample interval of 200 m; the Cyprus deposits could be detected by halos of Zn and Co enrichment and Cu depletion in lavas sampled at the same interval.

PART IV. SUMMARY AND CONCLUSIONS

Chapter 13

SYNTHESIS OF GEOCHEMICAL RESPONSES AND OPERATIONAL CONCLUSIONS

INTRODUCTION

An obvious conclusion from the data presented in the preceding chapters is that the host rocks of all mineral deposits show some geochemical response for some elements at some scale. In terms of planning geochemical exploration programmes this is a singularly useless item of information unless it is possible to also state which elements respond and on what scale. In this chapter an attempt is made to provide at least a partial answer to these questions.

There are considerable limitations on making the ideal comparative survey required to answer the questions "what element?" and "what scale of response?" because of the nature of the data available. A wide variety of rock types have been sampled and analyzed for a bewildering variety of elements, the choice of elements in any particular case may have been dictated by the whim of the investigator, and the availability of analytical techniques and the objective of the investigations have varied from case to case. Furthermore, a particular element response is influenced in many cases by such factors as the analytical sensitivity of the technique used or the textural variety of the rock sampled.

The nature of the geochemical response of a mineral deposit is fundamentally controlled by the geochemical processes associated with the mineralizing event and the geochemical nature of the host rock. The general factors that will influence the geochemical signature in any particular case include the following:

— the genesis of the ore deposit, especially whether it is syngenetic or epigenetic;

— whether the deposit is related to igneous intrusive rocks; if it is, whether it lies within the intrusion or within surrounding rocks host to the intrusion;

— whether the deposit is within volcanic rocks and due to volcanic processes;

— whether the deposit is within sedimentary rocks and is related to sedimentary processes;

332

— the size, composition, and grade of the deposit;
— the effect of post-ore processes, e.g., metamorphism, weathering;
— the fundamental laws of element distribution, e.g., the control of major element chemistry on the distribution of trace elements in crystal structures;
— the determination of the *relevant* background.

In exploration situations geological information relevant to the above factors is generally inadequate to accurately forecast all details of geochemical response. Even in the detailed case histories of massive sulphides (whose geological character and genesis are now quite well understood) in areas where the level of geological information is high there is no immediate explanation of the apparently capricious behaviour of K — which in some cases is enriched, in some cases is depleted, and in some cases shows no variation at all.

Moreover, regardless of how comprehensive the geological data may be, there are commonly practical difficulties in separating, in terms of element concentration, the geochemical consequences of several geological events. This is because the analytically measured total element concentration in a sample of rock reflects the total geological history of the rock — which includes original composition, metamorphism (and diagenesis in the case of clastic rocks), and weathering, as well as any local mineralizing events. The geochemical results of various phases of the history of a rock frequently merge. For example, around syngenetic deposits there may be regional geochemical anomalies in host rocks upon which are superimposed local scale anomalies; if the local scale anomalies are of the same general geochemical character as the regional anomalies, the actual location of mineralization may be difficult to determine. Furthermore, the same type of *geological* process may have quite different geochemical results in different places. Thus, hydrothermal alteration may increase the content of Mg due to the formation of chlorite, or decrease the content of Mg due to destruction of ferromagnesian minerals. Since many trace elements in plutonic and volcanic rocks reside in the lattice of ferromagnesian minerals, variation in the amount (and kind) of these minerals will also have a profound effect on trace element distributions.

Notwithstanding these many constraints, there are now enough data to derive at least some general guides. The primary division of material in this book has been the scale of response; secondary subdivisions have been based largely on the deposit type. As was pointed out in the Introduction, this classification was adopted in accord with normal operational procedures. It has resulted, however, in similar deposits being discussed in widely separated chapters (for example, geochemical responses for vein-type deposits have been covered in Chapters 4, 6, and 9). In this review chapter essentially the reverse procedure has been adopted. Mineral deposits are divided here into three main groups, regardless of the scale of the geochemical response. These are:

— recognition of productive intrusions that have genetically associated mineralization (i.e., tin, porphyry, and some vein and replacement deposits) and halos around porphyry deposits;

— vein and replacement deposits (which include those deposits with and without igneous associations);

— deposits of volcanic and sedimentary associations.

RECOGNITION OF PRODUCTIVE INTRUSIONS AND OF HALOS AROUND PORPHYRY DEPOSITS

Regional scale

Particular types of mineral deposits are associated with particular petrological varieties of intrusions. At the simplest level it is obvious that nickel sulphide deposits are associated with basic rather than with acid intrusions. Limited data indicate that mineralized basic intrusions may be distinguished from barren basic intrusions by enhanced levels of S, Ni, and probably Cu and Co (Table 13-I); basic intrusions are not discussed further here. Within the large group of felsic intrusive rocks particular deposit types are associated with particular types of felsic rocks — e.g., tin deposits occur in the more felsic granitoids and do not occur in granodiorites; porphyry deposits occur in granodiorite—quartz monzonite—quartz diorite intrusions. Beyond this gross petrological indication of deposit type there is the problem of discriminating between productive and barren intrusions.

The geochemical responses recorded by various workers for porphyry deposits, tin deposits, and various base metal and precious metal vein and replacement deposits are summarized in Table 13—I (the lack of an indicated response for many elements is simply due to the fact that the element was not determined). Felsic intrusions that are host to mineralization, or which have genetically associated mineralization in surrounding rocks, are characterized by an enrichment in K and Rb and a depletion of Ca and Sr (these patterns also appear to generally hold at the local scale around the deposits). Although case history data are limited, it is probable that productive plutons are also depleted in Ba and enriched in Li (data in Table 13-I clearly show that intrusions associated with tin deposits are enriched in Li); Mg is likely to be seriously depleted in intrusions that are metasomatically altered. Broadly, highly fractionated intrusions are genetically associated with mineralization; they are characterized by high Rb:Sr ratios and low K:Rb, Ba:Rb, and Mg:Li ratios. With the possible exception of tin-bearing granites, potentially productive plutons can only be recognized by high or low ratios within a particular region; i.e., no universal background value can be assigned to them. The tin granites appear to have absolutely low K:Rb and high Rb:Sr ratios that distinguish them from both barren intrusions and intrusions with other types of mineralization.

TABLE 13-I

Summary of characteristic geochemical responses in recognition of productive intrusions, and local scale responses for porphyry deposits (derived from data in Chapters 4, 5, and 8)

Mineralization	Mine or region	Sample type	Geochemical response																	comments	Halo dimensions
			Cu	Pb	Zn	Sn	Au	Ag	Mo	Mn	S	Ba	Li	K	Na	Ca	Mg	Rb	Sr		
Cu porphyry	Guichon (Highland Valley)	R	+		+																intrusion
	Copper Canyon	R	+					+													intrusion
	Coed-y-Brenin	R	+		−				+					+	−	−		+	−		intrusion
	Ely	Bt	+											+	−	−				<350 ppm Zn	intrusion
	Bingham	Bt	+																	<350 ppm Zn	intrustion
	Esperanza, Sierrita	Bt	+																		intrusion
	Lights Creek	R	+		−*			*	−*	−*	+	−		+		−*	−	+	−*	+FeS + CuS	2−3 km
	Valley Copper	R	+		−*			*	−*	−*	+			+		−*	−	+	−*	+Hg, +Cl,	≃500 m
	Bethlehem JA	R	+																	+CuS, +FeS	
	Lornex	R	+		−*			+	−	−									−	+Bi	≃500 m
	Highmont	R	+		−*			+												+B	≃500 m
	Kalamazoo—San Manuel	R	+		−*				−*	−*	+							−		+Co, −Tl	≃500 m
	Mineral Butte	R	+	−	+		+													zoned	>900 m
	Copper Mt., Ingerbelle	R	+		+									−	+	+		−	+	−F	500−1000 m
	El Teniente																				500−800 m
	Rio Blanco	R												+	−	−		+	−		400−1000 m
	Los Bronces																				
	Copper Canyon	R	+	+	+		+	+	+		+								+	zoned	200−500 m
	Ely	Rf	+	+	+		+	+	+	+	+									+Te, +As, +Sb, +Bi	
Mo—Cu porphyry	Valley Copper, Highmont	Bt	+																	zoned	1.5−7 km
Mo porphyry	Valley Copper, Highmont	Mt	+	−	−																≃500 m
	Setting Net Lake	R	+	+				+													500−1000 m
	Nojal Peak	R	+					+													400 m
W-disseminated and skarn	Canadian Cordillera	R	+	+																form of W frq. dist; base metal residuals	intrusion

								Notes	Host
Sn deposits									
Anchor	R		+				−		intrusion
Finlayson	R		+	+	+		−		intrusion
Erzgebirge	R		+	+	+		−		intrusion
S. Mountain	R		+	+	+		−		intrusion
and Mo, W, Pb, Cu									
Elizabeth Creek	R		+	+	+		−		intrusion
and W, Cu									
Mareeba	R		+		+		−		intrusion
and Au									
Esmeralda	R		+	+					intrusion
and Cu, Pb, Zn, Ag									
Almaden	R		+		+		−		intrusion
and W, Mo, Bi									
Mt. Pleasant	R	+	+				−		intrusion
and base metals									
Cornwall	R	+	+				−		intrusion
and base metals									
Cornwall	Fr	+	+	+	+		−		intrusion
and base metals									
Cornwall	Bt	+	+	+	+	−	−		intrusion
and base metals									
Cornwall	M	+	+	+	+				intrusion
Ni—Cu deposits									
Canada	R	+S	+					+NiS, +CoS	intrusion
U.S.S.R.	R		+					+Ni, +Co, +H$_2$O	
Finland	R	+	+					+Ni	intrusion
Vein, replacement base metal deposits									
and Ag, Au									
Sardinia	R	+	+	+				<350 ppm Zn	intrusion
Silver City	Bt	+	+					<350 ppm Zn	intrusion
Clayton Peak	Bt	+	+					<350 ppm Zn	intrusion
San Francisco	Bt	+	+					<350 ppm Zn	intrusion
Gold Hill	Bt	+	+					<350 ppm Zn	intrusion
Santa Rita Mts.	Bt	+							intrusion

+ = enrichment; − = depletion; * = peripheral positive anomaly; +S, NiS = sulphide-held element; R = whole rock; Rf = fault or fissure material, jasperoid; Rw = weathered rock; Bt = biotite; Fr = feldspar; M = muscovite; Mt = magnetite.

In terms of the ore elements themselves, intrusions that are host to copper porphyry deposits are enriched in Cu, although the Highland Valley deposits in British Columbia occur in granitic phases that are *depleted* in Cu relative to the batholith as a whole (which is enriched in Cu). Tin is generally, but not always, enriched in intrusions associated with tin deposits.

Productive intrusions may also be recognized by the form of frequency distribution of elements. Intuitively it may be expected that ore elements will have a high variance in productive plutons, and this has been demonstrated to be the case for Sn, Cu, and Co in the U.S.S.R. and for base metals and W in the Canadian Cordillera.

In most cases it is difficult to determine whether anomalous element concentrations in productive plutons are a primary magmatic feature or are a result of metasomatic alteration. This is, of course, largely irrelevant for exploration purposes.

Discrete halos

Around copper porphyry deposits (Table 13-I) halos are generally 500—1000 m in extent and are, naturally, strongly influenced by the mineralogical alteration --- high K and Rb in zones of potassic alteration, and relatively high Ca and Sr in the outer propylitic zone (broadly, K and Rb are enriched and Ca and Sr are depleted). In all cases for which data are available Cu and S have strong positive anomalies that generally persist furthest from mineralization. There is a pronounced zoning of element dispersion:

-- Cu has peak values over the centre of deposits;

-- S has peak values at the periphery of ore zones;

-- Zn and Mn are generally depleted over the ore zone, but they have positive anomalies at the periphery;

-- Mo may have a positive anomaly over the ore zone or only at the periphery.

A generally applicable zoning sequence, in terms of peak positive (+) or negative (−) anomalies, from the centre of mineralization outwards is: (+Cu, +Rb, +K, −Zn, −Mn, −Sr, −Ca)—(+S)—(+Zn, +Mn, +Ca, +Sr). It should be noted that K and Rb, and especially Cu and S, have positive anomalies throughout this sequence and decline in value with increasing distance from the ore zone.

Mineral separates and halogens

Despite considerable work, the measurement of total and water-soluble C1 and F in both whole rock and mineral separates has yielded largely negative results. The abundance of these elements appears to be specific to each intrusion (and varies with petrology), regardless of whether the intrusion is mineralized or barren. The distribution of ore elements Cu, Pb, Zn, and Sn

in mineral separates (chiefly biotite, with fewer studies on feldspar, muscovite, and magnetite) has yielded better results. As with the halogens, however, it appears that each intrusion has its own characteristic abundance. In
cases where comparative whole rock data are available it has been demonstrated that there is no advantage in using mineral separates for either
regional or local scale exploration.

There is potential value in the determination of the mineralogical site of
trace elements. Little work has been done on this, but it is probable that the
distribution of trace elements (especially the ore elements) within silicate
minerals and the partition of ore elements between silicate and sulphide
minerals varies markedly between high background and mineralized situations. It is also probable — as the work on uranium in the U.K. discussed
in Chapter 4 has demonstrated — that the mineralogical site of ore elements
will vary according to the type and genesis of mineralization.

VEIN AND REPLACEMENT DEPOSITS

Regional scale

The recognition of intrusions that have associated vein and replacement
deposits was included in the last section. In addition to these types of regional
anomalies, extensive halos of more than 1 km have been detected in a variety
of non-carbonate rocks (including intrusive rocks). In all examples of geochemical halos cited in this book (summarized in Table 13-II) the mobile
elements Te or As, or both, are included in the anomalous elements; Sb and
Bi (other mobile elements) are also prominent (where none of these four
elements appear, they were not determined). At least one of the ore elements (i.e., Cu, Pb, Zn, Au, Ag) are also anomalous. Tellurium consistently
appears to give the largest halos, although its precise relation to mineralization
is not always clear.

In a number of investigations vein material, fracture fillings, and jasperoid
were sampled rather than country rock. Since such material represents the
"plumbing" system of a mineralizing event, the logic and success of this
approach is obvious. Even where the country rock was sampled, these large
halos are clearly spatially related to structural features in some cases, and to
widespread hydrothermal alteration in other cases.

The implication is clear — detection of vein-type mineralization on a
regional scale depends on obtaining samples that reflect the passage of
mineralizing solutions. Rocks that had high permeability are therefore likely
to have the widest halo. Reactive carbonate rocks are likely to restrict the
extent of halos, but this can be overcome by sampling vein and fracture
material; indeed, it is probable that in all situations the widest halos will be
obtained from samples of the "plumbing" system. Apart from the variations

338

TABLE 13-II

Summary of characteristic geochemical responses around vein and replacement deposits (derived from data in Chapters 6 and 9)

Mineralization	Mine or Region	Sample type	Cu	Pb	Zn	Sn	Au	Ag	Mo	Mn	S	Te	As	Sb	Hg	comments	Halo dimensions
Base metal deposits																	
and Ag	Mykonos	R	+	+	+			+					+	+	+	zoned partly secondary	3 km
	Goosly, Bradina	R	+	+	+								+				4—6 km
	Tres Hermanas	R	+	+	+												1—3 km
	Hanover—Fiero	R	+		+												1—3 km
	Granite Mt.	R	+		+												1—3 km
	Lordsburg	R	+		(+)						+						1—3 km
and Ag, Au	Gortdrum	Rf	+	(+)	(+)		+						+	+	+		1—3 km
	Santa Rita Mts.	Rf	+	+	+		+	+			+	+	+	+	+	+Bi	1—3 km
	Mykonos	R	+	+	+			+				+	+	+	(+)	wallrock, granite	20 m
	Tintic, Swansea	R		+	+									+		wallrock, monzonite	15—30 m
	Tintic, Carisa	R	+	+	+											wallrock, dolomite	1 m
	Tintic, Eureka Hill	R		+	+											wallrock, dolomite	3 m
and Ag	Park City	R	+	+	+			+								wallrock, sst.	12—24 m
	Nenthead	R	+	+	+					+						wallrock, lst.	5—10 m
and Ag	Wisconsin	R	+	+	+			+								wallrock, lst.	5—50 m
Au, Ag deposits																	
Ag(Pb, Zn)	Coer d'Alene	R	+	+				+				+	+			+Cd	>6 km
Ag(Pb, Zn)	Montezuma	R						+				+	+	+			>6 km
Au—Ag—Cu—Pb—Zn	Crater Creek	R					+					+					>6 km
Au—Ag	Cripple Creek	R					+					+					>6 km
Ag(Sb, Au, Cu, Pb, Zn)	Taylor	Rf	+	+	+		+	+			+	+	+	+	+		3 km
Cu—Ag—Au—Mo	Detroit	Rf	+	+			+	+	+						+	+Bi	3 km
Au, Ag	Marysville	R					+										1—3 km
Au, Ag	Marysville	Bt				+	+										1—3 km
	Marysville	H					+										1—3 km
Au	Cortez, Carlin	Rf					+						+	+		+W	1—3 km
Au	Ogofau	R								+			+			+Rb, +Li	200—500 m
Au	Costerfield	Rw											+	+			20—200 m
Au	Vuda	Rw					+										10—30 m
Ag	Cobalt	R						+					+	+		+Co, +Ni; wallrock greywacke greenstones diabase	30—45 m 15—16 m 1 m
Au—Ag	Searchlight	R					+	+					+	+		wallrock, andesite	30—45 m
Ag	Eureka	R		+			+	+					+			wallrock, lst.	20—30 m

+ = enrichment; − = depletion; (+),(−) = minor enrichment or depletion; R = whole rock; Rf = fault or fissure material, jasperoid; Rw = weathered rock; Bt = biotite; H = hornblende.

due to the nature of the rock, large mineral districts appear to give rise to large halos.

Wallrock halos

Halos in the wallrock around individual veins have a well-defined logarithmic decay pattern of limited extent — generally 20—40 m. In all cases discussed in this book anomalous elements include the ore elements (see Table 13-II). It is presumed that wallrock halos are diffusion-controlled, and that the larger halos occur in the more permeable rocks. The halos in the reactive carbonate rock are generally the smallest (in some cases less than 1 m). An exception to this was demonstrated around lead—zinc veins in Wisconsin (see Chapter 9) where, after correction for background Zn in clay material in limestone, it was shown that the extent of Zn halos was related to the width of the vein; halos of up to 50 m in extent were detected by this technique.

VOLCANIC-SEDIMENTARY MASSIVE SULPHIDES

As with most exploration studies, difficulties arise in making comparisons of responses between different deposits because of the variety of approaches and range of different elements determined by different investigators. The data for massive sulphides are, however, more comparable than for other types of deposits, and the geochemical responses show a remarkable consistency for deposits of all ages (see Table 13-III). This is, of course, a reflection of their broadly similar genesis. Such differences that do occur between deposits are attributable to differences in host rocks (degree of fractionation of lavas, relative proportions of volcanic and sedimentary rock), differences in ore composition, and whether deposits are proximal or distal to the primary source of mineralizing fluids. Responses also vary as a function of the texture of volcanic rock as it reflects its initial permeability.

Although regional scale studies are not as numerous as those on a local and mine scale, some broad generalizations are possible. There is some uncertainty whether the Archaean deposits of the Canadian Shield occur in calc-alkaline or tholeiitic lavas, but there is general agreement that the productive volcanic cycles that are host to zinc—copper deposits are distinguished from barren cycles by enhanced levels of Zn, Fe, and Mg (and probably Mn) and are depleted in Na. These differences are more clearly evident if the element contents are normalized against SiO_2 to compensate for variations in the degree of fractionation.

Massive sulphide deposits of other types and ages, including the essentially sedimentary McArthur River and Tynagh deposits and the cupriferous pyrite deposits of Cyprus also exhibit regional scale Zn enrichment in their host

TABLE 13-III

Summary of characteristic geochemical responses around massive sulphide deposits (derived from data in Chapters 7, 10, 11 and 12). All results are for whole rock samples

Mineralization	Mine or area and age	Geochemical response														Halo dimensions
		Cu	Pb	Zn	Ni	Co	Fe	Mn	Na	K	Ca	Mg	H₂O	S	comments	
Zn–Cu	Abitibi, A								−			+				regional
Zn–Cu	Noranda, A			+										+		regional
Zn–Cu	High Lake, A			+												regional
Zn–Cu–Pb	Mattabi, A			+			+	+	−			+				regional
Zn–Cu–Pb	Sturgeon Lake, A			+			+	+	−			+				regional
Zn–Cu	Uchi-Birch Lake, A			+		+	+									regional
Cu	Ingladhal, A	−				+										regional
Zn–Cu–Pb	East Tuva, Pa	−	−	−	−	−		+							+Ba	regional
Zn–Pb–Cu	Buchans, Pa		+	+											+Ba	3 km
Pb–Zn–Cu—																
Ag	Tynagh, Pa			+			+	+								7 km
Pb–Zn–Ba	Meggan, Pa						+	+								5 km
Zn–Pb–Cu	McArthur River, Pz		+	+			+	+								15—23 km
Zn–Pb–Cu	New Brunswick, Pa	−													Zn:Cu, Zn: Pb, Pb:Cu ratios	regional
Cu	Cyprus, M	−					+	+	−							regional
Cu–Pb–Zn	Red Sea, R	+		+	(−)	+	+	+							+Hg	9—10 km
Zn–Cu	Millenbach, A	+		+			+	+	−	±	−	+				>200 m
Zn–Cu–Pb	Mattabi, A	+		+			+	+	−	±	−	(+)		(+)	(+CO₂)	400—800 m
Zn–Cu	Fox, A	+		+					−			+				100—250 m
Zn–Cu	Jay, A	+		+					−							500 m
Cu–Pb–Zn	Hanson, A	+														60 m
Zn–Cu	East Waite, A						+		−	−	−	+				600 m
Zn–Cu	South Bay, A						+		−	−	−	+				100 m
Cu–Zn	Norbec, A						+	+	−	−	−	+				450—700 m
Cu	Louvem, A						+		−	−	−	+	+			200—400 m
Zn–Cu	Detour, A						+		−	−	−	+	+	+		200 m
Cu {	Boliden, outer, Pz						−	−	−	−	−	−	+	+	+Si, +Al	} >300 m
Cu {	Boliden, inner, Pz	(+)					(+)	+	−	+	−	−	+	+	−Sr, +Si,	} >300 m
Pb–Zn	Broken Hill, Pz	(+)	+				+	+	−			−		+	+Rb, (+Ti)	500 m

Type	Deposit, age										Notes	Dimensions
Zn–Pb–Cu	Brunswick No. 12, Pa	±	+	+	+	+	±	±		+	+Cr, +P, –Sr	450 × 915 m
Zn–Pb–Cu	Heath Steele ACD, Pa	+	+	+	+	+	±	+	–	+	+Cr, +P, –Sr	>100 × >1140 m
Zn–Pb–Cu	Heath Steele B, Pa	+	+	+	+	±	+		–	+		450 × 120 m
Zn–Pb–Cu	Caribou, Pa	+	+	+	+	+	+			+		200 × 800 m
Zn–Pb–Cu	Key Anacon, Pa	+	+					+	–	+		205 × 1800 m
Zn–Pb–Cu	Woodlawn, Pa		+	+	+				–	+	+Si, –Sr	400–800 m
Cu–Au	Mt. Morgan, Pa	+	+	+	+				–	+	+Ba, +Ag	>1400 m
Cu	Prince Lyell, P		(+)	+	+	(+)		–	(+)	(+)	–CO₂, +Mo, –Sr, (+Rb)	
Zn–Pb	Limerick, Pa		+	+	+			+		+	+Ba	>300 × 700 m
Zn–Cu–Pb	East Tuva, Pa		+	+	+				+	+	+Ba	400 m
Cu	Cyprus, M	–	+	(–)				+				?100 m
Cu	Ergani-Maden, M	(±)	+	+	–							>1000 m
Cu–Zn–Pb	Kuroko, C		–				–		±	+	+Si	1000 m
Zn–Cu	Wainaleka, C	+	+	+	–				–	+	–Sr, –Rb	300–1200 m

A = Archaean; Pz = Proterozoic; Pa = Palaeozoic; M = Mesozoic; C = Cenozoic; R = Recent; + = enrichment; + = enrichment or depletion; ± = enrichment and depletion in different parts of halo. — = depletion; (+), (–) = minor enrichment or depletion; (±) = enrichment and depletion in different parts of halo.

rocks (in Table 13-III only the eastern Tuva example from the U.S.S.R. is an exception in showing a depletion of Zn). The copper deposits, as well as the zinc—lead deposits of New Brunswick are characterized by regional scale Cu depletion in the host rocks.

Enrichment of Mn and regional scale Mn halos are common. As may be expected from the geochemistry and genesis of the massive sulphides, large-scale Mn halos are likely to be best developed in sedimentary and, especially, exhalative rocks at the stratagraphic ore horizons.

There are considerably more data on local and mine scale exploration for massive sulphides, and it is in these studies that the remarkable consistency of response — especially for major elements — becomes evident (see Table 13-III). In virtually all deposits for which there are data Fe and Mg are enriched and Na and Ca are depleted. The exceptions are:

— a Na enrichment at Caribou and in surface weathered rocks at the small Limerick deposit;

— a depletion of Mg at Broken Hill in Australia and at the Detour deposit in Quebec (although at the Detour deposit this may be a function of the part of the alteration zone sampled).

Particular elements can be both enriched and depleted at the same deposit, depending upon the distance from the ore zone; thus, Mg and Fe are depleted close to the Boliden ore zone and are enriched further away. Potassium commonly shows depletion near the sulphides and enrichment more distant from the deposits. Manganese is similarly variable; in the New Brunswick deposits, for example, it is generally depleted in the footwall and in the hanging wall immediately over the sulphides, but is enriched in the hanging wall lateral to the deposits.

Notwithstanding the local variations, extensive anomalies of hundreds of metres to more than 1000 m are defined by some simple combinations of a few elements — Fe, Mg, Na, K, Ca, and Mn. These geochemical halos are 2—10 times larger than the mineralogical alteration zones around the deposits. In many cases the halos are readily defined by a single element or by a simple element ratio; however, they may be smoothed and enhanced by various normalization procedures to compensate for petrological variation.

Trace and minor element halos are generally far less distinct and well defined. Single-element halos of the ore elements (Cu, Pb, Zn) are normally restricted to a few tens of metres from the sulphides, but, when combined by multivariate statistical techniques, the elements allow halos to be detected hundreds of metres from the deposits.

The ore elements (especially Zn) are normally enriched, although complications occur. Copper is depleted (except in the immediate wallrocks) around the Cyprus deposits; it is enriched relative to Pb in footwall rocks and is depleted relative to Pb in hanging-wall rocks in the New Brunswick deposits. The variance of the ore elements characteristically increases towards mineralization. Less comprehensive data indicate that Co, H_2O, and Rb are enriched and Ni and Sr are depleted.

Significant anomalies have not been reported in hanging-wall rocks of Archaean deposits of the Canadian Shield, but extensive anomalies occur in the hanging wall of other deposits.

OPERATIONAL CONCLUSIONS

In detail, geochemical halos around particular deposits or mineral districts vary in extent and intensity as a function of the type of deposit, the nature of the rock sampled, the elements determined, and the analytical techniques used. Despite the large scope for variation in response, the conclusion from the data presented in this book is that geochemical signatures of mineralization are, in fact, fairly consistent and can be defined by relatively few elements.

The most useful elements for different targets and various scales of exploration are given in Table 13-IV. The data show that exploration rock geochemical surveys for all the most common types of mineral deposits can be undertaken with no more than about two dozen elements. Many other elements have proved to be useful in individual cases described throughout the book, but those listed in Table 13-IV are the most consistently applicable.

The fewer the number of elements that are required to *reliably* detect mineralization, the more economic and useful rock geochemistry will be. Whereas efforts should certainly be made to reduce the number of elements required for a particular target, considerably more experimentation is needed — both in research and in actual field programmes. In particular, elements that have been shown to be very effective in the case of a few deposits (e.g., Rb and Sr for massive sulphides) should be tested for universal applicability. Moreover, development of techniques using volatile elements may show them to be indispensible and thus extend the list of useful elements.

Pronounced zoning of elements in geochemical halos has been described for a number of case studies. Recognition of this zoning offers the potential to locate deeply buried deposits and also to assess their composition and size; zoning patterns should also be very useful in the interpretation of drill core. Little work has been published on this aspect of rock geochemistry outside of the U.S.S.R., but it clearly requires further investigation — even although development of a successful zoning sequence will inevitably increase the number of trace and minor elements that must be determined.

Sample densities (see Table 13-IV) are small in relation to the amount of information that can be derived from a rock survey; this compensates, in part, for the greater costs involved in having to collect larger samples than in other types of geochemical surveys — generally the sample should not be less than 1 kg — and the greater costs of having to crush and grind samples rather than the simple sieving of soil or sediment samples. Moreover, if bedrock is available for sampling, exploration information is potentially available that cannot possibly be derived from soils or stream sediments.

TABLE 13-IV

Summary of elements to be determined and surface sampling density for different targets in regional, and local and mine scale exploration. Elements in parentheses have been shown to be useful in some cases but have uncertain status; elements with asterisk are expected to be useful but there are little data. Si should be determined in all cases where petrological variation is expected to cause variations to the content of other elements

Scale	Target	Elements		Sampling density
		non-ore	ore	
Regional	identification of productive plutons	K, Rb, Sr, Ba, Li, Na*, Ca*	e.g. Cu, Pb, Zn, Sn, W, Mo, U, Ni	min. 30/intrusion but see Appendix 3
	massive sulphides	Fe, Na, Mg, Mn, Na, (K), (Ca), (Ba)	Cu, Zn, (Pb)	$0.2-5/km^2$
	vein and replacement	As, Sb, Ta, Bi*	e.g. Cu, Pb, Zn, Au, Ag	$1-10/km^2$
Local and Mine	porphyry	K, Ca, Rb, Sr, Mn, (Mg)	Cu, Zn, Mo, S	$2-30/km^2$
	massive sulphides	Fe, Mn, Na, K, Ca, Mg, (H_2O), (Rb), (Sr)	Cu, Pb, Zn, (S)	150—200 m interval
	vein and replacement		e.g. Cu, Pb, Zn, Au, Ag	5—10 m interval

The data presented in this book should be adequate to demonstrate that primary halos can be detected around all economic mineral deposits. This being the case, it should be obvious that interpretation of drilling data can be immeasurably improved if all drill core is geochemically analyzed for elements appropriate to the target rather than simply assayed for the ore elements.

The ultimate limitation on the use of rock geochemistry in mineral exploration is the availability of exposed rock. Lack of exposure can be overcome by surface drilling through overburden, but this adds to the already relatively high cost of using rock geochemical techniques.

The limitations imposed by surface weathering have probably been exaggerated in some of the literature. Surface rocks in Cyprus (Chapters 7 and 12), Mykonos (Chapters 6 and 9) and around the Mount Morgan deposit in Australia (Chapter 12) are all quite badly weathered, but extensive and intensive halos have been identified in surface rocks; indeed, in Mykonos, weathering appears to have enhanced the geochemical response. The effect of weathering on geochemical dispersion will necessarily be a function of the degree of weathering which has taken place and the geochemical characteristics of the individual elements.

The analytical techniques used for a particular element can significantly affect the usefulness of that element in exploration. Poor analytical sensitivity can result in small halos or even no response at all. Similarly, as the data given in Chapter 2 show, the physical size of the sample collected for analysis can be an important factor in geochemical response. Anomalies for low-concentration trace elements (Au is an extreme example) can quickly disappear into a noisy background even with the 1-kg sample size recommended throughout this book simply because of the sparse and random distribution of the element in the rock. The theoretical solution to the problem is to increase the sample size; however, this is rarely practical. An alternative approach is to undertake some physical concentration or separation of the minerals containing the trace element of interest prior to analysis or to use mineral-selective digestions.

Some of the problems of discriminating between element content due to a mineralizing event and background conditions can also be overcome by selective extraction techniques. Sulphide-selective leaches have been successfully used in a number of studies reported here. Limited data from the New Brunswick work (Chapter 11) indicate that partial digestions (even water leaches) can be effective in anomaly detection.

Each of the three major instrumental techniques used in rock geochemistry — AAS, XRF, and emission spectroscopy — has both advantages and disadvantages (these are fully discussed in Fletcher, 1981). For the limited suite of elements proposed for general purpose exploration (see Table 13-IV) flame and flameless AAS is at present the most useful technique in terms of element detection limits; AAS also has the advantage that it can be used with different extraction techniques and the capital equipment costs are relatively low. The technique, however, is not appropriate for some elements of interest, and sequential element determination is a disadvantage. Both XRF and emission spectroscopy offer the advantages of making a range of simultaneous multi-element determinations, but emission spectroscopy, in particular, has relatively poor detection limits and precision for many of the elements of interest (although these disadvantages have been overcome in the Soviet work by the determination of a very large number of elements and the use of element ratios). The development of inductively coupled plasma emission spectroscopy offers the potential of both simultaneous multi-element analysis and adequate detection limits; this instrumentation will undoubtedly provide a major impetus for the use of exploration rock geochemistry as it becomes more widely available.

In conclusion, there are a number of operating principles — most of which are applicable in varying degrees to all geochemical surveys — that should be considered in the planning, execution, and interpretation of rock geochemical surveys. These are:

— sampling density should be dictated by the desired scale of geochemical response and the particular phase of the exploration sequence;

— where a choice of sample material is available, the decision on which material to sample should be guided by an understanding of the likely dispersion processes on the scale of response sought;

— the choice of elements to be determined should be dictated by the nature of the target and the mineralizing processes;

— the analytical technique must be appropriate to the concentration level anticipated and the mineralogical site of the element concerned;

— even minor variations in lithological composition or rocks necessarily cause variations in chemical composition, particularly in the level of trace and minor elements; geochemical anomalies are generally enhanced by appropriate compensation for lithological variations (usually achieved by regression against a major element and the calculation of a residual value);

— the difficulty of determining a valid absolute background value for a particular element must be appreciated and more reliance placed on recognizing spatial patterns, relative enrichment and depletion, and the use of multi-element interpretative techniques.

This last point is absolutely fundamental to successful interpretation of all geochemical data — and especially rock geochemical data. Apart from the problems of establishing even a local background for many elements, a number of the more subtle, but extremely useful, anomalies lie *within* so-called background ranges. Identification of these types of anomalies is generally only possible through the recognition of spatial patterns of relative enrichment or depletion. Much of the successful application of the conclusions reached in this book depends upon the understanding by geochemists, both in the field and in the head office, of the importance of looking for and recognizing these spatial patterns.

APPENDIX 1 — CRUSTAL ABUNDANCE, DISTRIBUTION, AND CRYSTAL CHEMISTRY OF THE ELEMENTS

P.C. RICKWOOD

School of Applied Geology, University of New South Wales, Kensington, N.S.W. 2033 (Australia)

METHODS USED IN CRUSTAL ABUNDANCE ESTIMATES

There is no known way of *accurately* ascertaining the abundance of any element within the earth's crust because that material is so extremely variable, and so incompletely exposed, that there is no satisfactory method of sampling it for the purpose of determining element abundance. Nevertheless, Soviet geologists in particular (e.g., Perel'man, 1977) have made extensive use of these data for which Fersman (1923; in Beus, 1976) coined the unit the "clarke"; the clarke value of a particular element is its abundance in the lithosphere.

At least 24 authors have attempted to compile tables of crustal abundance of the elements, and several have published more than one estimate. Thirty-six sets of these estimates are listed in Table A-I, and the various methods by which they were derived are identified by an alphanumeric code used in the following descriptions.

Three distinct methods have been used to derive the estimates of crustal abundance of the elements.

— Geochemical computations (1). In this method one of three techniques is used to estimate the average volume proportions of the lithosphere that are occupied by rocks of known type. Then, by ascertaining the densities of these rocks and their average chemical compositions — from compilations such as those of Nockolds (1954), Vinogradov (1962), Beus (1976) or Le Maitre (1976) — it is possible to calculate the contribution that each rock type makes to the entire lithospheric composition. Consider an element i; let the number of rock types in a lithospheric model be j; let V_j be the volume percentage of the lithosphere occupied by rock type j which has a density of $d_j \, \mathrm{g/cm^3}$. Let $D \, \mathrm{g/cm^3}$ be the average density of the lithosphere, and let w_{ij} be the weight % of element i in rock type j. By the algebra given by Rickwood (1966), the crustal abundance of element i, expressed as a weight percentage is:

$$W_i = \sum_{i=1}^{i=k} V_j \frac{d_j}{D} \frac{w_{ij}}{100} \%$$

The problem with this approach is determining V_j, and the following methods have been used:

    — Geophysical-geochemical methods (1a): evaluation of geophysical data to construct a "typical" or "average" cross section through the lithosphere (e.g., Poldervaart, 1955; Ronov and Yaroshevsky, 1969; Beus, 1976, p. 35; Wedepohl, 1969, p. 244).

    — Cartographic-geochemical methods (1b): estimation of relative proportions of rock types by the areas that they occupy on map projections of certain regions, usually Precambrian Shields. Daly (1910, 1914, 1933), Moore (1959) and Rudman et al. (1965) used this method for the U.S.; it was used by Sederholm (1925) for Finland; by Lodochnikov (1927, in Wedepohl, 1971) for the U.S.S.R.; by Vogt (1931) and Barth (1961) for Norway; by Grout (1938) and Shaw et al. (1967) for the Canadian Shield; and by Beus (1976) for Malagasy. Knopf (1916) used the estimate of Daly; Wedepohl (1971) averaged the results of Daly, Moore, Grout, and Sederholm and then applied an arbitrary correction for weathering.

    — Simplistic model-geochemical methods (1c): for example, use of equal mixtures of granite and basalt (S.R. Taylor, 1964). This proportion was also adopted by Reeves and Brooks (1978) who used analyses of the U.S. Geological Survey rock standards G-1 and W-1 which were never intended for such use but for which there are extensive analytical data (Fairbairn et al., 1951; S.R. Taylor, 1964, Fleischer, 1953, 1965, 1969; and Fleischer and Stevens, 1962). Other variations on granite-basalt mixtures include a ratio of 2:1 granite to basalt (Vinogradov, 1962) and a ratio of 1:2 of "granitic" and "basaltic" layers (Beus, 1976). Other artificial mixtures, usually of igneous rocks, have been used by Noddack and Noddack (1930; in Rösler and Lange, 1972), Hevesy (1930, 1932; in Rösler and Lange, 1972); Vogt (1931); and Goldschmidt (1937). Additional problems have arisen because some workers have included parts of the hydrosphere and atmosphere (Clarke and Washington, 1924) and even the biosphere (Vernadskii, 1954; in Rösler and Lange, 1972).

    — Averages of analyses of igneous rocks (2). An assumption is made that the crust is essentially equivalent to igneous rock (sedimentary and metamorphic rocks are assumed to be derivatives), and *all* available analyses of a set standard of quality are averaged (Clarke et al., 1915; Clarke, 1916). Clarke and Washington (1924) averaged 5159 such analyses in one data set, and Le Maitre (1976) averaged 25,924 analyses (although he did not infer that the results represented the average lithospheric composition). The problem with this method is that some rocks of particular scientific, aesthetic, commercial, or logistic significance get oversampled compared to their proportion in the lithosphere (see Solovyov, 1952; in Beus, 1976).

    — Averages of analyses of sedimentary rocks (3). An assumption is made that "A good chemical average of the Upper Crust exposed throughout Earth's history can probably be derived from its sediment layers" (Wedepohl,

1969, p. 244). The chosen sediments must have experienced little or no chemical alteration; therefore Goldschmidt (1933, 1954) used 78 analyses of glacial and post-glacial loams from southern Norway. However, Wedepohl (1971) based his calculations on 70 analyses of greywackes obtained from several continents "Because these rocks, the debris of mountain ranges, are transported only short distances, chemical alterations are few" (pp. 61—62).

PUBLISHED ESTIMATES OF CRUSTAL ABUNDANCE

The data sets in Table A-I are tabulated in ascending order of atomic number, and the abundances are all expressed as ppm by weight (ppm, μg/g or g/tonne). Recalculations from oxide percentages were made by using the IUPAC (1971) atomic weights; for some of the older data sets this has resulted in the elemental abundances shown in Table A-I differing slightly from those published by the originators of the data (e.g., the data of Clarke and Washington, 1924 as presented by Washington, 1925). The data sets are listed in chronological order of publication; it would be unwise, however, to assume that the most recent estimate is necessarily the most reliable.

The data in Table A-I are summarized in Table A-II where the abundances listed as "approximate concensus" are estimates made to indicate an order of magnitude; statistical analysis of the data is not justified. The succeeding columns give the range of the estimates and the percentage of the highest value compared to the lowest value; the latter figure is indicative of the concordance of the various models. The data in the last column of Table A-II is shown in different form in Table A-III.

It is evident from these summaries that the use of various models has resulted in estimates of vastly different concentrations for most of the elements and that knowledge of the crustal abundance of most elements is still poor. There is greater disagreement for the major elements with low concentrations than for those with high concentrations as is illustrated in Fig. A-1. In general, this trend continues into the lower concentration ranges of the trace elements.

In nearly all cases the data sets have purported to represent the average lithospheric composition, whereas it is obvious that this is not so. Some authors have only estimated the composition of the average igneous rock (e.g. Clarke et al., 1915; Clarke, 1916; Clarke and Washington, 1924; Vogt, 1931; Le Maitre, 1976), whereas some estimates are supposedly of the composition of the upper crust (usually of certain continental areas, although if these are large enough they should be reasonable samples of the continental crust of the whole earth; see for example Knopf, 1916; Sederholm, 1925; Grout, 1938; Goldschmidt, 1954; Shaw et al., 1967; Wedepohl, 1969, 1971). Other workers have claimed to estimate the composition of the entire crust down to the Mohorovicic discontinuity (e.g., Poldervaart, 1955; Berry and

TABLE A-I

Various estimates of crustal abundance (ppm) of elements (for explanation see p. 363)

Data source	Method	Element (atomic number)						
		H (1)	He (2)	Li (3)	Be (4)	B (5)	C (6)	N (7)
1	2	1409.8						
2	1b	1611.2						
3	1c, 2	8800		40	10	10	870	300
4	1c, 2	1300						
5	2	1286.7		37.2			275.6	
6	1b	883.9					327.5	
7	1b, 1c							
8	1b, 1c							
9		10000	0.01	50	4	50	3500	400
10	1c			65	6	3		
11	1b	883.9					1064.4	
12		1500		65	6	3	1000	100
13	3	3379.1					1473.8	
14	3			65	6	10	320	
15	1a							
16	1c	1400	0.003	30	2	3	320	46
17	1c			32	3.8	12	230	19
18	2		0.003	29	2.5	3		46.3
19	1c			20	2.8	10	200	20
20				22	1.5	10	1200	19
21				21	1.3	13	2800	18
22	1a							
23	1b	883.9		22	1.3		3693.4	
24	3	2573.5						
25	1b	671.3						
26	1b	700	0.003	30	2	9	320	20
27	1a	1711.9					3992.5	
28	1a	1544.1					4902.3	
29	1a	1532.9					3820.9	
30	1a	671.3						
31	1a							
32	1a	1200						
33	1a, 1c	1000						
34	1a, 1c	1000						
35	2	1253.2					409.4	
36	1c	1400		18	1.9	10	200	15

O (8)	F (9)	Ne (10)	Na (11)	Mg (12)	Al (13)	Si (14)	P (15)
473652.4			25223.2	17910.3	83145.2	288130.0	1134.7
469837.7			25297.4	23217.0	79758.0	283661	1265.6
495200	270		26400	19400	75100	257500	1200
465900			28500	20900	81300	277200	
466218.6	300		28487.4	21046.1	81187.0	276444.0	1304.9
480680.8			22775.1	10191.4	77588.1	315896.0	480.1
469725.6			26039.3	16101.2	83145.2	299301.8	785.5
472757.1			25445.8	13447.8	81557.4	307248.3	741.9
491300	800	0.005	24000	23500	74500	260000	1200
494000	270		28300	21000	88200	276000	786
476113.1			27003.7	10734.1	88649.4	294861.2	960.1
470000	270		26400	21000	80000	276000	800
477620.6			15208.1	19900.3	83727.4	276677.7	960.1
466800	800	7×10^{-5}	28300	20900	81300	277200	1200
452341			21513.9	31358.1	80975.3	258026.9	1309.3
466000	700		28300	20900	81300	277200	1180
470000	660		25000	18700	80500	295000	930
		7×10^{-5}	23800*				
464000	625		23600	23300	82300	281500	1050
460000	470		25000	21000	83000	290000	1200
460000	450		23000	28000	83000	270000	1200
460250.9			22255.8	33770.2	80446.0	270180.3	1309.3
479120.9	500		25668.4	13508.1	77429.3	303508.8	654.6
489887.6			23739.5	12663.8	76212.0	313652.2	872.8
473842.4			24481.4	13869.9	78329.0	305705.8	872.8
472500	720		24500	13900	78300	305400	810
481305.0			22700.9	13327.2	80340.2	298881.1	872.8
474054.0			22033.2	18453	80340.2	281492.4	1047.4
483736.6			22255.8	24121.6	84150.8	277191.9	872.8
478750.4			26707.0	13266.9	78858.3	310380	785.6
471156.8			25223.2	19297.3	82563	292150	785.6
480000			22000	12000	80000	308000	800
466000			23000	24000	81000	277000	1000
463000			23000	30000	81000	267000	800
461248.1			25965.0	30513.6	79493.5	266580.7	1134.7
464000	500		24000	23000	82000	282000	1050

TABLE A-I (*continued*)

Data source	Method	Element (atomic number)					
		S (16)	Cl (17)	Ar (18)	K (19)	Ca (20)	Sc (21)
1	2				21999.0	36163.8	
2	1b				25153.5	34877.4	
3	1c, 2	480	1900		24000	33900	0.x
4	1c, 2				26000	36300	
5	2	520	480		25983.7	36306.8	
6	1b				29553.3	24299.8	
7	1b, 1c				29221.3	33019.1	
8	1b, 1c				31462.7	28659.5	
9		1000	2000	4	23500	32500	6
10	1c	500	480		25900	36300	5
11	1b	1200			25485.6	29088.3	
12		500	450		26000	36000	6
13	3	1020.4			32624.9	21941.3	
14	3	520	480		25900	36300	5
15	1a				15772.9	62893.6	
16	1c	520	200	0.04	25900	36300	5
17	1c	470	170		25000	29600	10
18	2		150	0.04			20
19	1c	260	130		20900	41500	22
20		330	100		20000	41000	17
21		400	280		17000	52000	18
22	1a				16603.0	53602.5	
23	1b	600	100		25734.7	29445.6	
24	3				15772.9	16438.1	
25	1b				28225.1	28588.0	
26	1b	310	320	0.04	28200	28700	14
27	1a	400	500		27311.9	28445.1	
28	1a	400	500		23742.3	39380.0	
29	1a				19923.6	51458.4	
30	1a				27395.0	27158.6	
31	1a				19093.5	42882	
32	1a				27000	25000	
33	1a, 1c				18000	43000	
34	1a, 1c				16000	50000	
35	2				20338.6	43811.2	
36	1c	260	150		21000	41000	20

Ti (22)	V (23)	Cr (24)	Mn (25)	Fe (26)	Co (27)	Ni (28)	Cu (29)
4376.4			1239.1	45616.0			
4796.1			774.5	45950.6			
5800	160	330	800	47000	100	180	100
				50100			
6294.9	176.7	286.0	960.3	51080.2		196.5	100
2458.0			309.8	34456.3			
3597.1				36063.9			
3273.3				32566.2			
6100	200	300	1000	42000	20	200	100
6300	100	200	930	51000	40	100	100
4856.0			154.9	39266.2			
6000	150	200	900	51000	30	80	100
4736.1			851.9	51678.3			
4400	150	200	1000	50000	40	100	70
9592.2			1548.9	64668.0			
4400	110	200	1000	50000	23	80	45
4500	90	83	1000	46500	18	58	47
					32	90	38
5700	135	100	950	56300	25	75	55
5300	120	77	1000	48000	18	61	50
6400	140	110	1300	58000	25	89	63
7194.1			1548.9	58838.9			
3117.5	53	99	526.6	30888.3	21	23	14
3597.1			774.5	41816.0			
4196.6			774.5	35443.3			
4700	95	70	690	35400	12	44	30
3417.2			774.5	36219.7			
4376.4			1084.2	46650.5			
5395.6			774.5	52464.7			
4196.6			619.6	33810.8			
4076.7			1006.8	42665.2			
3300			700	35000			
6000			900	57000			
6000			900	60000			
6294.8			1084.2	56273.1			
5700	180	71	1000	56000	26	75	60

TABLE A-I (*continued*)

Data source	Method	Element (atomic number)					
		Zn (30)	Ga (31)	Ge (32)	As (33)	Se (34)	Br (35)
1	2						
2	1b						
3	1c, 2	40	$\times 10^{-5}$	$\times 10^{-5}$	x	0.0x	x
4	1c,2						
5	2	40					
6	1b						
7	1b, 1c						
8	1b, 1c						
9		200	1	4	5	0.8	10
10	1c	40	15	7	5	0.6	
11	1b						
12		50	15	7	5	0.6	1.6
13	3						
14	3	80	15	7	5	0.09	2.5
15	1a						
16	1c	65	15	2	2	0.09	3
17	1c	83	19	1.4	1.7	0.05	2.1
18	2	79	19	1.3	2		3.1
19	1c	70	15	1.5	1.8	0.05	2.5
20		81	18	1.4	1.7	0.059	4.0
21		94	18	1.4	2.2	0.075	4.4
22	1a						
23	1b		14				
24	3						
25	1b						
26	1b	60	17	1.3	1.7	0.09	2.9
27	1a						
28	1a						
29	1a						
30	1a						
31	1a						
32	1a						
33	1a, 1c						
34	1a, 1c						
35	2						
36	1c	70	17	1.4	1.9	0.1	0.26

Kr (36)	Rb (37)	Sr (38)	Y (39)	Zr (40)	Nb (41)	Mo (42)	Tc (43)
	x	170		230		x	
		186.0		288.7			
0.0002	80	350	50	250	0.32	10	0.001
	310*	420	51	190	15	15*	
	300	400	28	200	10	3	
	280	150	28.1	220	20	2.3	
	120	450	40	160	24	1	
	150	340	29	170	20	1.1	
	120	450	40	156	24	1	
	90	375	33	165	20	1.5	
	90	470	27	140	20	1.1	
	78	480	24	130	19	1.3	
	118	340		400			
	120	290	34	160	20	1	
	125	215	19	155	12	2	

356

TABLE A-I (*continued*)

Data source	Method	Ru (44)	Rh (45)	Pd (46)	Ag (47)	Cd (48)	In (49)
1	2						
2	1b						
3	1c, 2	$\times 10^{-5}$	$\times 10^{-5}$	$\times 10^{-5}$	0.0x	0.x	$\times 10^{-5}$
4	1c, 2						
5	2						
6	1b						
7	1b, 1c						
8	1b, 1c						
9		0.05	0.01	0.05	0.01	5	0.1
10	1c		0.001	0.01	0.1	0.5	0.1
11	1b						
12		0.005	0.001	0.01	0.1	5	0.1
13	3						
14	3		0.001	0.01	0.2	0.18	0.1
15	1a						
16	1c	0.001	0.001	0.01	0.1	0.2	0.1
17	1c			0.013	0.07	0.13	0.25
18	2						0.11
19	1c				0.07	0.2	0.1
20				0.0084	0.065	0.15	0.11
21				0.01	0.075	0.15	0.14
22	1a						
23	1b						
24	3						
25	1b						
26	1b	0.001	0.001	0.01	0.06	0.1	0.07
27	1a						
28	1a						
29	1a						
30	1a						
31	1a						
32	1a						
33	1a, 1c						
34	1a, 1c						
35	2						
36	1c	0.0001	0.005	0.01	0.05	0.18	0.05

Sn (50)	Sb (51)	Te (52)	I (53)	Xe (54)	Cs (55)	Ba (56)	La (57)
x	0.x	0.00x	0.x		0.00x	470	
						492.6	
80	0.5	0.01	10	3×10^{-5}	10	500	6.5
40	1		0.3		7	390	19
						179.1	
40	0.4	0.01	0.5		7	500	18
40	1	0.0018	0.3		3.2	430	18.3
3	0.2	0.002	0.3		1	400	18
2.5	0.5	0.001	0.4		3.7	650	29
2	0.15		0.521		2.7	400	
2	0.2		0.5		3	425	30
1.6	0.45	0.00036	0.51		1.6	400	39
1.7	0.62	0.00055	0.56		1.4	390	39
						1070	
3	0.2	0.002	0.5		2.7	590	44
4	0.7	0.01	0.05		1.3	600	100

TABLE A-I (*continued*)

Data source	Method	Element (atomic number)					
		Ce (58)	Pr (59)	Nd (60)	Pm (61)	Sm (62)	Eu (63)
1	2						
2	1b						
3	1c, 2						
4	1c, 2						
5	2						
6	1b						
7	1b, 1c						
8	1b, 1c						
9		29	4.5	17		7	0.2
10	1c	44	5.6	24		6.5	1.0
11	1b						
12		45	7	25		7	1.2
13	3						
14	3	46.1	5.53	23.9		6.47	1.06
15	1a						
16	1c	46	6	24		7	1
17	1c	70	9	37		8	1.3
18	2						
19	1c	60	8.2	28		6	1.2
20		57	6.5	28		7.6	1.2
21		43	5.7	26		6.7	1.2
22	1a						
23	1b						
24	3						
25	1b						
26	1b	75	7.6	30		6.6	1.4
27	1a						
28	1a						
29	1a						
30	1a						
31	1a						
32	1a						
33	1a, 1c						
34	1a, 1c						
35	2						
36	1c	96	10	36		7	1.2

Gd (64)	Tb (65)	Dy (66)	Ho (67)	Er (68)	Tm (69)	Yb (70)	Lu (71)
7.5	1	7.5	1	6.5	1	8	17
6.3	1.0	4.3	1.2	2.4	0.3	2.6	0.7
10	1.5	4.5	1.3	4	0.8	3	1
6.36	0.91	4.47	1.15	2.47	0.20	2.66	0.75
6	0.9	5	1	3	0.2	3	0.9
8	4.3	5	1.7	3.3	0.27	0.33	0.8
5.4	0.9	3	1.2	2.8	0.48	3	0.5
7.6	1.2	5.0	1.5	3.0	0.28	3.0	0.93
6.7	1.1	4.1	1.4	2.7	0.25	2.7	0.82
8.8	1.4	6.1	1.8	3.4	0.6	3.4	0.6
5	0.7	3	0.7	3	0.3	2	0.27

TABLE A-I (*continued*)

Data source	Method	Hf (72)	Ta (73)	W (74)	Re (75)	Os (76)	Ir (77)	Pt (78)
1	2							
2	1b							
3	1c, 2	30		50		0.000x	0.000x	0.00x
4	1c, 2							
5	2							
6	1b							
7	1b, 1c							
8	1b, 1c							
9		4	0.24	70	0.001	0.05	0.01	0.2
10	1c	3.2		69	0.001		0.001	0.005
11	1b							
12		3.2	2	1	0.001	0.05	0.001	0.005
13	3							
14	3	4.5	2.1	1	0.001		0.001	0.005
15	1a							
16	1c	5	2	1	0.001	0.001	0.001	0.005
17	1c	1	2.5	1.3	0.0007			
18	2	3.9			0.05			
19	1c	3	2	1.5				
20		1.8	2.3	1.2	0.00042			0.028
21		1.5	1.6	1.1	0.00047			0.046
22	1a							
23	1b							
24	3							
25	1b							
26	1b	3	3.4	1.3	0.001	0.001	0.001	0.005
27	1a							
28	1a							
29	1a							
30	1a							
31	1a							
32	1a							
33	1a, 1c							
34	1a, 1c							
35	2							
36	1c	4	1.2	0.4	0.0005	0.0002	0.001	0.01

Au (79)	Hg (80)	Tl (81)	Pb (82)	Bi (83)	Po (84)	At (85)	Rn (86)	Fr (87)
0.00x	0.0x	0.000x	20	0.0x				
			20					
0.005	0.05	0.1	16	0.1	0.05			
0.005	0.5	0.3	16	0.2				
0.005	0.07	3	16	0.2	2×10^{-10}		7×10^{-12}	
0.001	0.05	0.3	16	0.2				
0.005	0.05	1	16	0.2				
0.0043	0.083	1	16	0.009				
	0.077	1.3	15		3×10^{-10}			
0.004	0.08	0.45	12.5	0.17				
0.0035	0.08	0.56	13	0.0029	0.00082			
0.0035	0.089	0.48	12	0.0043	0.001			
0.004	0.03	1.3	15	0.2				
0.005	0.17	0.7	14	0.2				

TABLE A-I (*continued*)

Data source	Method	Element (atomic number)					
		Ra (88)	Ac (89)	Th (90)	Pa (91)	U (92)	Sum
1	2						999999.9
2	1b						995200.0
3	1c, 2	10^{-6}		20		80	1000960
4	1c, 2						987500
5	2						999999.9
6	1b						999900.1
7	1b, 1c						997000.1
8	1b, 1c						997160.0
9		2×10^{-6}		10	7×10^{-7}	4	1000000.1
10	1c			11		4	1032295.72
11	1b						1000500.0
12		10^{-6}	$x \times 10^{-10}$	8	10^{-6}	3	1000299.1
13	3						991800.1
14	3			11.5		4	996706.7
15	1a						1000000.1
16	1c			7		2	997607.75
17	1c			13		2.5	1000323.1
18	2	1.3×10^{-6}	3×10^{-11}	10	8×10^{-7}	2.4	—
19	1c			9.6		2.7	1000168.9
20				6.8		2.2	999444.04
21				5.8		1.7	1005741.03
22	1a						1000000.6
23	1b						997555.4
24	3						998000.0
25	1b						995000.1
26	1b			11		3.5	997330.129
27	1a						1000200.0
28	1a						1000000.0
29	1a						1027700.1
30	1a						1002600.3
31	1a						1000900.1
32	1a						995000
33	1a, 1c						997900
34	1a, 1c						998700
35	2						994400.1
36	1c			10		2	1005262.7

NOTES TO TABLE A-I

The methods used by the various authors are indicated by number and letter as follows: 1a = geophysical-geochemical, 1b = cartographic-geochemical, 1c = simplistic model-geochemical; 2 = averaging available analyses of igneous rocks; 3 = averaging analyses of sediments.

The sources of data are:

1 = Clarke (1916, p. 26);
2 = Knopf (1916, p. 622);
3 = Clarke and Washington (1924) — earth's crust including lithosphere, hydrosphere and atmosphere;
4 = Clarke and Washington (1924) — earth's crust, magmatites only;
5 = Clarke (1924, p. 29);
6 = Sederholm (1925);
7 = Vogt (1931; in Rösler and Lange, 1972, p. 241) — granite 50%;
8 = Vogt (1931; in Rösler and Lange, 1972, p. 240) — granite 60%;
9 = Fersman (1933—1939; in Rösler and Lange, 1972, p. 240);
10 = Goldschmidt (1937, p. 656);
11 = Grout (1938, p. 502);
12 = Vinogradov (1949; in Rösler and Lange, 1972, pp. 230—231);
13 = Goldschmidt (1954, p. 54) — Norwegian post-glacial loams;
14 = Goldschmidt (1954, pp. 74—75);
15 = Poldervaart (1955, p. 133);
16 = Berry and Mason (1959, p. 212);
17 = Vinogradov (1962, pp. 647—648);
18 = Rankama (1963, p. 123);
19 = S.R. Taylor (1964, p. 1280);
20 = Lee Tan and Yao Chi-lung (1965, p. 782) — continental crust;
21 = Lee Tan and Yao Chi-lung (1965, p. 782) — earth's crust;
22 = Pakiser and Robinson (1967);
23 = Shaw et al. (1967, p. 848);
24 = Wedepohl (1971, p. 61) — 70 greywackes;
25 = Wedepohl (1971, p. 61) — igneous rocks in the upper crust corrected for sediments;
26 = Wedepohl (1971, p. 65) — igneous rocks in the upper crust;
27 = Ronov and Yarovshevsky (1969, p. 47) — granitic;
28 = Ronov and Yarovshevsky (1969, p. 47) — total continental crust;
29 = Ronov and Yarovshevsky (1969, p. 55) — total crust;
30 = Wedepohl (1969, p. 247) — standard column of intrusions;
31 = Holland and Lambert (1972, p. 676) — average continental crust;
32 = Beus (1976, p. 329) — granitic;
33 = Beus (1976, p. 329) — continental lithosphere;
34 = Beus (1976, p. 329) — total crust;
35 = Le Maitre (1976, p. 599) — average of igneous rock analyses;
36 = Reeves and Brooks (1978, pp. 79—81).

* Possible copying error.

364

TABLE A-II

Summary of estimates of crustal abundances

Atomic No.	Element	Approximate consensus (ppm)	Lowest (L) (ppm)	Highest (H) (ppm)	$H/L \times 100$ (%)
1	H	1400	671.3	10,000	1490
2	He		0.003	0.01	
3	Li	30	18	65	361
4	Be	3	1.3	10	769
5	B	10	3	50	1667
6	C		200	4902	2451
7	N	20?	15	400	2667
8	O	470,000	452,341	495,200	109
9	F	500	270	800	296
10	Ne		7×10^{-5}	0.005	7143
11	Na	24,000	15,208	28,500	187
12	Mg	21,000	10,191	33,770	331
13	Al	80,000	74,500	88,649	119
14	Si	270,000	257,500	315,896	123
15	P	1000	480	1309	273
16	S	500	260	1200	462
17	Cl	500?	100	2000	2000
18	Ar		0.04	4	10,000
19	K	26,000	15,773	32,625	207
20	Ca	30,000	16,438	62,894	383
21	Sc		5	22	440
22	Ti	5000	2458	9592	390
23	V	150	53	200	377
24	Cr	200	70	330	471
25	Mn	900	155	1549	1000
26	Fe	50,000	30,888	64,668	209
27	Co	25	12	100	833
28	Ni	80	23	200	870
29	Cu	60?	14	100	714
30	Zn	70	40	200	500
31	Ga	17	1	19	1900
32	Ge	15?	1.3	7	538
33	As	2	1.7	5	294
34	Se	0.09	0.05	0.8	1600
35	Br	3	0.26	10	3846
36	Kr		0.0002		
37	Rb	120	78	310	397
38	Sr	350	150	480	320
39	Y	30	19	50	263
40	Zr	160	130	400	308
41	Nb	20	0.32	24	7500
42	Mo	2	1	15	1500
43	Tc		0.001		
44	Ru		0.0001	0.05	50,000
45	Rh	0.001	0.001	0.01	1000
46	Pd	0.01	0.0084	0.05	595

TABLE A-II (*continued*)

Atomic No.	Element	Approximate consensus (ppm)	Lowest (L) (ppm)	Highest (H) (ppm)	$H/L \times 100$ (%)
47	Ag	0.07	0.02	0.1	500
48	Cd	0.18	0.1	5	5000
49	In	0.1	0.05	0.25	500
50	Sn	2.5	2	80	4000
51	Sb		0.15	1	667
52	Te		0.00036	0.01	2778
53	I	0.5	0.05	10	20,000
54	Xe		3×10^{-5}		
55	Cs	3	1	10	1000
56	Ba	430	179	1070	598
57	La		6.5	100	1538
58	Ce	45	29	96	331
59	Pr	6	4.5	10	222
60	Nd	25	17	37	218
61	Pm				
62	Sm	7	6.5	8	123
63	Eu	1.2	0.2	1.4	700
64	Gd	7	5	10	200
65	Tb	1	0.7	4.3	614
66	Dy	4.5	3	7.5	250
67	Ho	1.2	0.7	1.8	257
68	Er	3	2.4	6.5	271
69	Tm	0.3	0.2	1	500
70	Yb	3	0.33	8	2424
71	Lu	0.9	0.27	1.7	630
72	Hf	3	1	30	3000
73	Ta	2	0.24	3.4	1417
74	W	1	0.4	70	35,000
75	Re	0.001	0.00042	0.001	238
76	Os		0.0002	0.05	25,000
77	Ir	0.001	0.001	0.01	1000
78	Pt	0.005	0.005	0.2	4000
79	Au	0.004	0.001	0.005	500
80	Hg	0.08	0.03	0.5	1667
81	Tl	1	0.1	3	3000
82	Pb	16	12	20	167
83	Bi	0.2	0.0029	0.2	6897
84	Po		2×10^{-10}	0.05	2.5×10^{10}
85	At				
86	Rn		7×10^{-12}		
87	Fr				
88	Ra		1×10^{-6}	2×10^{-6}	200
89	Ac		3×10^{-11}		
90	Th	10	5.8	20	345
91	Pa		7×10^{-7}	1×10^{-6}	143
92	U	3	1.7	80	4706

366

TABLE A-III

Range (highest/lowest × 100) of estimates of crustal abundance of elements

Range (%)	Elements
>3000	Ne, Ar, Br, Nb, Ru, Cd, Sn, I, Hf, W, Os, Pt, Tl, Bi, Po, U
2500—2999	N, Te
2000—2499	C, Cl, Yb
1500—1999	B, Ga, Se, Mo, La, Hg
1100—1499	H, Ta
1050—1099	
1000—1049	Mn, Rb, Cs, Ir
950—999	
900—949	
850—899	Ni
800—849	Co
750—799	Be
700—749	Eu, Cu
650—699	Sb
600—649	Tb, Lu
550—599	Pd, Ba
500—549	Zn, Ag, In, Tm, Au, Ge
450—499	S, Cr
400—449	Sc
350—399	Li, V, Ca, Rb, Ti
300—349	Sr, Mg, Zr, Ce, Th
250—299	Dy, Ho, Y, Er, P, As, F
200—249	Gd, Re, Ra, K, Fe, Nd, Pr
150—199	Pb, Na
100—149	O, Al, Si, Sm, Pa

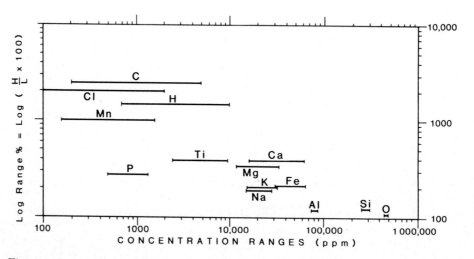

Fig. A-1. Percentage variation in estimates of crustal abundance of selected elements as a function of the range of reported concentrations.

Mason, 1959; Ronov and Yaroshevsky, 1969; Beus, 1976; Reeves and Brooks, 1978).

For the purpose of economic assessment the data should relate to accessible regions — i.e., the continental upper crust — because ocean mining has not yet attained economic significance, and mining to the Mohorovičić discontinuity is impractical. Moreover, although world-wide estimates are desirable for the general situation, those confined to certain continents or countries may be preferable for most types of exploration surveys.

Few of the data sets relate to the world-wide continental upper crust. Those that *seem* to fit the requirements are Wedepohl's (1971) two data sets based on analyses of greywackes or average regional rock abundance; Ronov and Yaroshevsky's (1969) data for the average composition of the granitic layer (although the total continental crust differs only slightly); and Beus' (1976) average of the "granitic" layer. The results of a reappraisal of these estimates are given in Table A-IV (this list is attenuated compared to Table A-II as only Wedepohl reported values for the trace elements). The data show that the range of values drops to less than 180% for all elements except H, C, and Cl; however, the greywacke model of Wedepohl gives an extreme value for 9 of the elements, but its exclusion causes significant changes only in the ranges for H, K, and Ca (see Fig. A-2), and the values of H, C and Cl remain excessively large.

Our knowledge of the crustal abundance of most elements is still poor and advances in analytical procedures are unlikely to bring much improvement. The models used to estimate the composition of the upper continental crust have yielded results that are in reasonable agreement for O, Na, Mg, Al, Si, P, Ti, Mn, and Fe; the data for K and Ca is reasonably good if the greywacke model is excluded. There is, however, only one set of data for most of the elements of interest to the economic geologist (Wedepohl, 1971), but for the major elements this data set provides extreme values for C, O, Na, Mg, Cl, K, Ca, Mn, and even Al if the greywacke model is excluded. Thus, the model is itself atypical, although it is not necessarily wrong; nevertheless, caution is needed in placing reliance on the trace element data.

It is evident (Table A-III) that no average crustal abundance (or clarke value) can be assigned for many of the economically important elements. Thus, the Soviet practice (e.g., Perel'man, 1977) of deducing genetic factors from clarke of concentration values (cc) — the ratio of element abundance in an ore to its lithospheric abundance — is not therefore tenable for most of the economically interesting elements, although perhaps it can be applied to Al, Pb, Fe, Zn, and P.

On the basis of the available data, the economically valuable elements can be *broadly* grouped by concentration as follows:

— those with average abundances less than 800 ppm (0.08%): Li, Be, C, S, V, Cr, Co, Ni, Cu, Zn, Zr, Mo, Ag, Cd, Sn, Sb, Ta, W, Pt, Au, Pb, U;

— those with average abundances greater than 800 ppm but less than

TABLE A-IV

A summary of the estimates of elemental abundances in the continental upper crust

Atomic No.	Symbol	Concentrations (ppm) lowest (L)	highest (H)	Range (%) (H/L × 100)	Concentrations (ppm) excluding those derived from greywackes lowest (L)	highest (H)	Range (%) (H/L × 100)
1	H	671	2574	384	671	1678	250
6	C	320	4183	1307			
8	O	472,500	489,888	104	472,500	480,976	102
11	Na	22,000	24,500	111			
12	Mg	12,000	13,900	116			
13	Al	76,212	80,446	106	78,300	80,446	103
14	Si	298,694	313,652	105	298,694	308,000	103
15	P	800	872.8	109	800	872.8	109
17	Cl	320	1000	313			
19	K	15,773	28,200	179	27,000	28,200	104
20	Ca	16,438	28,700	175	25,000	28,700	115
22	Ti	3300	4700	142			
25	Mn	690	774.5	112	690	774.5	112
26	Fe	35,000	41,816	119	35,000	36,531	104

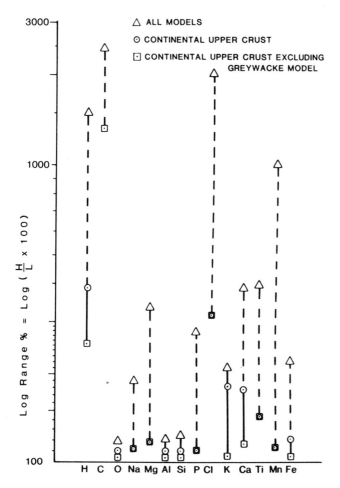

Fig. A-2. Percentage range in estimates of crustal abundance for selected elements for various models.

20,000 ppm (0.2%): P, Ti, Mn;
   — those with average abundances greater than 20,000 ppm: Al, Fe.

With the possible exception of U, Pt, and Au, the elements in the first two categories have to be very highly concentrated by natural processes to create economic deposits.

CLASSIFICATION OF ELEMENTS, IONIC SITE, AND COORDINATION NUMBERS

Goldschmidt's classification and size restrictions on lattice site.

Goldschmidt (1922; in Brownlow, 1979) subdivided the naturally occur-ring elements into four categories based on their distribution in the phases that exist in meteorites and the products of ore smelting. These categories are:

(1) *siderophile* — elements that concentrate in metallic phases (dominantly metallic bonding);

(2) *chalcophile* — elements that concentrate in sulphide phases (dominantly covalent bonding);

(3) *lithophile* — elements that concentrate in silicate phases (dominantly ionic bonding);

(4) *atmophile* — elements that concentrate in the atmosphere.

Some elements possess dual character; the character that they display depends on such factors as the relative stability of competing phases, temperature, pressure, and the nature of coexisting elements. The classification shown in Table A-V is only qualitative, but it does indicate the tendencies of the elements to occur in metallic, covalent, or ionically bonded structures, and it is adequate for the present purpose. It should be noted, however, that quantification of bond character has been attempted by Gordy and Thomas (1956), Pauling (1960), Ringwood (1955), and others, and the topic has been well reviewed by Ahrens (1964).

The continental upper crust is dominantly silicate in composition; therefore, the elements that are most likely to occur in the ionic minerals are lithophile in character. It is not possible, however, for each of the lithophile elements to enter into any particular lattice site because the size of the site imposes physical limits which some ions exceed.

The geometrical constraints of stacking spheres of radii r_A around one of radius r_C dictate the final arrangement, and the number of anions stacked against such a cation (at approximately equal distance; see Zemann, 1969, p. 17) is the so-called coordination number of that cation in that particular arrangement. The geometry is simply, and generally, described by letting $r_A = 1$ arbitrary unit. The range of units of r_C for a given coordination number is shown in Table A-VI together with the approximate shape of the lattice site.

Cations smaller than the lower value of r_C would lead to a condition known as "rattling", i.e., excessive distance between cations and anions would exist. Cations larger than the upper values of r_C lead to excessive distortion of lattice structure. For a given size of r_A there is only one value of r_C that gives a regular coordination polyhedron for each coordination number. Generally, some distortion exists and shapes of coordination polyhedra can only be roughly described; it is preferable to refer to the specific cation coordination number.

Although until relatively recently it was generally held that each cation had a set radius, the work of Shannon and Prewitt (1969, 1970), modified by Whittaker and Muntus (1970), demonstrated that the radius of an ion varies with the number and type of surrounding ions. Thus, to apply the geometric concept it is necessary to use the appropriate ionic radius for any particular coordination number.

TABLE A-V

Goldschmidt's (1937) classification of the elements. Italics indicate primary character, normal type indicates secondary character. Single underlining indicates primary character according to Berry and Mason (1959); double underlining indicates primary character according to Brownlow (1979)

Period	Class and geochemical character			
	siderophile (metallic)	chalcophile (sulphides)	lithophile (silicates)	atmophile (atmosphere)
I			H	H He
II	C	O	Li Be B C O F	C N O Ne
III	P	S	Na Mg Al Si P S Cl	Cl Ar
IV	Fe Co Ni Ge As Se	Cr Mn Fe Co Ni Cu Zn Ga Ge As Se	K Ca Sc Ti V Cr Mn Fe Co Ni Cu Zn Ga Ge As Se Br	Br Kr
V	Nb Mo Ru Pd Sn Te	Mo Ru Rh Pd Ag Cd In Sn Sb Te	Rb Sr Y Zr Nb Mo Sn Sb Te I	I Xe
VI	Ta W Re Os Ir Pt Au	Re Pt Au Hg Tl Pb Bi	Cs Ba La REE Hf Ta W Tl Pb Bi	
VII			Th U	

372

TABLE A-VI

Range of cation radii (r_C) compatible with different coordination numbers ($r_A = 1$ arbitrary unit of length)

r_C	Coordination No.	Arrangement of anions
0.155—0.225	III	apices of equilateral triangle
0.225—0.414	IV	apices of tetrahedron
0.414—0.732	VI	apices of octahedron
0.732—1	VIII	apices of cube
>1	XII	midpoints of cube edges

Coordination with oxygen (VI)

The convention of showing coordination numbers in superscript Roman numerals is from Shannon and Prewitt (1969). It prevents confusion with charge numbers.

The most abundant element in the lithosphere is undoubtedly oxygen; it constitutes about 47% by weight (see Table A-II above). More significantly, it occupies about 90% by volume, and, as Barth (1948) demonstrated, most lithospheric minerals can be considered to be meshes of oxygen anions held together by cations. As oxygen is commonly in six-fold coordination in many silicate structures, the value of r_A can be taken as 1.32 Å (see Whittaker and Muntus, 1970). Boundary values of r_C for different arrangements with oxygen are shown in Table A-VII.

Fig. A-3 illustrates the Whittaker-Muntus ionic radii to allow a rapid assessment of possible coordination with ^{VI}O (only the more common valence states are shown). Cations in coordination II with ^{VI}O are unlikely. Of cations with coordination III, only Be^2 has a radius suitable for this arrangement with ^{VI}O, but B^3 is known to occur with this coordination. Ions that can have IV, VI, VIII, and XII coordination with ^{VI}O are given in Table A-VIII; ions that have radii close to, but outside of, the limiting values for a particular coordination may fit with extra distortion of the coordination polyhedra.

Ten cations have ionic radii appropriate for coordination IV with ^{VI}O:

TABLE A-VII

Range of cation radii (r_C) for different coordinations with $^{VI}O^{2+} = 1.32$ Å

r_C (Å)	Coordination No.	Approximate shape of polyhedron
0.205—0.297	III	triangle
0.297—0.546	IV	tetrahedron
0.546—0.966	VI	octahedron
0.966—1.32	VIII	cube — ^{VI}O at apices
1.32	XII	cube — ^{VI}O at midpoints of edges

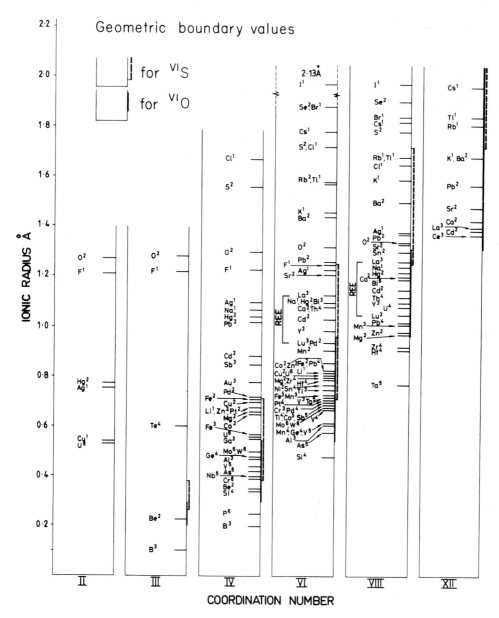

Fig. A-3. Whittaker-Muntus ionic radii for various co-ordination numbers.

two (Al and Si) are major constituents of the lithosphere. The hexavalent ions are unlikely to enter many silicate structures because they differ too greatly in charge from the major elements. Phosphorus is a minor element, and it is more probable that it will be diadochic with the hexavalent ions

TABLE A-VIII

Ions suitable for IV, VI, VIII, and XII coordination with ^{VI}O. Ions in brackets have radii close to, but outside of, the limiting values for the coordination (trace elements are shown in italics)

Coordination	Charge					
	1+	2+	3+	4+	5+	6+
IV		*Be*	Al (Fe) (*B*) (*Ga*)	Si (*Ge*)	(P) *V* *As* *Nb*	*Cr* *Mo* *W* (*U*)
VI	*Li*	Mg Mn Fe *Co* *Ni* *Cu* *Zn* *Pd*	Al Ti Mn Fe *Sc* *V* *Cr* *Co* *Ga* (*Y*) (*REE-high Z*) *Ru* *Rh* *In* *Ir*	(Si) Ti Mn *V* *Ge* *Zr* *Ru* *Rh* *Pd* *Sn* *Hf* *Re* *Os* *Ir* *Pt* *Pb*	*V* *As* *Nb* *Sb* *Ta*	*Mo* *W* *Re* *U*
VIII	Na (*Ag*)	Mg Ca Mn Fe *Zn* (*Sr*) *Cd* *Sn* *Hg* (*Pb*)	(*Sc*) *Y* *REE* *In* *Bi*	(*Zr*) (*Hf*) *Pb* *Th* *U*		
XII	K *Rb* *Cs* *Tl*	Ca Sr *Cd* Ba *Pb*	*La* *Ce*			

(elements that can occupy the same lattice site in a structure are said to be diadochic in that structure; see Rösler and Lange, 1972; and Burns, 1973). Phosphorus has only a minor lithophile character, as do Fe, Ga, As, and Mo.

Magnesium, Al, Ti, Fe, and Mn are major elements with appropriate radii for coordination VI with ^{VI}O; opportunity for diadochic relationships is common. ^{VI}Si only occurs in the rare high-pressure phase stishovite. Iron occurs in two valence states in minerals, Mn occurs in three, and the role of Ti is not resolved. The main valence state of Ti is Ti^4 but Ti^3 is believed to exist in some silicates, e.g., diopside (Schröpfer, 1968), garnets (Ito and Frondel, 1967). Ions that are non-lithophile in character are Ru, Rh, Pd, In, Re, Os, Ir, Pt; the lithophile role is said to be minor for Fe, Co, Ni, Cu, Zn, Ga, As, Mo, Sn, Sb, and Pb.

Relatively few ions have radii appropriate for coordination VIII with ^{VI}O; however, they include the major elements Mg, Ca, Na, and Fe^2 (note that Whittaker and Muntus (1970) did not give a radius for $^{VIII}Fe^2$). Silver, Cd, Hg, and In are essentially non-lithophile; Zn, Sn, Pb, and Bi favour other roles. Tetravalent ions have a much greater charge than the major elements in this coordination with ^{VI}O, and diadochic relationships are unlikely.

The very large, and so-called incompatible ions, tend to occur in oxygen lattice minerals in coordination XII. These are predominantly alkali and alkaline earth elements of Groups I and II. Of these elements only Cd has a non-lithophile character, but this is a lesser role for Tl and Pb.

Conclusions on the distribution of elements in phase types

On the basis of spatial considerations, substitution into oxygen lattices seems possible for ^{VI}(Ru, Rh, Pd, In, Re, Os, Ir, Pt), ^{VIII}(Ag, Cd, In, Hg), and ^{XII}(Cd); however, these elements have no lithophile character and they will normally be located in metallic and sulphide phases. Similarly, ^{IV}(P, Fe, Ga, As, Mo), ^{VI}(Fe, Co, Ni, Cu, Zn, Ga, As, Mo, Sn, Sb, Pb), ^{VIII}(Zn, Sn, Pb, Bi), and ^{XII}(Tl, Pb) are less likely to occur in silicate or oxide minerals than in sulphides.

Of those elements of economic importance, Ag, Cd, Pt, and Au should be sought in metallic and sulphide phases. Iron, Co, Ni, Cu, Zn, Mo, Sn, Sb, and Pb can occur in oxygen lattice minerals, but they prefer those which are metallic or sulphide. Lithium, Be, Al, Ti, V, Cr, Zr, U, Ta, and W have significant lithophile character and readily enter many silicate and oxide structures.

Sampling plans for geochemical exploration programmes depend on the type of host mineral for the element of interest. Metallic and sulphide phases are minor constituents of the continental upper crust; accordingly, it is difficult to adequately sample rocks that contain them. Some oxides and silicates (e.g., tantalite and zircon) may present equally difficult sampling problems. Phases that occupy large volume fractions of rocks, and which might contain elements of interest, require far less stringent sampling criteria.

INCORPORATION OF IONS IN MINERALS

Rock-forming minerals

The cation coordination numbers for some of the common rock-forming mineral groups are shown in Table A-IX based on data given by Deer et al. (1962—1963) and Wells (1962). Very many of these mineral groups have cations in IV and VI coordination; nesosilicates, sorosilicates, inosilicates, and tectosilicates have VIII-fold coordinated cations. The very large XII coordinated ions can be incorporated into phyllosilicates, tectosilicates, and

TABLE A-IX

Cation coordination with ^{VI}O in some groups of rock-forming minerals

Mineral group	Cation coordination numbers							
	IV	V	VI	VII	VIII	IX	X	XII
Nesosilicates								
Olivine	X		X					
Zircon	X				X			
Sphene	X		X	X				
Garnet	X		X		X			
Vesuvianite	X		X		X			
Al$_2$SiO$_5$	X	(X)	X					
Sorosilicates								
Epidote	X		X		X			
Melilite	X				X			
Cyclosilicates								
Beryl	X		X					
Cordierite	X		X					
Tourmaline	X		X			?		
Inosilicates								
Pyroxene	X		X		X			
Wollastonite	X		X					
Rhodonite	X		X					
Amphibole	X		X		X		X	
Phyllosilicates								
Mica	X		X					X
Talc	X		X					
Chlorite	X		X					
Serpentine	X		X					
Clays								
kandites	X		X					
illites	X		X					X
smectites	X		X					X

TABLE A-IX (*continued*)

Mineral group	Cation coordination numbers							
	IV	V	VI	VII	VIII	IX	X	XII
Tectosilicates								
Feldspars								
alkali	X					X		
plagioclase	X				X			
Silica	X							
Feldspathoids								
nepheline	X				X	X		
leucite	X							X
Analcite	X		X					
Zeolites	X		X	X	X			
Oxides								
Cassiterite*			X					
Corundum			X					
Hematite			X					
Ilmenite			X					
Rutile*			X					
Perovskite			X					
Spinels**	X		X					

* Oxygen coordination number III.
** Oxygen coordination number IV.

perovskite, but $^{XII}Pb^2$ is the only cation of economic importance that may occur in these minerals. Odd numbered coordination is uncommon. Tauson (1974) asserted that if conditions of crystallization favour the growth of two or more phases that can accommodate a particular ion, that ion will tend to preferentially enter the phase with the largest lattice site. Thus, Rb and Tl favour XII coordination in biotite over IX coordination in potash feldspars.

Rules for predicting the order of incorporation of ions in minerals

Elements that occur in minerals in trace quantities (i.e., less than 0.1%) are often described as "substituting" for a major element. Use of this word is misleading as it might convey a two-way diffusion process whereby ions of the trace element enter the crystal lattice to occupy sites recently vacated by ions of the major element. More commonly the ions of the trace and major elements compete for available lattice sites while crystals are growing. Indeed, the relative proportions of these ions in the resultant minerals can be dependent on temperature, and pressure, or both, thereby providing methods for geothermometry and geobarometry. Thus, the distribution of Ni between olivine and pyroxene can be used for geothermometry (see Häkli and Wright, 1967). It was Goldschmidt (1937) who first suggested rules that indicate preferential incorporation of one ion over another in minerals. In essence, he stated that for a suitable lattice site in a crystal the following rules apply:

— *Rule 1:* Ions of equal radius and charge have equal chance of incorporation. A major element is said to *camouflage* a trace element in these circumstances, e.g., VI-coordinated Zr^{4+} (0.80 Å) and Hf^{4+} (0.79 Å) or IV-coordinated W^{6+} (0.50 Å) and Mo^{6+} (0.50 Å) for the minerals zircon and scheelite, respectively.

— *Rule 2:* For two ions with different radii but the same charge the smaller ion is preferentially incorporated into crystals formed at higher temperatures, for example in potash feldspars:

$$^{XII}K^+ \text{ (1.68 Å)} - Rb^+ \text{ (1.81 Å)} - Tl^+ \text{ (1.84 Å)} - Cs^+ \text{ (1.96 Å)}$$

high temperature $\longleftarrow$ $\longrightarrow$ low temperature

Although a pair of ionic radii may be appropriate for a particular lattice site, they may be excessively different from each other. Ringwood (1955) noted that for two ions to be able to replace one another in a crystal structure, the ionic radii must not differ by more than 15%. There are several exceptions to this limit; Sobolev and Soboleva (1948; in Tauson, 1965) believed it to extend even to 40% at crystallization temperatures; Ahrens (1964), however, placed the upper limit at about 15—20%.

— *Rule 3:* For two ions with appropriate radii for a particular lattice site but with different charges, the ion with the higher charge is preferentially incorporated into crystals formed at higher temperatures.

A major element is said to *capture* a trace element of higher charge and to *admit* one of lower charge. For example:

$$^{VI}Li^+ \text{ (0.82 Å)} - Mg^{2+} \text{ (0.80 Å)} - Sc^{3+} \text{ (0.83 Å)}$$

(admitted)   (captured)

lower temperatures $\longleftarrow$ $\longrightarrow$ higher temperatures

For such processes to occur with maintenance of charge neutrality in the resultant crystal it is a necessary consequence that capture in one lattice site must be accompanied by admittance in another. This raises a difficulty because Goldschmidt's rule can only apply to one of the processes; as Shaw (1953) has observed "There is no reason why attention should be focused only on one cation, to the neglect of the other" (p. 147). For Goldschmidt's rule to function it seems necessary to add the corollary that it be applied to the larger of the two lattice sites. Generally these events do not occur if the charge difference is greater than 1, so that Mg^{2+} (0.80 Å) does not capture Zr^{4+} (0.80 Å) in sites of coordination number VI (see Berry and Mason, 1959).

Goldschmidt's (1937) rules effectively mean that ions that lead to the strongest bond are preferentially incorporated because both smaller radius and higher charge give an increase in bond strength. Ringwood (1955) found it necessary to introduce a fourth rule to account for exceptions to

Goldschmidt's rules which he believed to be largely due to the degree of covalency in the bonding.

— *Rule 4:* For ions with the same charge and similar radii, the ion with the lowest electronegativity is preferentially incorporated, provided that values differ by more than 0.1 unit. This rule was intended to account for the behaviour of ion pairs that seemingly contravened either Goldschmidt's second rule relating to ionic size or his third rule relating to ion charge. The former is largely a geometrical problem; the latter, the problem of bonding, is the one to which Ringwood's rule is most closely applicable.

When Ringwood (1955) introduced his rule the tables of ionic radii then in in use where those of Ahrens (1952); with the availability of the Whittaker-

TABLE A-X

Application of Goldschmidt (1937) rules with the Whittaker-Muntus (1970) radii to ion pairs selected by Ringwood (1955) for which the Ahrens (1952) ionic radii gave incorrect predictions

Trace ions		Major ions		Predicted preference		Observed preference
ion	Whittaker-Muntus radii (Å)	ion	Whittaker Muntus radii* (Å)	Whittaker-Muntus radii	Ahrens radii	
Cu^2	$^{VI}(0.81)$	Fe^2	$^{VI}_L(0.69)$	Fe^2	Cu^2	Fe^2
			$^{VI}_H(0.86)$	Cu^2		
Zn^2	$^{VI}(0.83)$	Fe^2	$^{VI}_L(0.69)$	Fe^2	equal	Fe^2
			$^{VI}_H(0.86)$	Zn^2		
Sn^2	$^{VIII}(1.30)$	Ca^2	$^{VIII}(1.20)$	Ca^2	Sn^2	Ca^2
V^3	$^{VI}(0.72)$	Fe^3	$^{VI}_L(0.63)$	Fe^3	Fe^3	V^3
			$^{VI}_H(0.73)$	V^3		
Cr^3	$^{VI}(0.70)$	Fe^3	$^{VI}_L(0.63)$	$_L Fe^3$	equal	Cr^3
			$^{VI}_H(0.73)$	$_H Cr^3$		
Ag^1	$^{VIII}(1.38)$	K^1	$^{VIII}(1.59)$	Ag^1	Ag^1	K^1
Cd^2	$^{VIII}(1.15)$	Ca^2	$^{VIII}(1.20)$	Cd^2	Cd^2	Ca^2
Ni^2	$^{VI}(0.77)$	Mg^2	$^{VI}(0.80)$	Ni^2	Mg^2	Ni^2

*H = high spin; L = low spin.

Muntus (1970) ionic radii it is appropriate to re-examine the ion pairs that Ringwood cited as being exceptions to Goldschmidt's second rule. Details of the re-examination are given in Table A-X. The behaviour of Ca^2/Sn^2, V^3/Fe^3, and Cr^3/Fe^3 accord with Goldschmidt's rule. In the last two cases Fe^3 is in the high-spin state (maximum number of unpaired electrons), which Burns and Fyfe (1967) indicated to be the case for most oxide and silicate structures; this accounts for their paramagnetism (or ferromagnetism, as in the case of magnetite). The ion pairs Fe^2/Cu^2 and Fe^2/Zn^2 can only accord with Goldschmidt's rules if Fe^2 is in the low-spin state; consequently, these ion pairs should be considered exceptions, as are K^1/Ag^1 and Ca^2/Cd^2. The ion pair Na^1/Cu^1 could not be examined as Whittaker and Muntus (1970) did not give ionic radii for Cu^1 in coordinations other than II.

Shaw (1953) listed 17 pairs of silicate and oxide phases, together with their melting points. The melting order is correctly predicted by Goldschmidt's rules (Whittaker-Muntus ionic radii) for 13 pairs; the exceptions are shown in Table A-XI. Ringwood's rule satisfactorily accounts for the oxides, and possibly for potassium and sodium feldspars, but does not account for their strontium and calcium analogues. It is one thing to invoke Ringwood's rule in hindsight, but in attempting to predict element behaviour the opposing predictions from Goldschmidt's and Ringwood's rules lead to the dilemma of deciding which to use.

The ions Mg^2 and Ni^2 have a seemingly curious behaviour. In the Skaergaard intrusion Ni^2 is concentrated in early differentiates, and many workers have interpreted this as incorporation into structures at the expense of Mg^2. Ringwood (1955) discussed this phenomenon at length and concluded that Ni^2 is actually camouflaged by Fe^2. The data in Table A-XII show that Ringwood's rule completely fails to account for the order of incorporation

TABLE A-XI

Some exceptions to Goldschmidt's rules (electronegativity values from Gordy and Thomas, 1956)

Compound	Melting point (°C)	Significant ion	Electronegativity
$SrAl_2Si_2O_8$	>1700	Sr^2	1.0
$CaAl_2Si_2O_8$	1550	Ca^2	1.0
$KAlSi_3O_8$	1170	K^1	0.8
$NaAlSi_3O_8$	1118	Na^1	0.9
MgO	2800	Mg^2	1.2
NiO	1990	Ni^2	1.8
MnO	1785	Mn^2	1.4
FeO	1370	Fe^2	1.7

TABLE A-XIl

Some ionic parameters for Mg, Fe and Ni

Ion	Electronegativity (Gordy and Thomas, 1956, p. 440)	Ionic Radii (Å) (Ahrens, 1952)	Ionic radii (Å) (Whittaker and Muntus, 1970, pp. 952—953)	Octahedral site preference energy (kcal/mole) (Burns and Fyfe, 1967, p. 265)
$^{VI}Ni^2$	1.8	0.69	0.77	20.6
$^{VI}Mg^2$	1.2	0.66	0.80	
$^{VI}_H Fe^2$	1.7	0.74	0.86	4.0
Predicted order of incorporation				
1	Mg^2	Mg^2	Ni^2	Ni^2
2	Fe^2	Ni^2	Mg^2	Fe^2
3	Ni^2	Fe^2	Fe^2	

of the ions, and Goldschmidt's rule with Ahrens' radii is only partially successful. Burns and Fyfe (1967) applied crystal field theory to this problem and concluded that as Ni^2 has higher crystal field stabilization energy than Fe^2 it should be preferentially admitted into mineral structures. They noted an apparent paradox that Ni^2 enters olivine ahead of Mg^2, and yet Ni_2SiO_4 has a lower melting point (1640°C) than Mg_2SiO_4 (1890°C), forsterite. The order is, however, correctly predicted by the Goldschmidt rules using Whittaker-Muntus ionic radii.

ORDER OF INCORPORATION OF IONS DURING FRACTIONAL CRYSTALLIZATION

The degree of correlation between the actual and predicted orders of uptake of ions during crystallization of a magma has been tested using the Spearman rank correlation coefficient r_s (Siegel, 1956). Predictions derived from four different rules on the order of ion uptake are compared to the observed order for the Skaergaard and Thingmuli intrusions in Table A-XIII. Values of r_s (Table A-XIV) are not directly comparable unless the number of data pairs are the same, so emphasis is directed to values of probabilities (p) of correctness of the null hypothesis that the two sequences are unrelated. It is most important to note that the degree of association of sequences reported for Skaergaard and Thingmuli is not significant at $p = 0.05$. Thus, the probability of the sequences being different is $\gg 5\%$, and it is conventional to reject the alternative hypothesis of similarity. Significant disagreements in the order of uptake of ions in nature therefore exists. In the prediction of the sequence of ion uptake at Skaergaard there is little to choose between

TABLE A-XIII

Observed and predicted orders of ion incorporation into octahedral lattice sites in ^{VI}O structures (brackets indicate that an ion is subject to Jahn-Teller distortion)

Skaergaard (Williams, 1959)	Thingmuli (Carmichael et al., 1974)	Crystal field stabilization energies (C.D. Curtis, 1964)	Octahedral site preference energies (Burns and Fyfe, 1964, 1967)	Whittaker-Muntus ionic radii — Goldschmidt	
				rules 2 + 3	rule 2
Ni^2	$Cr^3=Ni^2$	Cr^3	Cr^3	Ti^4	Al^3
Cr^3	Cu^2	V^3	Ni^2	Al^3	Ti^4
Co^2	Co^2	Ni^2	(Cu^2)	$Cr^3=Ga^3$	$Cr^3=Ga^3$
V^3	V^3	Co^2	V^3	V^3	V^3
$Mg^2=Al^3$	$Fe^{2,3}$	Fe^2	Co^2	Fe^3	Fe^3
$Ca^2=Sc^3$	Ti^4	$Fe^3=Mn^2=Zn^2$	Fe^2	Sc^3	Ni^2
Ga^3	Mn^2	Cu^2	$Ti^4=Sc^3=Fe^3=$	Y^3	Mg^2
Mn^2	Zn^2		$=Ga^3=Ca^2=$	La^3	Cu^2
Fe^2			$=Mn^2=Zn^2$	Ni^2	$Sc^3=Zn^2=Co^2$
Cu^2				Mg^2	Fe^2
$La^3=Y^3$				Cu^2	Mn^2
Ba^2				$Zn^2=Co^2$	Y^3
Fe^3				Fe^2	Ca^2
				Mn^2	La^3
				Ca^2	Ba^2
				Ba^2	

TABLE A-XIV

Spearman correlation coefficients (r_s) between observed and predicted order of ion incorporation in the Skaergaard and Thingmuli intrusions. The observed order for Skaergaard has been derived by combining the data in figs. 1 and 2 in Williams (1959). The observed order for Thingmuli has been derived from the position of the maxima in fig. 3-1 in Carmichael et al. (1967); iron has been disregarded because data for Fe^2 and Fe^3 were combined. The number of data pairs is given in brackets (n)

	All ions		Divalent ions		Trivalent ions
	Skaergaard	Thingmuli	Skaergaard	Thingmuli	Skaergaard
Octahedral crystal field stabilization energies (C.D. Curtis, 1964)	0.802* (8)	0.455 (7)	0.900* (5)	0.462 (5)	0.894 (5)
Octahedral site preference energies (Burns and Fyfe, 1967)	0.535* (11)	0.945** (8)	0.232 (6)	0.975* (5)	
Goldschmidt rules 2 and 3, Whittaker-Muntus radii	0.260 (16)	0.410 (8)			
Goldschmidt rule 2, Whittaker-Muntus radii	0.523* (16)	0.410 (8)	0.619 (8)	0.821 (5)	0.619 (8)
Thingmuli (Carmichael et al., 1967)	0.638 (6)	—	0.400 (4)		
Octahedral site preference energies less Jahn-Teller distortions	0.759** (10)	0.916** (7)	0.667 (5)	0.949 (4)	

Critical values of r_s corrected for tied rankings:

n	$p = 0.05$*	$p = 0.01$**	n	$p = 0.05$*	$p = 0.01$**
4	1.0	—	8	0.643	0.833
5	0.900	—	10	0.564	0.746
6	0.829	0.943	11	0.521	—
7	0.714	0.893	12	0.506	0.712
			16	0.425	0.601

use of octahedral crystal field stabilization energies (C.D. Curtis, 1964); octahedral site preference energies (Burns and Fyfe, 1964, 1977); or Gold-schmidt's second rule based on the size of ions. The second of these methods of prediction is marginally less successful than the others, and Goldschmidt's second rule would be the most favoured for it has been tested on the greatest number of ions (16).

When applied to the Thingmuli data, only the Burns and Fyfe (1964, 1967) prediction is successful and that highly so. However, these tests were carried out without consideration of Jahn-Teller distortions which Burns and Fyfe (1967, p. 278) claimed to be sufficient to warrant removal of Cr^2, Cu^2 and Mn^3 from the sequences predicted. Of these, only Cu^2 is involved here, and its removal from the prediction sequence results in variations of r_s that bring about only a minor change in the conclusions.

Some authors (e.g., Williams, 1959; Burns and Fyfe, 1964, 1967; and C.D. Curtis, 1964) have separately discussed sequences of divalent and trivalent ions. This tends to make their predictions more presentable, but in all cases the ions are competing for octahedral lattice sites and should be considered as one sequence irrespective of the need for similar competition to be occur-ring elsewhere in the lattice so as to maintain charge balance. There are not adequate data to warrant correlation tests on trivalent ions for Thingmuli, nor for comparison of trivalent ions at Skaergaard with C.D. Curtis' (1964) prediction. The other data pairs have been tested; the results are given in Table A-XIV. The correlation between observed ion incorporation at Skaer-gaard and Thingmuli is extremely poor, and only two predictions correlate significantly $(0.05 \geqslant p \geqslant 0.01)$ with reported sequences.

This examination should disturb geochemists — either there is little similarity between ion behaviour in different environments (Skaergaard, Greenland versus Thingmuli, Iceland), or true sequences of ion uptake by developing crystals have not been established for these locations. Testing methods of prediction has little relevance until this dilemma is resolved. Skaergaard is the better studied of the two areas, and based on these studies, predictions dependent on mere ion size are approximately equally success-ful compared to those based on lattice site preference energies. Moreover, the latter relate only to transition elements, and the former relate to all elements.

That there should be concordance between crystal field theory and Gold-schmidt's (Whittaker-Muntus radii) rules is to be expected. The crystal field splitting parameter is:

$$\Delta \propto Q/(r_A + r_C)^5$$

where Q = ligand charge, r_A = anion radius, and r_C = cation radius (Burns and Fyfe, 1967). It follows that r_C is dependent both on charge and the crystal field splitting parameter, and the Whittaker-Muntus ionic radii have automatically taken account of these features. Indeed, these authors tabulated radii for both high- and low-spin conditions for a number of ions, and, as predicted by Burns and Fyfe (1967), the low-spin ions have smaller radii.

It appears, therefore, that the order of incorporation of ions during crystal growth can *largely* be predicted by Goldschmidt's rules, provided that Whittaker-Muntus ionic radii are used. There are exceptions — most of which do not follow the charge rule — and, in hindsight, these can often be shown to follow Ringwood's rule and can be explained by crystal field theory.

Goldschmidt's rules are simplistic, and yet their success is amazing. Their application necessitates the use of tables of ionic radii which are based on *average* interatomic distances within a particular lattice site. Zemann (1969) has indicated that quite large variations of interatomic distance can occur at a single lattice site, so that even the choice of coordination numbers can be problematical. Indeed, he noted that it is usual for interatomic distances within the range (minimum) to (minimum + 15%) to be counted for this purpose. These variations result in distorted coordination polyhedra; in addition, the ions themselves are seldom spherical. Burns (1973) has stressed the need to consider both fluid and solid structures when determining preferences of ion incorporation in crystals, but Goldschmidt's rules ignore the liquid phase. Despite these failings, his empirical rules work well.

INCORPORATION OF IONS OF ECONOMIC SIGNIFICANCE IN MINERALS

From the preceding discussion it may be concluded that prediction of the location of ions in mineral types can reasonably be made from Goldschmidt's classification of elements combined with his second rule (and Whittaker-Muntus ionic radii) and radius ratio relationships. These principles have been applied to a number of elements of economic importance. Table A-XV lists the mineral type in which these elements can be expected and ion charge and coordination number for silicate or oxide phases are also given. This latter information can be used in conjunction with Table A-IX to ascertain probable rock-forming minerals that might host these ions. In addition, the sequence of incorporation of these ions in oxide structures has been predicted from Goldschmidt's second rule and Whittaker-Muntus ionic radii. In Table A-XVI these sequences are given for some ions of economic importance together with major elements with which they might be diadochically related. The sequences for individual coordination numbers are not necessarily related but are shown here linked at major elements so as to give common reference points. Ions at the top of the table will tend to be preferentially incorporated in crystals and thus tend to occur in those phases crystallizing early from a melt and therefore at the higher temperatures for the system. For any ion of interest it is possible to ascertain its position,

TABLE A-XV

Probable locations of ions of economic importance in mineral types (locations of lesser importance indicated by question mark; arabic numerals indicate charge number)

Element	Metals	Sulphides	Coordination numbers in silicates/oxides				
			III	IV	VI	VIII	XII
Li	—	—	—	—	1	—	—
Be	—	—	2	2	—	—	—
C	X	—	?	4	—	—	—
Al	—	—	—	3	3	—	—
P	X	—	? —	5	—	—	—
S	—	X	—	6	—	—	—
Ti	—	—	—	4	2, 3, 4	—	—
V	—	—	—	5	2, 3, 4, 5	—	—
Cr	—	?	—	5, 6	2, 3	—	—
Mn	—	?	—	—	2	2	—
Fe	X	X	? —	3	2, 3	—	—
Co	X	?	? —	—	2, 3	—	—
Ni	X	?	? —	—	2	—	—
Cu	—	X	? —	—	2	—	—
Zn	—	X	? —	—	2	2	—
Zr	—	—	—	—	4	(4)	—
Mo	X	X	? —	6	2, 3, 6	—	—
Ag	—	X	—	—	—	—	—
Cd	—	X	—	—	—	—	—
Sn	X	X	? —	—	4	—	—
Sb	—	X	? —	—	5	—	—
Ta	X	—	—	—	5	—	—
W	?	—	—	6	6	—	—
Pt	X	?	—	—	—	—	—
Au	X	?	—	—	—	—	—
Pb	—	X	? —	—	4	4	2
U	—	—	—	—	5	4	—

relative to its competitors for the same lattice site, and thus to deduce whether it is likely to occur in high-, medium-, or low-temperature phases and rocks. In turn this gives guidance as to whether to seek the ion in basic, intermediate, or acidic rocks. A complication is that ions of high charge may not necessarily be able to enter common rock-forming minerals because the charge difference (2 or more units) from the major constituent that normally occupies the same site has to be compensated elsewhere in the structure, and this is difficult to achieve. Thus for IV and VI coordination sites, pentavalent and hexavalent ions usually will be located in minor phases. Likewise in VIII coordination of the quadrivalent ions Zr^4, Pb^4, and U^4 also can be expected in accessory minerals where they are major components. These accessory minerals tend to form late in a crystallization sequence.

TABLE A-XVI

Order of incorporation of some ions of economic importance in silicate/oxide structures as predicted by Goldschmidt's second rule (brackets indicate major element in continental upper crust; underlining indicates stronger non-lithophile character)

Within coordination number group ranking of temperature of incorporation	Coordination number with ^{VI}O			
	IV	VI	VIII	XII
Higher	(P^5)			
↑	(Si^4)			
	Be^2			
	Cr^6			
	V^5			
	(Al^3)	(Al^3)		
		$V^5 = (Mn^4)$		
		V^4		
	$\underline{Mo^6} = \underline{W^6}$	$\underline{Mo^6} = \underline{W^6}$		
		$(Ti^4) = Sb^5$		
	U^6	Cr^3		
		$V^3 = Ta^5$		
	(Fe^3)	$(\underline{Fe}^3) = (Mn^3)$		
		$Ni^2 = Sn^4$	Zr^4	
		$Zr^4 = (Mg^2)$	(Mg^2)	
		$Cu^2 = U^6$		
		Li^1		
		$\underline{Co}^2 = \underline{Zn}^2$	$\underline{Zn}^2$	
		$(\underline{Fe}^2) = \underline{Pb}^4$		
		(Mn^2)	(Mn^2)	
			$\underline{Pb}^4$	
			U^4	
			(Ca^2)	(Ca^2)
			(Na^1)	
			Sn^2	
			$\underline{Pb}^2$	$\underline{Pb}^2$
Lower				(K^1)

APPENDIX 2 — ABUNDANCE OF ELEMENTS IN COMMON ROCK TYPES

The following table gives abundances in the main rock types of selected elements used in exploration rock geochemistry. Values for international standards (PCC-1, BCR-1, AGV-1, G-2, SCo-1, and SDC-1) are from Flanagan (1976) and are the most likely values *for those particular samples*. They should not, however, be regarded as necessarily representative of the class to which they belong. The average abundances are from Vinogradov (1962). The ranges are from tables in Wedepohl (1969–1978) and are the ranges of *average* values (i.e., the actual range for individual reported values is up to several orders of magnitude greater in some cases); some extreme values have been ignored in this compilation.

These data are included as a guide to the *approximate* range of concentrations to be expected. Not only is there variation in concentration of elements in similar rocks in different environments, but, as demonstrated in the main text of this book, there is considerable regional variation in similar rocks in similar geological environments. Neither these data, nor similar data from other sources, can be used as background; background values must be established in each region and area as part of the exploration programme.

Abundance of elements in common rock types

Element	Symbol	Atomic number	Atomic weight	Ultramafic
				peridotite PCC-1
Major and minor elements (wt. %)				
Aluminium	Al	13	26.98154	0.39
Calcium	Ca	20	40.08	0.36
Iron	Fe	26	55.847	5.84
Magnesium	Mg	12	24.305	26.04
Manganese	Mn	25	54.9380	0.09
Phosphorus	P	15	30.97376	0.0009
Potassium	K	19	39.098	0.003
Silicon	Si	14	28.086	19.59
Sodium	Na	11	22.98977	0.004
Titanium	Ti	22	47.90	0.009
Trace elements (ppm)				
Antimony	Sb	51	121.75	1.4
Arsenic	As	33	74.9216	0.05
Barium	Ba	56	137.34	1.2
Beryllium	Be	4	9.01218	
Bismuth	Bi	83	208.9804	0.013
Boron	B	5	10.81	6.0
Cadmium	Cd	48	112.40	0.1
Chlorine	Cl	17	35.453	60
Chromium	Cr	24	51.996	2730
Cobalt	Co	27	58.9332	112
Copper	Cu	29	63.546	11.3
Fluorine	F	9	18.9984	15
Gold	Au	79	196.9665	0.0016
Lead	Pb	82	207.2	13.3
Lithium	Li	3	6.941	2
Mercury	Hg	80	200.59	0.0072
Molybdenum	Mo	42	95.94	0.2
Nickel	Ni	28	58.71	2339
Niobium	Nb	41	92.9096	<2
Rubidium	Rb	37	85.4678	0.063
Selenium	Se	34	78.96	<0.18
Silver	Ag	47	107.868	0.005
Strontium	Sr	38	87.62	0.41
Sulphur	S	16	32.06	<10
Tantalum	Ta	73	180.9479	<0.1
Tellurium	Te	52	127.60	<1
Thallium	Tl	81	204.37	0.0008
Thorium	Th	90	232.0381	0.01
Tin	Sn	50	118.69	1
Tungsten	W	74	183.85	0.06
Uranium	U	92	238.029	0.005
Vanadium	V	23	50.9414	30
Zinc	Zn	30	65.38	36

average	range	Mafic basalt BCR-1	average	range
0.45		7.20	8.76	3.01—12.16
0.7		4.95	6.72	
9.85		9.37	8.56	
25.9	9.15—27.80	2.09	4.5	2.07—4.86
0.15	0.096—0.106	0.14	0.2	0.124—0.147
0.017	0.001—0.05	0.16	0.14	0.011—0.597
0.03	0.0004—0.0125	1.41	0.83	0.09—1.89
19.0	18.79—23.61	25.48	24.0	22.62—23.89
0.57	0.011—0.57	2.43	1.94	1.34—2.87
0.03	0.10—1.70	1.32	0.9	0.56—1.80
0.1		0.69	1.0	0.1—1.4
0.5	0.3—10	0.70	2	0.28—2.5
1.0	8.8—23	675	300	14.5—613
0.2	0.15—0.8	1.7	0.4	0.3—2.6
0.001		0.050	0.007	0.008—0.22
0.1	1—210	5.0	5.0	3.2—700
0.05	0.001 < 2	0.12	0.19	0.006—0.6
50	18—600	50	50	70—750
2000	2410—4080	17.6	200	100—400
200	88—128	38	45	37—50
20	7.3—61	18.4	100	47—160
100	19—252	470	370	280—730
0.005	0.0017—0.023	0.00095	0.004	0.0003—0.018
0.1		17.6	8.0	0.76—10.2
0.5	2.0—15	12.8	15	10—18
0.01		0.0107	0.09	
0.2	0.2—0.4	1.1	1.4	0.6—2.8
2000	1390—1450	15.8	160	69—530
1.0	0.7—5.5	13.5	20	3—20
2.0	0.072—2.42	46.6	45	6—110
0.05		0.1	0.05	0.1—0.2
0.05		0.036	0.1	
10	1.54—76	330	440	97—1400
100	190—2650	392	300	40—1600
0.018	<0.1—0.55	0.91	0.48	0.55—2.0
0.001		<1	0.001	
0.01	0.07—1.04	0.30	0.2	0.06—0.54
0.005		6.0	3.0	0.19—3.8
0.5	0.1—1.3	2.6	1.5	1.1—8.7
0.1	0.1—0.77	0.4	1.0	0.27—6.3
0.003		1.74	0.5	0.15—0.99
40	17—300	399	200	50—400
30	41—60	120	130	57—154

Abundance of elements in common rock types (*continued*)

Element	Intermediate			Felsic	
	andesite AGV-1	average	range	granite G-2	average
Major and minor elements (wt. %)					
Aluminium	9.13	8.85	5.22—15.55	8.15	7.7
Calcium	3.50	4.05		1.39	1.58
Iron	4.73	5.85		1.85	2.7
Magnesium	0.92	2.18		0.46	0.56
Manganese	0.08	0.12	0.100—0.132	0.03	0.06
Phosphorus	0.21	0.16	0.067—0.276	0.06	0.07
Potassium	2.40	2.3	0.86—2.03	3.74	3.34
Silicon	27.58	26.0	25.33—27.81	32.30	32.3
Sodium	3.16	3.0	2.36—2.84	3.02	2.77
Titanium	0.62	0.8	0.34—1.56	0.30	0.23
Trace elements (ppm)					
Antimony	4.5	0.2		0.1	0.26
Arsenic	0.8	2.4	1.2—2.8	0.25	1.5
Barium	1208	650	703—1379	1870	830
Beryllium	3.8	18	<0.8—14	2.6	5.5
Bismuth	0.057	0.01	0.03—0.12	0.043	0.01
Boron	5.0	15	4.8—40	2.0	15
Cadmium	0.09		0.017—0.32	0.039	0.1
Chlorine	110	100	140—285	50	240
Chromium	12.2	50	11—103	7.0	25
Cobalt	14.1	10	13—28	5.5	5
Copper	59.7	35	34—67	11.7	20
Fluorine	435	500	210—505	1290	800
Gold	0.0006		0.0006—0.017	0.001	0.0045
Lead	35.1	15	1.8—27.9	31.2	20
Lithium	12	21	6—25	34.8	41
Mercury	0.015			0.039	0.08
Molybdenum	2.3	0.9	0.7—1.0	0.36	1.0
Nickel	18.5	55	3—28	5.1	8
Niobium	15	20	0.3—49	13.5	20
Rubidium	67	100	20—88	168	200
Selenium	<0.14	0.05		<0.7	0.05
Silver	0.11	0.07		0.049	0.05
Strontium	657	800	332—703	479	300
Sulphur	<10	200	130—395	24	400
Tantalum	0.9	0.7		0.91	3.5
Tellurium	<1	0.001		<1	0.001
Thallium	1	0.5	0.21—1.8	1	1.5
Thorium	6.41	7.0		24.2	18
Tin	4.2		0.01—2.0	1.5	3.0
Tungsten	0.55	1.0	1.1—2.7	0.1	1.5
Uranium	1.88	1.8		2.0	3.5
Vanadium	125	100	8—70	35.4	40
Zinc	84	72	60—90	85	60

range	Shales			Metamorphic
	shale SCo-1	average	range	mica-schist SDC-1
	7.09	10.45		8.63
	1.92	2.53		1.00
	3.57	3.33		4.90
0.17—3.97	1.62	1.34		0.96
0.016—0.059	0.04	0.067	0.013—0.082	0.09
0.005—0.157	0.19	0.07	0.039—0.709	0.08
3.28—4.26	2.32	2.28	0.008—7.06	2.66
	29.91	23.8		30.80
2.29—2.58	0.72	0.66	0.1—9.2	1.56
0.08—0.58	0.50	0.45	0.42—0.64	0.59
0.1—0.3		2.0	0.1—3.0	
1.0—4.8		6.6	3.2—18	
	300	800	83—866	1000
1.5—24	1.0	3.0	0.2—7.0	3
0.15—2.0		0.01		
1.2—900	70	100	25—800	30
0.001—0.2		0.3	<0.3—2.6	
43—507		100	72—265	
1—25	70	100	35—185	70
1—7	10	20	11—22	20
8—34	30	57	21—67	30
520—2950		500	50—5950	
0.0002—0.0061		0.001	0.001—0.057	
14.9—56.1	15	20	17.6—28	30
30—40		61	4—400	
		0.4		
0.6—3.3		2.0	0.7—2.0	
4.5—16	30	95	20—90	50
0.9—39	15	20	13.3—20.1	15
170—910		200	20—663	
0.1—0.2		0.06	0.1—0.7	
		0.1		0.3
55—252	200	450	20—360	200
70—1530		3000	2400—69000	
2.1—2.5		3.5		
		0.01		
0.6—6.9		1.0	0.36—1.4	
9.7—56		11	10.2—13.1	
<1.0—8.8		10	2.5—19	3
1.4—3.7		2.0	1.5—3.8	
2.2—7.6		3.2	2.0—8.0	
<5—70	100	130	98—260	70
23—89		80	46—197	

APPENDIX 3 — PROBABILITY TABLE FOR REQUIRED NUMBER OF SAMPLES

The table below is an abbreviated form of the tables given in Garrett (1982) to determine the number of samples required in a sample set (for a probability of 0.95) to ensure that a specified number (r) of samples have values greater (or less) than threshold for various a priori probabilities (P) of the occurrence of values greater (or less) than threshold. Some examples of how to use the table are given below.

Example 1. Suppose it has been determined that 20% of samples from tin-bearing granites have Ba:Rb ratios equal to or less than 2.0, i.e., the a priori probability of occurrence is 0.20. Calculate the number of samples required to ensure that at least 5 samples from a tin-bearing granite will have Ba:Rb ratios equal to or less than 2.0. The solution is given by finding the number of samples in the table that correspond to $P = 0.2$ and $r = 5$, i.e., 44 samples.

Example 2. Suppose it has been determined that 10% of samples from tin-bearing granites have a Sn content equal to or greater than 20 ppm, i.e., the a priori probability of occurrence is 0.1. Calculate the number of samples required to ensure that at least 3 samples from a tin-bearing granite will have a Sn content equal to, or greater than, 20 ppm. The solution is given by finding the number of samples in the table that correspond to $P = 0.1$ and $r = 3$, i.e., 61 samples.

Required number of samples in sample set for a probability of 0.95 that a specified number (r) of samples have values greater (or less) than threshold for various a priori probabilities (P) of occurrence of values greater (or less) than threshold

Number of samples, r, greater (or less) than threshold	Probability of occurrence, P									
	0.01	0.03	0.05	0.08	0.1	0.14	0.20	0.3	0.4	0.5
1	299	99	59	36	29	20	14	9	6	5
3	628	208	124	77	61	43	30	19	14	11
5	913	303	181	112	89	63	44	28	21	16
7	1182	392	234	146	116	82	57	37	27	21
10	1568	521	311	193	154	109	76	49	36	28

APPENDIX 4 — LIST OF DEPOSITS AND MINERALIZED DISTRICTS AND SOURCE OF REFERENCES FOR GEOCHEMICAL DATA

Note: this list gives the main references only and excludes occurrences in the U.S.S.R.; for complete data see index and text.

Deposit or mineralized district, and country	Ore minerals	Reference
Above Rocks, Jamaica	Cu	Kesler et al., 1975b
Agricola Lake, Canada	Pb, Zn, (Cu)	Cameron, 1975
Agrokypia, Cyprus	Cu	Govett and Pantazis, 1971
		Govett, 1972
		Constantinou and Govett, 1973
Ajax, Australia	Zn, Cu	Large, 1980
Ajo, U.S.A.	Cu	Davis and Guilbert, 1973
Akita district, Japan	Cu, Zn, Pb	Tono, 1974
		H. Ishikawa et al., 1962
Algonquin (Philipsberg Stock), U.S.A.	Mn	Mohsen and Brownlow, 1971
Almaden, Australia	Cu, Pb, Ag, Zn	Sheraton and Black, 1973
Anchor Mines, Australia	Sn	Groves, 1972
Annan River Tinfield (Finlayson granite), Australia	Sn	Sheraton and Black, 1973
Belmont (Marysville Stock), U.S.A.	Au, Ag	Mantei and Brownlow, 1967
		Mantei et al., 1970
		Tilling et al., 1973
Bethlehem JA, Canada	Cu	Olade and Fletcher, 1975
		Olade and Fletcher, 1976a
		Olade and Fletcher, 1976b
		Olade, 1977
Bingham, U.S.A.	Cu	Slawson and Nackowski, 1959
		Parry and Nackowski, 1963
		Parry and Jacobs, 1975
		Kesler et al., 1975c
Bode Lake, Canada	Cu	Garrett, 1975
Boliden, Sweden	Cu	Nilsson, 1968
Bootstrap, U.S.A.	Au	Erickson et al., 1964b
Bradina, Canada	Zn, Cu	Church et al., 1976
Breckenridge Mining District, Wirepatch Mine, U.S.A.	Zn, Pb, (Au, Ag) Mo(?)	Price et al., 1979

Source of reference for geochemical data (*continued*)

Deposit or mineralized district, and country	Ore minerals	Reference
Broken Hill, Australia	Pb, Zn	Plimer, 1979
		Plimer and Elliott, 1979
Brunswick Mining and Smelting No. 12, Canada	Zn, Pb, Cu	Goodfellow, 1975a
		Goodfellow, 1975b
		Govett and Goodfellow, 1975
		Govett, 1976b
		Goodfellow and Wahl, 1976
Buchans, Canada	Zn, Pb, Cu	Thurlow et al., 1975
Butte, U.S.A.	Cu	Kesler et al., 1975c
Canadian Shield, Canada	massive sulphides, general	Sakrison, 1971
		Descarreaux, 1973
		Bennett and Rose, 1973
		Cameron, 1975
		Roscoe, 1965
		Larson and Webber, 1977
		Dumitriu et al., 1979
		Sopuck et al., 1980
	Ni	Cameron et al., 1971
Cantung, Canada	W	Garrett, 1971a
		Garrett, 1971b
		Garrett, 1973
		Garrett, 1974
Caribou, Canada	Zn, Pb, Cu	Gandhi, 1978
Carlin, U.S.A.	Au	Erickson et al., 1964b
Cerro Colorado, Panama	Cu	Kesler et al., 1975b
Clayton Peak, U.S.A.	Cu, Pb, Zn, Ag, Au	Slawson and Nackowski, 1959
		Parry and Nackowski, 1963
		Parry and Jacobs, 1975
Cobalt District, Canada	Ag, Ni, Co	Dass et al., 1973
Coed-y-Brenin, U.K.	Cu	Allen et al., 1976
Coeur d'Alene, U.S.A.	Ag, (Pb, Zn)	Gott and Botbol, 1973
		Botbol et al., 1977
		Watterson et al., 1977
Coppermine River, Canada	Cu	Cameron and Baragar, 1971
Copper Canyon, U.S.A.	Cu	Nash and Theodore, 1971
		Theodore and Nash, 1973
		Theodore et al., 1973
		Theodore and Blake, 1975
Copper Mountain, Canada	Cu	Gunton and Nichol, 1975
Cornwall, U.K.	Sn, Cu	Bradshaw, 1967
		Edwards, 1976
		Ahmad, 1977
		Wilson and Jackson, 1977

Source of reference for geochemical data (*continued*)

Deposit or mineralized district, and country	Ore minerals	Reference
Cornwall, U.K.	U	Simpson et al., 1977
Cortez, U.S.A.	Au	Erickson et al., 1964a
		Erickson et al., 1966
		Wells et al., 1969
Costerfield, Australia	Sb, Au, As	Hill, 1980
Crater Creek, U.S.A.	Cu, Au, Pb, Ag, Zn	Watterson et al., 1977
Cripple Creek, U.S.A.	Au, Ag	Watterson et al., 1977
Cruse (Marysville stock), U.S.A.	Au, Ag	Mantei and Brownlow, 1967
		Mantei et al., 1970
		Tilling et al., 1973
Cuyon, Puerto Rico	Cu	Kesler et al., 1975b
Cyprus	massive sulphides, general	Govett and Pantazis, 1971
		Govett, 1972
		Constantinou and Govett, 1972
		Constantinou and Govett, 1973
		Govett, 1976b
Detour, Canada	Zn, Cu	I.G.L. Sinclair, 1977
Detroit Mining District, U.S.A.	Cu, Au, Ag	Lovering and McCarthy, 1978
Drumlummon (Marysville stock), U.S.A.	Au, Ag	Mantei and Brownlow, 1967
		Mantei et al., 1970
		Tilling et al., 1973
East Waite, Canada	Zn, Cu	Nichol et al., 1977
El Teniente, Chile	Cu	Oyarzun, 1975
		Armbrust et al., 1977
Ely, U.S.A.	Cu	Slawson and Nackowski, 1959
		Parry and Nackowski, 1963
		Parry and Jacobs, 1975
		Watterson et al., 1977
		McCarthy and Gott, 1978
El Yunque, Puerto Rico	Cu	Kesler et al., 1975b
Ergani-Maden, Turkey	Cu	Erdogan, 1977
Esmeralda, Australia	Au, Sn	Sheraton and Black, 1973
Erzgebirge, East Germany	Sn	Tischendorf, 1973
Esperanza, U.S.A.	Cu	Lovering et al., 1970
Eureka Mining District, U.S.A.	Ag	J.S. Curtis, 1884
Finland	base metals, Ni	Häkli, 1970
Fox, Canada	Zn, Cu	Turek et al., 1976
Fukasawa, Japan	Cu, Zn, Pb	Y. Ishikawa et al., 1976
Getchell, U.S.A.	Au	Erickson et al., 1964b
Gloster (Marysville stock), U.S.A.	Au, Ag	Mantei and Brownlow, 1967
		Mantei et al., 1970
		Tilling et al., 1973

Source of reference for geochemical data (*continued*)

Deposit or mineralized district, and country	Ore minerals	Reference
Gold Acres, U.S.A.	Au	Erickson et al., 1964b
Gold Hill, U.S.A.	Cu, Pb, Zn, Au, Ag	Parry and Nackowski, 1963
		Parry and Jacobs, 1975
Gortdrum, Ireland	Cu	Steed and Tyler, 1979
Granisle, Canada	Cu	Kesler et al., 1975c
Granite-Bimetallic (Philipsberg stock), U.S.A.	Ag	Mohsen and Brownlow, 1971
Guichon Batholith, Canada	Cu	Warren and Delavault, 1960
		Brabec and White, 1971
		Kesler et al., 1975c
		Olade and Fletcher, 1976a
Hackett River, Canada	Pb, Zn, (Cu)	Cameron, 1975
Hanover-Fierro, U.S.A.	Zn, Cu	Belt, 1960
Hanson Lake, Canada	Cu, Pb, Zn	Fox, 1978
Heath Steele, Canada	Zn, Pb, Cu	Whitehead, 1973a
		Whitehead, 1973b
		Whitehead and Govett, 1974
		Govett, 1976b
		Goodfellow and Wahl, 1976
		Wahl, 1978
Herbert River, Australia	Cu, Pb, W	Sheraton and Black, 1973
Herberton Tinfield, (Elizabeth Creek granite), Australia	Sn	Sheraton and Black, 1973
High Lake, Canada	Zn, Cu	Cameron, 1975
Highmont, Canada	Cu	Olade and Fletcher, 1976a
		Olade, 1979
Ingladhal, India	Cu	Mookherjee and Philip, 1979
Jay Copper Zone, Canada	Zn, Cu	Descarreaux, 1973
Kalamazoo, U.S.A.	Cu	Chaffee, 1976b
Key Anacon, Canada	Zn, Pb, Cu	Wahl, 1978
Kuroko deposits, Japan	massive sulphides, general	H. Ishikawa et al., 1962
		Iijima, 1974
		Lambert and Sato, 1974
		Shirozu, 1974
Lights Creek, U.S.A. (Moonlight Valley, Copper Valley, Sulfide Ridge, Superior Mine, Engels Mine)	Cu	Putman, 1975
Limerick, Australia	Pb, Zn	Dunlop et al., 1979
Loma de Cabrera, Dominican Republic	Cu	Kesler et al., 1975b

Source of reference for geochemical data (*continued*)

Deposit or mineralized district, and country	Ore minerals	Reference
Lordsburg, U.S.A.	Cu, Zn	Belt, 1960
Lornex, Canada	Cu	Olade and Fletcher, 1976a
Los Bronces, Chile	Cu	Armbrust et al., 1977
Lou Lake, Canada	Co, As	Garrett, 1975
Louvem, Canada	Cu	Spitz and Darling, 1978
Magdalena (Granite Mountain), U.S.A.	Zn, Cu	Belt, 1960
Mathiati, Cyprus	Cu	Govett and Pantazis, 1971
		Govett, 1972
		Pantazis and Govett, 1973
		Govett and Goodfellow, 1975
Mattabi, Canada	Zn, Cu, Pb	Franklin et al., 1975
		Nichol et al., 1977
McArthur River, Australia	Zn, Pb, (Cu)	Lambert and Scott, 1973
Meggan, West Germany	Zn, Ba	Gwosdz and Krebs, 1977
Millenbach, Canada	Zn, Cu	Riverin and Hodgson, 1980
Minas de Oro, Honduras	Cu	Kesler et al., 1975b
Mineral Butte, U.S.A.	Cu	Chaffee, 1976a
Mineral Park, U.S.A.	Cu	Davis and Guilbert, 1973
Montezuma, U.S.A.	Ag, Pb, Zn	Watterson et al., 1977
Morenci, U.S.A.	Cu	Davis and Guilbert, 1973
Mt. Allen, Canada	W	Garrett, 1971a
		Garrett, 1971b
		Garrett, 1973
		Garrett, 1974
Mt. Carbine (Mareeba granite), Australia	W, Sn, Cu	Sheraton and Black, 1973
Mt. Lyell, Australia	Cu	Sheppard, 1981
Mt. Morgan, Australia	Cu, Au	Fedikow, 1982
Mt. Pleasant, Canada	W, Mo, Bi, Sn	Dagger, 1972
Mykonos, Greece	Ag, base metal	Lahti and Govett, 1981
Nantymwyn, U.K.	Pb, Zn	Al-Atia and Barnes, 1975
Nenthead, U.K.	Pb, Zn	Finlayson, 1910a
New Brunswick, Canada	massive sulphides, general	Pwa, 1978
		Govett and Pwa, 1981
New England district, Australia	Sn	Flinter et al., 1972
		Juniper and Kleeman, 1979
Nojal Peak (Rialto stock), U.S.A.	Mo, Au, Pb, Zn, Cu	Griswold and Missaghi, 1964
Noranda, Canada	Zn, Cu	Sopuck et al., 1980
Norbec, Canada	Cu, Zn	Pirie and Nichol, 1980
Normetal, Canada	Zn, Cu	Sopuck et al., 1980
Nova Scotia, Canada	Sn	Smith and Turek, 1976
Ogofau, U.K.	Au	Al-Atia and Barnes, 1975
		Steed et al., 1976

Source of reference for geochemical data (*continued*)

Deposit or mineralized district, and country	Ore minerals	Reference
Park City, U.S.A.	Pb, Zn, Cu, Ag	Bailey and McCormick, 1974
Petaquilla, Panama	Cu	Kesler et al., 1975b
Potato Hills, Canada	Au (alluvial)	Garrett, 1974
Providencia, Mexico	Cu	Stollery et al., 1971
Rio Blanco, Chile	Cu	Oyarzun, 1975
		Armbrust et al., 1977
Rio Pito, Panama	Cu	Kesler et al., 1975b
Rio Vivi, Puerto Rico	Cu	Kesler et al., 1975b
Sam Goosly, Canada	Cu	Church et al., 1976
Sanctuary Lake, Canada	Pb, Zn, (Cu)	Cameron, 1975
San Francisco, Honduras	Cu	Kesler et al., 1975b
San Francisco, U.S.A.	Cu, Pb, Zn, Au, Ag	Parry and Nackowski, 1963
		Parry and Jacobs, 1975
Santa Rita, U.S.A.	Cu	Lovering et al., 1970
		Drewes, 1973
		Davis and Guilbert, 1973
		Banks, 1974
		Jacobs and Parry, 1976
Sardinia, Italy	Cu, Pb	Hall, 1975
Scheelite Dome, Canada	Au (alluvial)	Garrett, 1974
Searchlight Mining District, U.S.A.	Au, Ag	Bolter and Al-Shaieb, 1971
Setting Net Lake, Canada	Mo, Cu	Wolfe, 1974
Sierrita, U.S.A.	Cu	Lovering et al., 1970
Skouriotissa, Cyprus	Cu	Govett and Pantazis, 1971
		Govett, 1972
		Constantinou and Govett, 1973
South Bay Mines Canada	Zn, Cu	Davenport and Nichol, 1973
		Nichol et al., 1977
		Sopuck et al., 1980
St. Louis (Marysville stock), U.S.A.	Au, Ag	Mantei and Brownlow, 1967
		Mantei et al., 1970
		Tilling et al., 1973
St. Martin, St. Martin	Cu	Kesler et al., 1975b
Sturgeon Lake, Canada	Zn, Cu	Nichol et al., 1977
		Sopuck et al., 1980
Taylor Mining District, U.S.A.	Ag, (Sb, Au, Cu, Pb, Zn)	Lovering and Heyl, 1974
Terra Mine, Canada	Ag, Cu	Garrett, 1975
Terre Neuve, Haiti	Cu	Kesler et al., 1975b
Tintic, U.S.A. (general)	Cu, Pb, Zn, Ag, Au	Slawson and Nackowski, 1959
		Parry and Nackowski, 1963
		Parry and Jacobs, 1975
Tintic (Swansea vein), U.S.A.	Pb, Zn	Morris and Lovering, 1952

Source of reference for geochemical data (*continued*)

Deposit or mineralized district, and country	Ore minerals	Reference
Tintic (Carisa), U.S.A.	Cu, Au	Morris and Lovering, 1952
Tintic (Eureka Hill), U.S.A.	Cu, Pb	Morris and Lovering, 1952
Tommie Lake, Canada	Cu	Garrett, 1975
Tres Hermanas, U.S.A.	Cu, Pb, Zn	Doraibabu and Proctor, 1973
True Fissure (Philipsberg stock), U.S.A.	Mn	Mohsen and Brownlow, 1971
Tynagh, Ireland	Pb, Zn, Cu, Ag	Russell, 1974 Russell, 1975
U.K., general	U	Simpson et al., 1977 Plant et al., 1980
Valley Copper, Canada	Cu	Olade and Fletcher, 1975 Olade and Fletcher, 1976a Olade and Fletcher, 1976b Olade, 1977 Olade, 1979
Virgin Islands	Cu	Kesler et al., 1975b
Vuda Valley, Fiji	Au	Govett et al., 1980
Wainaleka, Fiji	Zn, Cu	Rugless, 1981
West Ridge, Canada	Au (alluvial)	Garrett, 1974
Wisconsin lead-zinc district, U.S.A.	Pb, Zn	Lavery and Barnes, 1971 Barnes and Lavery, 1977
Woodlawn, Australia	Zn, Pb, Cu	Petersen and Lambert, 1979

REFERENCES

Ahmad, S.N., 1977. The geochemical distribution and source of copper in the metalliferous mining region of Southwest England. *Miner. Deposita*, 12: 1—21.

Ahrens, L.H., 1952. The use of the ionization potentials. I. Ionic radii of the elements. *Geochim. Cosmochim. Acta*, 2: 155—169.

Ahrens, L.H., 1954. The lognormal distribution of the elements. *Geochim. Cosmochim. Acta*, 5: 49—73.

Ahrens, L.H., 1964. The significance of the chemical bond for controlling the geochemical distribution of the elements, 1. *Phys. Chem. Earth*, 5: 1—54.

Al-Atia, M.J. and Barnes, J.W., 1975. Rubidium: a primary dispersion pathfinder at Ogofau Gold Mine, southern Wales. In: I.L. Elliott and W.K. Fletcher (Editors), *Geochemical Exploration 1974*. Elsevier, Amsterdam, pp. 342—352.

Allan, R. J., Cameron, E.M. and Durham, C.C., 1973. Lake geochemistry — a low sample density technique for reconnaissance geochemical exploration and mapping of the Canadian Shield. In: M.J. Jones (Editor), *Geochemical Exploration 1972*. Institution of Mining and Metallurgy, London, pp. 131—160.

Allen, P.M., Cooper, D.C., Fuge, R. and Rea, W.J., 1976. Geochemistry and relationships to mineralization of some igneous rocks from the Harlech Dome, Wales. *Inst. Min. Metall., Trans., Sect. B*, 85: 100—108.

Armbrust, G.A., Oyarzun, J. and Arias, J., 1977. Rubidium as a guide to ore in Chilean porphyry copper deposits. *Econ. Geol.*, 72: 1086—1100.

Armour-Brown, A. and Nichol, I., 1970. Regional geochemical reconnaissance and the location of metallogenic provinces. *Econ. Geol.*, 65: 312—330.

Aubrey, K.V., 1956. Frequency distributions of elements in igneous rocks. *Geochim. Cosmochim. Acta*, 8: 83—89.

Ayres, D.E., 1979. The mineralogy and chemical composition of the Woodlawn massive sulphide orebody. *J. Geol. Soc. Aust.*, 26: 155—168.

Bailey, G.B. and McCormick, G.R., 1974. Chemical halos as guides to lode deposit ore in the Park City District, Utah. *Econ. Geol.*, 69: 377—382.

Banks, N.G., 1974. Distribution of copper in biotite and biotite alteration products in intrusive rocks near two Arizona porphyry copper deposits. *J. Res. U.S. Geol. Surv.* 2: 195—211.

Barnes, H.L. and Czamanske, G.K., 1967. Solubilities and transport of ore minerals. In: H.L. Barnes (Editor), *Geochemistry of Hydrothermal Ore Deposits*. Holt, Rinehart and Winston, New York, N.Y., pp. 334—381.

Barnes, H.L. and Lavery, N.G., 1977. Use of primary dispersion for exploration of Mississippi Valley-type deposits. *J. Geochem. Explor.*, 8: 105—115.

Barsukov, V.L., 1957. The geochemistry of tin. *Geochemistry*, 1: 41—52.

406

Barsukov, V.L., 1967. Metallogenic specialization of granitoid intrusions. In: A.P. Vinogradov (Editor), *Chemistry of the Earth's Crust, Vol. II.* Israel Program for Scientific Translations, Jerusalem, pp. 211—231.

Barsukov, V.L., 1969. Genetic relationship between sulfide-cassiterite deposits and intrusions. In: N.I. Khitarov (Editor), *Problems of Geochemistry.* Israel Program for Scientific Translations, Jerusalem, pp. 225—233.

Barth, T.F.W., 1948. Oxygen in rocks; a basis for petrographic calculations. *J. Geol.*, 56: 50—61.

Barth, T.F.W., 1961. Abundance of the elements, areal averages and geochemical cycles. *Geochim. Cosmochim. Acta*, 23: 1—8.

Belt, C.B., 1960. Intrusion and ore deposition in New Mexico. *Econ. Geol.*, 55: 1244—1271.

Bennett, R.A. and Rose, W.I., 1973. Some compositional changes in Archean felsic volcanic rocks related to massive sulfide mineralization. *Econ. Geol.*, 68: 886—891.

Berman, B.I., and Agentov, V.B., 1965. Geochemical relations between pyritic-polymetallic mineralization and Lower Cambrian volcanism in Eastern Tuva. *Geochemistry*, 2: 213—222.

Berry, L.G. and Mason, B., 1959. *Mineralogy.* W.H. Freeman and Co., San Francisco, Calif., 630 pp.

Beus, A.A., 1969. Geochemical criteria for assessment of the mineral potential of the igneous rock series during reconnaissance exploration. In: F.C. Canney (Editor), *Proceedings, International Geochemical Exploration Symposium, Golden, Colorado, 1968. Q. Colo. Sch. Mines*, 64: 67—74.

Beus, A.A., 1976. *Geochemistry of the Lithosphere.* Mir Publishers, Moscow, 366 pp.

Beus, A.A. and Grigoryian, S.V., 1977. *Geochemical Exploration Methods for Mineral Deposits.* Applied Publishing Company, Wilmette, Ill., 287 pp.

Bignell, R.D., Cronan, D.S. and Tooms, J.S., 1976. Metal dispersion in the Red Sea as an aid to marine geochemical exploration. *Inst. Min. Metall., Trans., Sect. B*, 85: 274—278.

Bilibin, Yu.A., 1960. *Metallogenic Provinces and Metallogenic Epochs.* Gosgeoltekhizdat, Moscow (English transl. Secretary of State, Canada, 122 pp.).

Blaxland, A.B., 1971. Occurrence of zinc in granitic biotites. *Miner. Deposita*, 6: 313—320.

Bolotnikov, A.F. and Kravchenko, N.S., 1970. Criteria for recognition of tin-bearing granites. *Dokl. Akad. Nauk S.S.S.R.*, 191: 186—187.

Bolter, E. and Al-Shaieb, Z., 1971. Trace element anomalies in igneous wall rocks of hydrothermal veins. In: R.W. Boyle and J.I. McGerrigle (Editors), *Geochemical Exploration. Can. Inst. Min. Metall., Spec. Vol.*, 11: 289—290.

Bölviken, B., 1971. A statistical approach to the problem of interpretation in geochemical prospecting. In: R.W. Boyle and J.I. McGerrigle (Editors), *Geochemical Exploration. Can. Inst. Min. Metall., Spec. Vol.*, 11: 564—567.

Boström, K. and Peterson, M.N.A., 1966. Precipitates from hydrothermal exhalations on the East Pacific Rise. *Econ. Geol.*, 61: 1258—1265.

Botbol, J.M., Sinding-Larsen, R., McCammon, R.B. and Gott, G.B., 1977. Characteristic analysis of geochemical exploration data. *U.S. Geol. Surv. Open File Rep.*, 77—349: 55 pp.

Brabec, D. and White, W.H., 1971. Distribution of copper and zinc in rocks of the Guichon Creek Batholith, British Columbia. In: R.W. Boyle and J.I. McGerrigle (Editors), *Geochemical Exploration. Can. Inst. Min. Metall., Spec. Vol.*, 11: 291—297.

Bradshaw, P.M.D., 1967. Distribution of selected elements in feldspar, biotite and muscovite from British granites in relation to mineralization. *Inst. Min. Metall., Trans. Sect. B*, 76: 137—148.

Bradshaw, P.M.D., Clews, D.R. and Walker, J.L., 1972. *Exploration Geochemistry: A*

Series of Seven Articles Reprinted from Mining in Canada and Canadian Mining Journal. Barringer Research Ltd., Toronto, Ont., 49 pp.

Brownlow, A.H., 1979. *Geochemistry.* Prentice-Hall, Englewood Cliffs, N.J., 498 pp.

Buchanan, D.L., 1976. Identification of geological environments in basic and ultrabasic igneous rocks that favour the formation of sulphide segregations. *Inst. Min. Metall., Trans., Sect. B*, 85: 289—291.

Burns, R.G., 1973. The partitioning of trace transition elements in crystal structures: a provocative review with applications to mantle geochemistry. *Geochim. Cosmochim. Acta*, 37: 2395—2403.

Burns, R.G. and Fyfe, W.S., 1964. Site of preference energy and selective uptake of transition-metal ions from a magma. *Science*, 144: 1001—1003.

Burns, R.G. and Fyfe, W.S., 1967. Crystal-field theory and the geochemistry of transition elements. In: P.H. Abelson (Editor), *Researches in Geochemistry, Vol. 2.* John Wiley, New York, N.Y., pp. 259—285.

Butler, J.R., 1953. The geochemistry and mineralogy of rock weathering, 1. The Lizard area, Cornwall. *Geochim. Cosmochim. Acta*, 4: 157—178.

Butt, C.R.M. and Smith, R.E. (Compilers and Editors), 1980. Conceptual models in exploration geochemistry — Australia. *J. Geochem. Explor.*, 12: 365 pp.

Cameron, E.M., 1975. Geochemical methods of exploration for massive sulphide mineralization in the Canadian Shield. In: I.L. Elliott and W.K. Fletcher (Editors), *Geochemical Exploration 1974.* Elsevier, Amsterdam, pp. 21—49.

Cameron, E.M. and Baragar, W.R.A., 1971. Distribution of ore elements in rocks for evaluating ore potential: frequency distribution of copper in Coppermine River Group and Yellowknife Group volcanic rocks, N.W.T., Canada. In: R.W. Boyle and J.I. McGerrigle (Editors), *Geochemical Exploration. Can. Inst. Min. Metall., Spec. Vol.*, 11: 570—576.

Cameron, E.M., Siddeley, G., and Durham, C.C., 1971. Distribution of ore elements in rocks for evaluating ore potential: nickel, copper, cobalt, and sulphur in ultramafic rocks of the Canadian Shield. In: R.W. Boyle and J.I. McGerrigle (Editors), *Geochemical Exploration. Can. Inst. Min. Metall., Spec. Vol.*, 11: 298—313.

Carmichael, I.S.E., Turner, F.J. and Verhoogen, J., 1974. *Igneous Petrology.* McGraw Hill, New York, N.Y., 739 pp.

Chaffee, M.A., 1976a. Geochemical exploration techniques based on distribution of selected elements in rocks, soils, and plants, Mineral Butte Copper deposit, Pinal County, Arizona. *U.S. Geol. Surv. Bull.*, 1278-D: 55 pp.

Chaffee, M.A., 1976b. The zonal distribution of selected elements above the Kalamazoo porphyry copper deposit, San Manuel District, Pinal County, Arizona. *J. Geochem. Explor.*, 5: 145—165.

Chapman, R.P., 1976. Some consequences of applying lognormal theory to pseudo lognormal distributions. *J. Int. Assoc. Math. Geol.*, 8: 209—214.

Chayes, F., 1954. The lognormal distribution of elements: a discussion. *Geochim. Cosmochim. Acta*, 6: 119—120.

Chork, C.Y. and Govett, G.J.S., 1979. Interpretation of geochemical soil surveys by block averaging. *J. Geochem. Explor.*, 11: 53—71.

Church, B.N., Barakso, J.J. and Bowman, A.F., 1976. The endogenous distribution of minor elements in the Goosly-Owen Lake area of Central British Columbia. *Can. Inst. Min. Metall. Bull.*, 69(773): 88—95.

Clarke, F.W., 1889. The relative abundance of the chemical elements. *Philos. Soc. Washington Bull.*, 11: 131—142.

Clarke, F.W., 1916. *The Data of Geochemistry* (3rd edition). *U.S. Geol. Surv. Bull.*, 616: 821 pp.

Clarke, F.W., 1924. *The Data of Geochemistry* (5th edition). *U.S. Geol. Surv. Bull.*, 770: 841 pp.

Clarke, F.W. and Washington, H.S., 1924. The composition of the earth's crust. *U.S. Geol. Surv. Prof. Paper*, 127: 117 pp.

Clarke, F.W., Harker, A. and Washington, H.S., 1915. Analyses of rocks and minerals from the laboratory of the United States Geological Survey, 1880 to 1914. *U.S. Geol. Surv. Bull.*, 591: 376 pp.

Colley, H., 1976. Classification and exploration guide for Kuroko-type deposits based on occurrences in Fiji. *Inst. Min. Metall., Trans., Sect. B*, 85: 190—199.

Constantinou, G. and Govett, G.J.S., 1972. Genesis of sulphide deposits, ochre and umber of Cyprus. *Inst. Min. Metall., Trans., Sect B*, 81: 34—46.

Constantinou, G. and Govett, G.J.S., 1973. Geology, geochemistry and genesis of Cyprus sulphide deposits. *Econ. Geol.*, 68: 843—858.

Coope, J.A., 1977. Potential of lithogeochemistry in mineral exploration. Paper presented at AIME Annual General Meeting, 7 March 1977, Atlanta, Ga.

Coope, J.A. and Davidson, M.J., 1979. Some aspects of integrated exploration. In: P.J. Hood (Editor), *Geophysics and Geochemistry in the Search for Metallic Ores. Geol. Surv. Can., Econ. Geol. Rep.*, 31: 575—592.

Curtis, C.D., 1964. Application of the crystal-field theory to the inclusion of trace transition elements in minerals during magmatic differentiation. *Geochim. Cosmochim. Acta*, 28: 389—403.

Curtis, J.S., 1884. *Silver-Lead Deposits of Eureka, Nevada. U.S. Geol. Surv. Monogr.*, 7: 200 pp.

Cuturic, N., Kafol, N. and Karamata, S., 1968. Lead contents in K-feldspars of young igneous rocks of the Dinarides and neighbouring areas. In: L.H. Ahrens (Editor), *Origin and Distribution of the Elements*. Pergamon Press, Oxford, pp. 739—747.

Dagger, G.W., 1972. Genesis of the Mount Pleasant tungsten-molybdenum-bismuth deposit, New Brunswick, Canada. *Inst. Min. Metall., Trans., Sect. B*, 81: 73—102.

Daly, R.A., 1910. Average chemical composition of igneous rock types. *Am. Acad. Art Sci. Proc.*, 45: 211—240.

Daly, R.A., 1914. *Igneous Rocks and Their Origin*. McGraw-Hill, New York, N.Y., 563 pp.

Daly, R.A., 1933. *Igneous Rocks and the Depths of the Earth*. McGraw-Hill, New York, N.Y., 598 pp.

Dass, A.S., Boyle, R.W. and Tupper, W.M., 1973. Endogenic haloes of the native silver deposits, Cobalt, Ontario, Canada. In: M.J. Jones (Editor), *Geochemical Exploration 1972*. Institution of Mining and Metallurgy, London, pp. 25—35.

Davenport, P.H. and Nichol, I., 1973. Bedrock geochemistry as a guide to areas of base-metal potential in volcano-sedimentary belts of the Canadian Shield. In: M.J. Jones (Editor), *Geochemical Exploration 1972*. Institution of Mining and Metallurgy, London, pp. 45—57.

Davenport, P.H., Hornbrook, E.H.W. and Butler, A.J., 1975. Regional lake sediment geochemical survey for zinc mineralization in western Newfoundland. In: I.L. Elliott and W.K. Fletcher (Editors), *Geochemical Exploration 1974*. Elsevier, Amsterdam, pp. 555—578.

Davis, J.D. and Guilbert, J.M., 1973. Distribution of the radioelements potassium, uranium, and thorium in selected porphyry copper deposits. *Econ. Geol.*, 68: 145—160.

De Geoffroy, J. and Wignall, T.K., 1972. A statistical study of geological characteristics of porphyry-copper-molybdenum deposits in the Cordilleran Belt — application to the rating of porphyry prospects. *Econ. Geol.*, 67: 656—668.

Deer, W.A., Howie, R.A. and Zussman, J., 1962—1963. *Rock Forming Minerals, Vols. 1—5*. Longmans Green, London, Vol. 1: 333 pp.; Vol. 2: 379 pp.; Vol. 3: 270 pp.; Vol. 4: 435 pp.; Vol. 5: 371 pp.

De Launay, L., 1913. *Traité de Metallogenie, Vols. 1—3*. Libr. Polytechn. C. Beranger, Paris and Liege.

Descarreaux, J., 1973. A petrochemical study of the Abitibi volcanic belt and its bearing on the occurrences of massive sulphide ores. *Can. Inst. Min. Metall. Bull.*, 66(730): 61—69.

Divis, A.F. and Clark, J.R., 1979. Exploration for blind ore deposits and geothermal reservoirs by lithium isotope thermomtry-atomic absorption "mass spectrometry". In: J.R. Watterson and P.K. Theobald (Editors). *Geochemical Exploration 1978.* Association of Exploration Geochemists, Toronto, Ont., pp. 233—241.

Doraibabu, P. and Proctor, P.D., 1973. Trace base metals, petrography, and alteration, Tres Hermanas Stock, Luna County, New Mexico. *N.M., Bur. Mines Miner. Resour., Circ.*, 132: 29 pp.

Drewes, H.D., 1973. Geochemical reconnaissance of the Santa Rita Mountains, south east of Tucson, Arizona. *U.S. Geol. Surv. Bull.*, 1365: 67 pp.

Dumitriu, C., Webber, R. and David, M., 1979. Correspondance analysis applied to a comparison of some rhyolitic zones in the Noranda area (Quebec, Canada). *Math. Geol.*, 11: 299—307.

Dunham, K.C., 1973. Geological controls of metallogenic provinces. In: N.H. Fisher (Editor), Metallogenic Provinces and Mineral Deposits in the Southwest Pacific. *Bull. Aust. Bur. Miner. Resour., Geol. Geophys.*, 114: 1—12.

Dunlop, A.C., Ambler, E.P. and Avila, E.T., 1979. Surface lithogeochemical studies about a distal volcanogenic massive sulphide occurrence at Limerick, New South Wales. *J. Geochem. Explor.*, 11: 285—297.

Durovic, S., 1959. Contribution to the lognormal distribution of elements. *Geochim. Cosmochim. Acta*, 15: 330—336.

Edwards, R.P., 1976. Aspects of trace metal and ore distribution in Cornwall. *Inst. Min. Metall., Trans., Sect. B*, 85: 83—90.

Erdogan, B., 1977. *Geology, Geochemistry, and Genesis of the Sulphide Deposits of the Ergani-Maden Region, Southeast Turkey.* Ph.D. Thesis. University of New Brunswick, Fredericton, N.B. (unpublished).

Erickson, R.L., Masursky, H., Marranzino, A.P., Oda, U. and James, W.W., 1964a. Geochemical anomalies in the lower plate of the Roberts thrust, near Cortez, Nevada. *U.S. Geol. Surv. Prof. Paper*, 501-B: 92—94.

Erickson, R.L., Marranzino, A.P., Oda, U. and James, W.W., 1964b. Geochemical exploration near the Getchell Mine, Humboldt County, Nevada. *U.S. Geol. Surv. Bull.*, 1198—A: 26 pp.

Erickson, R.L., Van Sickle, G.H., Naicagawa, H.M., McCarthy, J.H. and Leong, K.W., 1966. Gold geochemical anomaly in the Cortez district, Nevada. *U.S. Geol. Surv. Circ.*, 534: 2 pp. plus tables and maps.

Exley, C.S., 1958. Magmatic differentiation and alteration in the St. Austell granite. *Q. J. Geol. Soc. London*, 114: 197—230.

Fairbairn, H.W., Schlecht, W.G., Stevens, R.E., Dennen, W.H., Ahrens, L.H. and Chayes, F., 1951. A cooperative investigation of precision and accuracy in chemical, spectrochemical and modal analysis of silicate rocks. *U.S. Geol. Surv. Bull.*, 980: 71 pp.

Fedikow, M., 1978. *Geochemical and Palaeomagnetic Studies at the Sullivan Mine, Kimberly, B.C.* M.Sc. Thesis, University of Windsor, Windsor, Ont. (unpublished).

Fedikow, M., 1982. *Rock Geochemical Exploration at Mount Morgan, Queensland, Australia.* Ph.D. Thesis, University of New South Wales, Kensington, N.S.W. (unpublished).

Finlayson, A.M., 1910a. The metallogeny of the British Isles. *Q. J. Geol. Soc. London*, 66(262): 281—298.

Finlayson, A.M., 1910b. Problems of ore deposition in the lead and zinc veins of Great Britain. *Q. J. Geol. Soc. London*, 66: 299—328.

410

Flanagan, F.J. (Editor), 1976. Descriptions and analyses of eight new U.S.G.S. rock standards. *U.S. Geol. Surv. Prof. Paper*, 840: 192 pp.

Fleischer, M., 1953. Recent estimates of the abundance of the elements in the earth's crust. *U.S. Geol. Surv. Circ.*, 285: 7 pp.

Fleischer, M., 1965. Summary of new data on rock samples G-1 and W-1, 1962—1965. *Geochim. Cosmochim. Acta*, 29: 1263—1283.

Fleischer, M., 1969. U.S. Geological Survey standards, I. Additional data on rocks G-1 and W-1, 1965—1967. *Geochim. Cosmochim. Acta*, 33: 65—79.

Fleischer, M. and Stevens, R.E., 1962. Summary of new data on rock samples G-1 and W-1. *Geochim. Cosmochim. Acta*, 26: 525—543.

Fletcher, W.K., 1981. *Analytical Methods in Geochemical Prospecting*. Elsevier, Amsterdam, 255 pp.

Flinter, B.H., 1971. Tin in acid granitoids: the search for a geochemical scheme of mineral exploration. In: R.W. Boyle and J.I. McGerrigle (Editors), *Geochemical Exploration*. *Can. Inst. Min. Metall. Spec. Vol.*, 11: 323—330.

Flinter, B.H., Hesp, W.R. and Rigby, D., 1972. Selected geochemical, mineralogical and petrological features of granitoids of the New England complex, Australia, and their relation to Sn, W, Mo and Cu mineralization. *Econ. Geol.*, 67: 1241—1262.

Fox, J.S., 1978. Interpretation of drill-hole geochemical data from the volcanic rocks of the Hanson Lake Mine, Saskatchewan. *Can. Inst. Min. Metall. Bull.*, 71(793): 111—116.

Fox, J.S., 1979. Host-rock geochemistry and massive volcanogenic sulphide ores. *Can. Inst. Min. Metall. Bull.*, 72(804): 127—134.

Franklin, J.M., Kasarda, J. and Poulsen, K.H., 1975. Petrology and chemistry of the alteration zone of the Mattabi massive sulfide deposit. *Econ. Geol.*, 70: 63—79.

Fuge, R. and Power, G.M., 1969. Chlorine and fluorine in granitic rocks from S.W. England. *Geochim. Cosmochim. Acta*, 33: 888—893.

Fyfe, W.S., 1964. *Geochemistry of Solids*. McGraw-Hill, New York, N.Y., 199 pp.

Gandhi, S.M., 1978. *Exploration Rock Geochemical Studies in and around the Caribou Sulphide Deposit, New Brunswick, Canada*. Ph.D. Thesis, University of New Brunswick, Fredericton, N.B. (unpublished).

Garrett, R.G., 1971a. Molybdenum, tungsten and uranium in acid plutonic rocks as a guide to regional exploration, S.E. Yukon. *Can. Min. J.*, 92: 37—40.

Garrett, R.G., 1971b. Molybdenum and tungsten in some acid plutonic rocks of southeast Yukon Territory. *Geol. Surv. Can. Open File Rep.*, 51: 5 pp.

Garrett, R.G., 1973. Regional geochemical study of Cretaceous acidic rocks in the Northern Canadian Cordillera as a tool for broad mineral exploration. In: M.J. Jones (Editor), *Geochemical Exploration 1972*. Institution of Mining and Metallurgy, London, pp. 203—219.

Garrett, R.G., 1974. Mercury in some granitoid rocks of the Yukon and its relation to gold-tungsten mineralization. *J. Geochem. Explor.*, 3: 277—289.

Garrett, R.G., 1975. Copper in Proterozoic acid volcanics as a guide to exploration in the Bear Province. In: I.L. Elliott and W.K. Fletcher (Editors), *Geochemical Exploration 1974*. Elsevier, Amsterdam, pp. 371—388.

Garrett, R.G., 1979. Sampling considerations for regional geochemical surveys. In: *Current Research, Part A. Geol. Surv. Can., Paper*, 79-1A: 197—205.

Garrett, R.G., 1982. Sampling methodology. In: R.J. Howarth (Editor), *Statistics and Data Analysis in Geochemical Exploration*. Elsevier, Amsterdam, pp. 83—110.

Garrett, R.G. and Nichol, I., 1967. Regional geochemical reconnaissance in eastern Sierra Leone. *Inst. Min. Metall., Trans., Sect. B*, 76: 97—112.

Goldschmidt, V.M., 1933. Grundlagen der quantitativen Geochemie. *Fortschr. Mineral., Kristallogr. Petrogr.*, 17: 112—156.

Goldschmidt, V.M., 1937. The principles of distribution of chemical elements in minerals and rocks. *J. Chem. Soc. London*, March, pp. 655—673.

Goldschmidt, V.M., 1954. *Geochemistry.* Clarendon Press, Oxford, 730 pp.

Goodfellow, W.D., 1975a. Major and minor element halos in volcanic rocks at Brunswick No. 12 sulphide deposit, N.B., Canada. In: I.L. Elliott and W.K. Fletcher (Editors), *Geochemical Exploration 1974.* Elsevier, Amsterdam, pp. 279—295.

Goodfellow, W.D., 1975b. *Rock Geochemical Exploration and Ore Genesis at Brunswick No. 12 massive sulphide deposit, N.B.* Ph.D. Thesis, University of New Brunswick, Fredericton, N.B. (unpublished).

Goodfellow, W.D. and Wahl, J.L., 1976. Water extracts of volcanic rocks — detection of anomalous halos at Brunswick No. 12 and Heath Steele B-zone massive sulphide deposits. *J. Geochem. Explor.*, 6: 35—59.

Gordy, W. and Thomas, W.J.O., 1956. Electronegativities of the elements. *J. Chem. Phys.*, 24: 439—444.

Gott, G.B. and Botbol, J.M., 1973. Zoning of major and minor metals in the Coeur d'Alene mining district, Idaho, U.S.A. In: M.J. Jones (Editor), *Geochemical Exploration 1972.* Institution of Mining and Metallurgy. London, pp. 1—12.

Govett, G.J.S., 1972. Interpretation of a rock geochemical survey in Cyprus — statistical and graphical techniques. *J. Geochem. Explor.*, 1: 77—102.

Govett, G.J.S., 1973. Differential secondary dispersion in transported soils and post-mineralization rocks: an electrochemical interpretation. In: M.J. Jones (Editor), *Geochemical Exploration 1972.* Institution of Mining and Metallurgy, London, pp. 81—91.

Govett, G.J.S., 1974. Exploration geochemistry in New Brunswick — discussion. *Can. Inst. Min. Metall. Bull.*, 67(743): 177—178.

Govett, G.J.S., 1975. Soil conductivities: assessment of an electrogeochemical technique. In: I.L. Elliott and W.K. Fletcher (Editors), *Geochemical Exploration 1974.* Elsevier, Amsterdam, pp. 101—118.

Govett, G.J.S., 1976a. Detection of deeply-buried and blind sulphide deposits by measurement of H[+] and conductivity of closely-spaced surface samples. *J. Geochem. Explor.*, 6: 359—382.

Govett, G.J.S., 1976b. The development of geochemical exploration methods and techniques. In: G.J.S. Govett and M.H. Govett (Editors), *World Mineral Supplies — Assessment and Perspective.* Elsevier, Amsterdam, pp. 343—376.

Govett, G.J.S., 1977. Presidential address, Annual General Meeting, Association of Exploration Geochemists, Vancouver, April, 1977. *J. Geochem. Explor.*, 8: 591—599.

Govett, G.J.S., 1978. Lithogeochemistry; letter to the editor. *J. Geochem. Explor.*, 9: 109—110.

Govett, G.J.S. and Chork, C.Y., 1977. Detection of deeply buried sulphide deposits by measurement of organic carbon, hydrogen ion, and conductance in surface soils. In: *Prospecting in Areas of Glaciated Terrain, Helsinki.* Institution of Mining and Metallurgy, London, pp. 49—55.

Govett, G.J.S. and Goodfellow, W.D., 1975. The use of rock geochemistry in detecting blind sulphide deposits — a discussion. *Inst. Min. Metall., Trans., Sect. B*, 84: 134—140.

Govett, G.J.S. and Hale, W.E., 1967. Geochemical orientation and exploration near a disseminated copper deposit, Luzon, Philippines. *Inst. Min. Metall., Trans., Sect. B*, 76: 190—201.

Govett, G.J.S. and Nichol, I., 1979. Lithogeochemistry in mineral exploration. In: P.J. Hood (Editor), *Geophysics and Geochemistry in the Search for Metallic Ores. Geol. Surv. Can., Econ. Geol. Rep.*, 31: 339—362.

Govett, G.J.S. and Pantazis, Th.M., 1971. Distribution of Cu, Zn, Ni, and Co in the Troodos Pillow Lava Series, Cyprus. *Inst. Min. Metall., Trans., Sect. B*, 80: 27—46.

Govett, G.J.S. and Pwa, A., 1981. Regional reconnaissance exploration rock geochemistry for massive sulphides, New Brunswick, Canada. In: A.W. Rose and H. Gundlach (Editors), *Geochemical Exploration 1980.* Elsevier, Amsterdam, pp. 139—158.

Govett, G.J.S., Goodfellow, W.D. Chapman, R.P. and Chork, C.Y., 1975. Exploration geochemistry — distribution of elements and the recognition of anomalies. *Math. Geol.*, 7: 415—446.

Govett, G.J.S., Lawrence, L.J. and Macdonald, E.H., 1980. *Summary Report of Status of the Vuda Prospect (Fiji) to October 1980*. Atherton Antimony NL (Sydney), Sydney, N.S.W., 10 pp.

Gregory, J.W., 1922. Ore deposits and their genesis in relation to geographical distribution. *J. Chem. Soc.*, 121/122(705): 750—772.

Grigoryan, S.V., 1974. Primary geochemical halos in prospecting and exploration of hydrothermal deposits. *Int. Geol. Rev.*, 16(1): 12—25.

Griswold, G.B. and Missaghi, F., 1964. Geology and geochemical survey of a molybdenum deposit near Nogal Peak, Lincoln County, New Mexico. *N. M. Inst. Min. Technol., Circ.*, 67: 24 pp.

Grout, F.F., 1938. Petrographic and chemical data on the Canadian Shield. *J. Geol.*, 46: 486—504.

Groves, D.I., 1972. The geochemical evolution of tin-bearing granites in the Blue Tier Batholith, Tasmania. *Econ. Geol.*, 67: 445—457.

Guilbert, J.M. and Lowell, J.D., 1974. Variations in zoning patterns in porphyry ore deposits. *Can. Inst. Min. Metall. Bull.*, 68(742): 99—109.

Guild, P.W., 1974. Distribution of metallogenic provinces in relation to major earth features. *Schriftenr. Erdwiss. Komm. Oesterr. Akad. Wiss.*, 1: 10—24.

Gummer, P.K., Guttenberg, R.V., Hattie, D., Pollock, D. and Thalenhorst, H., 1980. The Captain North extension lead-zinc-silver zone, Bathurst Camp, New Brunswick, Canada. Paper presented at 8th International Geochemical Exploration Symposium, Hanover, April, 1980.

Gunton, J.E. and Nichol, I., 1975. Chemical zoning associated with the Ingerbelle-Copper Mountain mineralization, Princeton, British Columbia. In: I.L. Elliott and W.K. Fletcher (Editors), *Geochemical Exploration 1974*. Elsevier, Amsterdam, pp. 297—312.

Gwosdz, W. and Krebs, W., 1977. Manganese halo surrounding Meggan ore deposit, Germany. *Inst. Min. Metall., Trans., Sect. B*, 86: 73—77.

Häkli, T.A., 1970. Factor analysis of the sulphide phase in mafic-ultramafic rocks in Finland. *Geol. Soc. Finl. Bull.*, 42: 109—118.

Häkli, T.A. and Wright, T.L., 1967. The fractionation of nickel between olivine and augite as a geothermometer. *Geochim. Cosmochim. Acta*, 31: 877—884.

Hall, A., 1975. Regional variation in the crustal abundance of minor elements: evidence from the granites of Sardinia. *Mineral. Mag.*, 40: 293—301.

Hawkes, H.E. and Webb, J.S., 1962. *Geochemistry in Mineral Exploration*. Harper and Row, New York, N.Y., 415 pp.

Helgeson, H.C., 1964. *Complexing and Hydrothermal Ore Deposition*. Macmillan, New York, N.Y., 128 pp.

Hesp, W.R., 1971. Correlations between the tin content of granitic rocks and their chemical and mineralogical composition. In: R.W. Boyle and J.I. McGerrigle (Editors), *Geochemical Exploration. Can. Inst. Min. Metall., Spec. Vol.*, 11: 341—353.

Hesp, W.R. and Rigby, D., 1974. Some geochemical aspects of tin mineralization in the Tasman Geosyncline. *Miner. Deposita*, 9: 49—60.

Hesp, W.R. and Rigby, D., 1975. Aspects of tin metallogenesis in the Tasman Geosyncline, eastern Australia, as reflected by cluster and factor analysis. *J. Geochem. Explor.*, 4: 331—347.

Hill, M., 1980. Costerfield Au—Sb prospect, Melbourne Trough, Victoria. *J. Geochem. Explor.*, 12: 304—306.

Holland, J.D. and Lambert, R. St. J., 1972. Major element composition of the shields and the continental crust. *Geochim. Cosmochim. Acta*, 36: 673—683.

Horn, M.K. and Adams, J.A.S., 1966. Computer-derived geochemical balances and element abundances. *Geochim. Cosmochim. Acta*, 30: 279—297.

Howarth, R.J., 1982. *Statistics and Data Analysis in Geochemical Prospecting*. Elsevier, Amsterdam.

Hutchinson, R.W., 1973. Volcanogenic sulphide deposits and their metallogenic significance. *Econ. Geol.*, 68: 1223—1246.

Iijima, A., 1974. Clay and zeolitic alteration zones surrounding Kuroko deposits in the Hokuroku district, northern Akita, as submarine hydrothermal-diagenetic alteration products. In: S. Ishihara (Editor), *Geology of Kuroko Deposits. Soc. Min. Geol. Jpn., Min. Geol. Spec. Issue*, 6: 267—289.

Ingamells, C.O. and Switzer, P., 1973. A proposed sampling constant for use in geochemical analysis. *Talanta*, 20: 547—568.

Ingerson, E., 1954. Nature of the ore-forming fluids at various stages — a suggested approach. *Econ. Geol.*, 49: 727—733.

Ishikawa, H., Rokuro, K. and Sudo, T., 1962. Minor elements in some altered zones of "kuroko" (black ore) deposits in Japan. *Econ. Geol.*, 57: 785—789.

Ishikawa, Y., Sawaguchi, T., Iwaya, S. and Horiuchi, M., 1976. Delineation of prospecting targets for Kuroko deposits based on modes of volcanism of underlying dacite and alteration halos. *Min. Geol.*, 26: 105—117 (in Japanese with English abstract). Results discussed in English in K. Hashimoto, The Kuroko deposits of Japan, geology and exploration strategies, Proc. XII Convencion National Asociacion de Inginieros de Minas Metalurgistas y Geologos de Mexico, 1977.

Ito, J. and Frondel, C., 1967. Synthetic zirconium and titanium garnets. *Am. Mineral.*, 52: 773—781.

IUPAC Commission on Atomic Weights, 1972. Atomic weights of the elements, 1971. *Pure Appl. Chem.*, 30: 637.

Ivanova, G.F., 1963. The content of tin, tungsten and molybdenum in granite enclosing tin and tungsten deposits. *Geochemistry*, 5: 492—500.

Izawa, E., Yoshida, T. and Saito, R., 1978. Geochemical characteristics of hydrothermal alteration around the Fukazawa Kuroko deposit, Akita, Japan. *Min. Geol.*, 28: 325—335.

Jacobs, D.C. and Parry, W.T., 1976. A comparison of the geochemistry of biotite from some Basin and Range stocks. *Econ. Geol.*, 71: 1029—1035.

Jambor, J.L., 1979. Mineralogical evaluation of proximal-distal features in New Brunswick massive sulphide deposits. *Can. Mineral.*, 17: 649—664.

James, C.H., 1967. The use of the terms "primary" and "secondary" dispersion in geochemical prospecting. *Econ. Geol.*, 62: 997—999.

Johns, W.D. and Huang, W.H., 1967. Distribution of chlorine in terrestial rocks. *Geochim. Cosmochim. Acta*, 31: 35—49.

Jonasson, I.R. and Boyle, R.W., 1972. Geochemistry of mercury and origins of natural contamination of the environment. *Can. Inst. Min. Metall. Bull.*, 65(717): 32—39.

Juniper, D.N. and Kleeman, J.B., 1979. Geochemical characterization of some tin-bearing granites of New South Wales. *J. Geochem. Explor.*, 11: 321—333.

Kesler, S.E., Van Loon, J.C. and Moore, C.M., 1973. Evaluation of ore potential of granodioritic rocks using water-extractable chloride and fluoride. *Can. Inst. Min. Metall. Bull.*, 66(730): 56—60.

Kesler, S.E., Jones, L.M. and Walker, R.L., 1975a. Intrusive rocks associated with porphyry copper mineralization in island arc areas. *Econ. Geol.*, 70: 515—526.

Kesler, S.E., Issigonis, M.I. and Van Loon, J.C., 1975b. An evaluation of the use of halogens and water abundances in efforts to distinguish mineralized and barren intrusive rocks *J. Geochem. Explor.*, 4: 235—245.

Kesler, S.E., Issigonis, M.J., Brownlow, A.H., Damon, P.E., Moore, W.J., Northcote, K.E.

414

and Preto, V.A., 1975c. Geochemistry of biotites from mineralized and barren intrusive systems. *Econ. Geol.*, 70: 559—567.

Kleeman, A.W., 1967. Sampling error in the chemical analysis of rocks. *J. Geol. Soc. Aust.*, 14: 43—47.

Knopf, A., 1916. The composition of the average igneous rock. *J. Geol.*, 24: 620—622.

Krauskopf, K.B., 1964. The possible role of volatile metal compounds in ore genesis. *Econ. Geol.*, 59: 22—45.

Krauskopf, K.B., 1967. *Introduction to Geochemistry*. McGraw-Hill, New York, N.Y., 721 pp.

Kuroda, Y., 1961. Minor elements in a metasomatic zone related to a copper-bearing pyrite deposit. *Econ. Geol.*, 56: 847—854.

Kwong, Y.T.J. and Crocket, J.H., 1978. Background and anomalous gold in rocks of an Archaean greenstone assemblage, Kakagi Lake area, northwestern Ontario. *Econ. Geol.*, 73: 50—63.

Lahti, H.R. and Govett, G.J.S., 1981. Primary and secondary halos in weathered and oxidized rocks — an exploration study from Mykonos. *J. Geochem. Explor.*, 16: 27—40.

Lambert, I.B., 1976. The McArthur zinc-lead-silver deposit; features, metallogenesis and comparisons with some other stratiform ores. In: K.H. Wolf (Editor), *Handbook of Strata-bound and Stratiform Ore Deposits, II. Regional Studies and Specific Deposits, Vol. 6, Cu, Zn, Pb, and Ag Deposits*. Elsevier, Amsterdam, pp. 535—585.

Lambert, I.B. and Heier, K.S., 1968. Geochemical investigations of deep-seated rocks in the Australian Shield. *Lithos*, 1: 30—53.

Lambert, I.B. and Sato, T., 1974. The Kuroko and associated ore deposits of Japan: a review of their features and metallogenesis. *Econ. Geol.*, 69: 1215—1236.

Lambert, I.B. and Scott, K.M., 1973. Implications of geochemical investigations of sedimentary rocks within and around the McArthur zinc-lead-silver deposit, Northern Territory. *J. Geochem. Explor.*, 2: 307—330.

Large, R.R., 1980. The Ajax Cu—Zn—Ag prospect, Yarrol Basin, Queensland. *J. Geochem. Explor.*, 12: 327—331.

Larson, L. and Webber, G.R., 1977. Chemical and petrographic variations in rhyolitic zones in the Noranda area, Quebec. *Can. Inst. Min. Metall. Bull.*, 70(784): 80—93.

Lavery, N.G. and Barnes, H.L., 1971. Zinc dispersion in the Wisconsin zinc-lead district. *Econ. Geol.*, 66: 226—242.

Lawrence, G., 1975. The use of rubidium/strontium ratios as a guide to mineralization in the Galway granite, Ireland. In: I.L. Elliott and W.K. Fletcher (Editors), *Geochemical Exploration 1974*. Elsevier, Amsterdam, pp. 353—370.

Lee Tan and Yao Chi-lung, 1970. Abundance of chemical elements in the earth's crust and its major tectonic units. *Int. Geol. Rev.*, 12: 778—786.

Le Maitre, R.W., 1976. The chemical variability of some common igneous rocks. *J. Petrol.*, 17: 589—637.

Lepeltier, C., 1969. A simplified statistical treatment of geochemical data by graphical representation. *Econ. Geol.*, 64: 538—550.

Levashev, G.B., Strizhkova, A.A. and Golubeva, E.D., 1973. Composition of biotite as a criterion for recognition of stanniferous granitoids and the factors responsible for tin mineralization. *Dokl. Akad. Nauk S.S.S.R.*, 202: 207—210.

Levinson, A.A., 1974. *Introduction to Exploration Geochemistry*. Applied Publishing Ltd., Calgary, Alta., 612 pp.

Lovering, T.G. and Heyl, A.V., 1974. Jasperoid as a guide to mineralization in the Taylor mining district and vicinity near Ely, Nevada. *Econ. Geol.*, 69: 46—58.

Lovering, T.G. and McCarthy, J.H., 1978. Detroit mining district. *J. Geochem. Explor.*, 9: 168—174.

Lovering, T.G., Cooper, J.R., Drewes, H. and Cone, G.C., 1970. Copper in biotite from igneous rocks in southern Arizona as an ore indicator. *U.S. Geol. Surv. Prof. Paper*, 700-B: 1—8.

Lowell, J.D. and Guilbert, J.M., 1970. Lateral and vertical alteration-mineralization zoning in porphyry ore deposits. *Econ. Geol.*, 65: 373—408.

Lynch, J.J., 1971. The determination of copper, nickel and cobalt in rocks by atomic absorption spectrometry using a cold leach. In: R.W. Boyle and J.I. McGerrigle (Editors), *Geochemical Exploration, Can. Inst. Min. Metall., Spec. Vol.*, 11: 313—314.

MacGeehan, P., 1977. A model for the formation of volcanogenic massive sulphide deposits in the Matagami Mining Camp, Quebec. *Geol. Assoc. Can. Annu. Meet., Progr.*, 2: 34 (abstract).

Malone, E.J., 1979. Nature, distribution and relationships of the mineralization at Wood-lawn, New South Wales. *J. Geol. Soc. Aust.*, 26: 141—153.

Mantei, E. and Brownlow, A.H., 1967. Variation in gold content of minerals of the Marysville quartz diorite stock, Montana. *Geochim. Cosmochim. Acta*, 31: 225—236.

Mantei, E., Bolter, E. and Al Shaieb, Z., 1970. Distribution of gold, silver, copper, lead and zinc in the productive Marysville stock, Montana. *Miner. Deposita*, 5: 184—190.

Mason, D.R., 1979. Chemical variations in ferromagnesian minerals: a new exploration tool to distinguish between mineralized and barren stocks in porphyry copper provinces. In: J.R. Watterson and P.K. Theobald (Editors), *Geochemical Exploration 1978*. Association of Exploration Geochemists, Toronto, Ont., pp. 243—249.

McCarthy, J.H. and Gott, G.B., 1978. Robinson (Ely) mining district near Ely, White Pine County, Nevada, *J. Geochem. Explor.*, 9: 225—232.

Miller, K.I. and Goldberg, E.D., 1955. The normal distribution in geochemistry. *Geochim. Cosmochim. Acta*, 8: 53—62.

Mikkola, A.K. and Niini, H., 1968. Structural position of ore-bearing areas in Finland. *Bull. Geol. Soc. Finl.*, 40: 17—33.

Mohsen, L.A. and Brownlow, A.H., 1971. Abundance and distribution of manganese in the western part of the Philipsburg Batholith, Montana. *Econ. Geol.*, 66: 611—617.

Mookherjee, H. and Philip, R., 1979. Distribution of copper, cobalt and nickel in ores and host rocks, Ingladhal, Kamataka, India. *Miner. Deposita*, 14: 33—55.

Moore, J.G., 1959. The quartz diorite boundary line in the western United States. *J. Geol.*, 67: 198—210.

Morris, H.T. and Lovering, T.S., 1952. Supergene and hydrothermal dispersion of heavy metals in wall rocks near ore bodies, Tintic District, Utah. *Econ. Geol.*, 47: 685—716.

Nash, J.T. and Theodore, T.G., 1971. Ore fluids in the porphyry copper deposit at Copper Canyon, Nevada. *Econ. Geol.*, 66: 385—399.

Ney. C.S., Anderson, J.M. and Panteleyev, A., 1972. Discovery, geologic setting and style of mineralization, Sam Goosly Deposit, B.C. *Can. Inst. Min. Metall. Bull.*, 65(723): 53—84.

Nichol, I., 1975. *Bedrock Composition as a Guide to Areas of Base Metal Potential in the Greenstone Belts of the Canadian Shield*. Queen's University, Kingston, Ont. 74 pp. (unpublished report).

Nichol, I., Bogle, E.W., Lavin, O.P., McConnell, J.W. and Sopuck, V.J., 1977. Lithogeo-chemistry as an aid in massive sulphide exploration. In: M.J. Jones (Editor), *Prospecting in Areas of Glaciated Terrain*. Institution of Mining and Metallurgy, London, pp. 63—71.

Nilsson, C.A., 1968. Wall rock alteration at the Boliden deposit, Sweden. *Econ. Geol.*, 63: 472—494.

Nockolds, S.R., 1954. Average chemical composition of some igneous rocks. *Geol. Soc. Am. Bull.*, 65: 1007—1032.

Olade, M.A., 1977. Major element halos in granitic wall rocks of porphyry copper deposits, Guichon Creek Batholith, British Columbia. *J. Geochem. Explor.*, 7: 59—71.

Olade, M.A., 1979. Copper and zinc in biotite, magnetite and feldspar from a porphyry copper environment, Highland Valley, British Columbia, Canada. *Min. Eng.*, September, pp. 1363—1369.

Olade, M.A. and Fletcher, W.K., 1974. Potassium chlorate-hydrochloric acid: a sulphide-selective leach for bedrock geochemistry. *J. Geochem. Explor.*, 3: 337—344.

Olade, M.A. and Fletcher, W.K., 1975. Primary dispersion of rubidium and strontium around porphyry copper deposits, Highland Valley, British Columbia. *Econ. Geol.*, 70: 15—21.

Olade, M.A. and Fletcher, W.K., 1976a. Trace element geochemistry of the Highland Valley and Guichon Creek Batholith in relation to porphyry copper mineralization. *Econ. Geol.*, 71: 733—748.

Olade, M.A. and Fletcher, W.K., 1976b. Distribution of sulphur, sulphide-iron, and copper in bedrock associated with porphyry copper deposits, Highland Valley, British Columbia. *J. Geochem. Explor.*, 5: 21—30.

Ovchinnikov, L.N. and Baranov, E.N., 1972. Endogenic geochemical halos of pyritic ore deposits. *Int. Geol. Rev.*, 14: 419—429.

Ovchinnikov, L.N. and Grigoryan, S.V., 1971. Primary halos in prospecting for sulphide deposits. In: R.W. Boyle and J.I. McGerrigle (Editors), *Geochemical Exploration. Can. Inst. Min. Metall., Spec. Vol.*, 11: 375—380.

Oyarzun, J.M., 1975. Rubidium and strontium as guides to copper mineralization emplaced in some Chilean andesitic rocks. In: I.L. Elliott and W.K. Fletcher (Editors), *Geochemical Exploration 1974*. Elsevier, Amsterdam, pp. 333—338.

Pakiser, L.C. and Robinson, R., 1967. Composition of the continental crust as estimated from seismic observations. In: J.S. Steinhart and T.J. Smith (Editors), *The Earth Beneath the Continents*. American Geophysical Union, Washington, D.C., pp. 620—626.

Pantazis, Th.M. and Govett, G.J.S., 1973. Interpretation of a detailed rock geochemical survey around Mathiati mine, Cyprus. *J. Geochem. Explor.*, 2: 25—36.

Parry, W.T., 1972. Chlorine in biotite from Basin and Range plutons. *Econ. Geol.*, 67: 972—975.

Parry, W.T. and Jacobs, D.C., 1975. Fluorine and chlorine in biotite from Basin and Range plutons. *Econ. Geol.*, 70: 554—558.

Parry, W.T. and Nackowski, M.P., 1963. Copper, lead, and zinc in biotites from Basin and Range quartz monzonites. *Econ. Geol.*, 58: 1126—1144.

Parslow, G.R., 1974. Determination of background and threshold in exploration geochemistry. *J. Geochem. Explor.*, 3: 319—336.

Pauling, L., 1960. *The Nature of the Chemical Bond and the Structure of Molecules and Crystals; An Introduction to Modern Structural Chemistry*. Cornell University Press, Ithaca, N.Y., 3rd ed., 644 pp.

Perel'man, A.I., 1977. *Geochemistry of Elements in the Supergene Zone*. Israel Program for Scientific Translations, Jerusalem, 266 pp.

Petersen, M.D. and Lambert, I.B., 1979. Mineralogical and chemical zonation around the Woodlawn Cu—Pb—Zn ore deposit, southeastern New South Wales. *J. Geol. Soc. Aust.*, 26: 169—186.

Petraschek, W.E., 1965. Typical features of metallogenic provinces. *Econ. Geol.*, 60: 1620—1634.

Pirie, I.D. and Nichol, I., 1980. Geochemical dispersion in wallrocks associated with the Norbec deposit, Noranda, Quebec. Paper presented at 8th International Geochemical Exploration Symposium, Hanover, 1980.

Plant, J., Brown, G.C., Simpson, P.R. and Smith, R.T., 1980. Signatures of metalliferous granites in the Scottish Caledonides. *Inst. Min. Metall., Trans., Sect. B*, 89: 198—210.

Plimer, I.R., 1979. Sulphide rock zonation and hydrothermal alteration at Broken Hill, Australia. *Inst. Min. Metall., Trans., Sect. B*, 88: 161—176.

Plimer, I.R. and Elliott, S.M., 1979. The use of Rb/Sr ratios as a guide to mineralization. *J. Geochem. Explor.*, 12: 21—34.

Poldervaart, A., 1955. Chemistry of the earth's crust. In: A. Poldervaart (Editor), *Crust of the Earth — A Symposium. Geol. Soc. Am. Spec. Paper*, 62: 119—144.

Polferov, D.V. and Suslova, S.I., 1966. Geochemical criteria of nickel mineralization in the mafic-ultramafic massifs. *Geochem. Int.*, 3: 487—496.

Pride, D.E., Timson, G.H. and Robinson, C.S., 1979. Use of selected elements to study hydrothermal alteration-mineralization of a porphyry molybdenum prospect, Breckenbridge mining district, Colorado. In: J.R. Watterson and P.K. Theobald (Editors), *Geochemical Exploration 1978*. Association of Exploration Geochemists, Toronto, Ont., pp. 251—267.

Putman, G.W., 1975. Base metal distribution in granitic rocks, II. Three-dimensional variation in the Lights Creek Stock, California. *Econ. Geol.*, 70: 1225—1241.

Putman, G.W. and Burnham, C.W., 1963. Trace elements in igneous rocks, northwestern and central Arizona. *Geochim. Cosmochim. Acta*, 27: 53—106.

Pwa, A., 1978. *Regional Rock Geochemical Exploration, Bathurst District*. Ph.D. Thesis, University of New Brunswick, Fredericton, N.B. (unpublished).

Radkevich, Ye.A., 1961. On the types of metallogenic provinces and ore districts. *Int. Geol. Rev.*, 3: 759—783.

Rankama, K., 1963. *Progress in Isotope Geology*. John Wiley, New York, N.Y., 705 pp.

Reeves, R.D. and Brooks, R.R., 1978. *Trace Element Analysis of Geological Materials*. John Wiley, New York, N.Y., 421 pp.

Rickwood, P.C. 1966. On the quality of representative mineral concentrates. *Geochim. Cosmochim. Acta*, 30: 545—551.

Rickwood, P.C., 1979. Sampling geological materials. In: *Proceedings, Symposium on Sampling and Sample Preparation*. Royal Australian Chemical Institute, pp. 2-1 to 2-12.

Ridler, R.H. and Shilts, W.W., 1973. Exploration for Archaean polymetallic sulphide deposits in permafrost terrains: an integrated geological-geochemical technique, District of Keewatin, Kaminak Lake Area. *Geol. Surv. Can., Open File Rep.*, 146: 45 pp.

Ringwood, A.E., 1955. The principles governing trace element distributions during magmatic crystallization. *Geochim. Cosmochim. Acta*, 7: 189—202.

Riverin, G. and Hodgson, C.J., 1980. Wall-rock alteration at the Millenbach Cu—Zn mine, Noranda, Quebec. *Econ. Geol.*, 75: 424—444.

Roedder, E., 1967. Fluid inclusions as samples of ore fluids. In: H.L. Barnes (Editor), *Geochemistry of Hydrothermal Ore Deposits*. Holt, Rinehart and Winston, New York, N.Y., pp. 515—574.

Roedder, E., 1977. Fluid inclusions as tools in mineral exploration. *Econ. Geol.*, 72: 503—525.

Ronov, A.B. and Yaroshevsky, A.A., 1969. Chemical composition of the earth's crust. In: P.J. Hart (Editor), *The Earth's Crust and Upper Mantle*. American Geophysical Union, Washington, D.C., pp. 37—57.

Roscoe, S.M., 1965. Geochemical and isotopic studies, Noranda and Matagami areas. *Can. Inst. Min. Metall. Bull.*, 58(641): 965—971.

Rose, A.W., 1970. Zonal relations of wallrock alteration and sulfide distribution at porphyry copper deposits. *Econ. Geol.*, 65: 920—936.

Rose, A.W., Hawkes, H.E. and Webb, J.S., 1979. *Geochemistry in Mineral Exploration*, Academic Press, London, 2nd ed., 657 pp.

Rösler, H.J. and Lange, H., 1972. *Geochemical Tables*. Elsevier, Amsterdam, 468 pp.

Rudman, A.J., Summerson, C.H. and Hinze, W.J., 1965. Geology of basement in midwestern United States. *Bull. Am. Assoc. Pet. Geol.*, 49: 894—904.

Rugless, C., 1981. *Geology and Geochemistry of the Wainaleka Zn—Cu deposit, Fiji.* Internal report for Ph.D. Thesis, School of Applied Geology, University of New South Wales, Kensington, N.S.W.

Russell, M.J., 1974. Manganese halo surrounding the Tynagh ore deposit, Ireland: a preliminary note. *Inst. Min. Metall., Trans., Sect. B,* 83: 665—666.

Russell, M.J., 1975. Lithogeochemical environment of the Tynagh base-metal deposit, Ireland, and its bearing on ore deposition. *Inst. Min. Metall., Trans., Sect. B,* 84: 128—133.

Saif, S.I., 1977. *Identification, Correlation and Origins of the Key-Anacon-Brunswick Mines Ore Horizon, Bathurst, New Brunswick.* Ph.D. Thesis, University of New Brunswick, Fredericton, N.B. (unpublished).

Saif, S.I., McAllister, A.L. and Murphy, W.L., 1978. Geology of the Key Anacon Mine Area, Bathurst, New Brunswick. *Can. Inst. Min. Metall. Bull.,* 71(791): 161—168.

Sakrison, H.C., 1971. Rock geochemistry — its current usefulness on the Canadian Shield. *Can. Inst. Min. Metall. Bull.,* 64(715): 28—31.

Sangster, D.F., 1972. Precambrian volcanogenic massive sulphide deposits in Canada — a review. *Can. Geol. Surv. Paper,* 72—22: 44 pp.

Sangster, D.F. and Scott, S.D., 1976. Precambrian, strata-bound, massive Cu—Zn—Pb sulfide ores of North America. In: K.H. Wolf (Editor), *Handbook of Strata-bound and Stratiform Ore Deposits, II. Regional Studies and Specific Deposits, Vol. 6. Cu, Zn, Pb, and Ag Deposits.* Elsevier, Amsterdam, pp. 129—222.

Sato, T., 1972. Behaviour of ore-forming solutions in seawater. *Min Geol.,* 22: 31—42.

Sato, T., 1973. A chloride complex model for Kuroko mineralization. *Geochem. J.,* 7: 245—270.

Schröpfer, L., 1968. Über den Einbau von Titan in Diopsid. *Neues Jahrb. Mineral., Monatsh.,* 11: 441—453.

Searle, D.L., 1972. Mode of occurrence of the cupriferous pyrite deposits of Cyprus. *Inst. Min. Metall., Trans., Sect. B,* 81: 189—197.

Sederholm, J.J., 1925. The average composition of the earth's crust in Finland. *Finl. Com. Geol. Bull.,* 12: 3—20.

Shannon, R.D. and Prewitt, C.T., 1969. Effective crystal radii in oxides and fluorides. *Acta Crystallogr., Sect. B,* 25: 925—946.

Shannon, R.D. and Prewitt, C.T., 1970. Revised values of effective ionic radii. *Acta Crystallogr., Sec. B,* 26: 1046—1048.

Shaw, D.M., 1953. The camouflage principle and trace-element distribution in magmatic minerals. *J. Geol.,* 61: 142—151.

Shaw, D.M., Reilly, G.A., Muysson, J.M., Pattenden, G.E. and Campbell, F.E., 1967. An estimate of the chemical composition of the Canadian Precambrian Shield. *Can. J. Earth Sci.,* 4: 829—853.

Sheppard, N.W., 1981. *Exploration Rock Geochemistry Studies around some Mt. Lyell Orebodies, Tasmania.* Internal report for Ph.D. Thesis, School of Applied Geology, University of New South Wales, Kensington, N.S.W.

Sheraton, J.W. and Black, L.P., 1973. Geochemistry of mineralized granitic rock of northeast Queensland. *J. Geochem. Explor.,* 2: 331—348.

Shergina, Yu.P. and Kaminskaya, A.B., 1965. On the possibility of utilizing natural variations in the isotopic composition of boron in geochemical prospecting. *Geochemistry,* 1: 64—67.

Shirozu, H., 1974. Clay minerals in altered wall rocks of the Kuroko-type deposits. In: S. Ishihara (Editor), *Geology of Kuroko Deposits. Soc. Min. Geol. Jpn., Min. Geol. Spec. Issue,* 6: 303—310.

Siegel, S., 1956. *Nonparametric Statistics for the Behavioral Sciences.* McGraw-Hill, New York, N.Y., 312 pp.

Sillitoe, R.H., 1972. Formation of certain massive sulphide deposits at sites of seafloor spreading. *Inst. Min. Metall., Trans., Sect. B,* 81: 141—148.

Sillitoe, R.H., 1973a. Environments of formation of volcanogenic massive sulfide deposits. *Econ. Geol.*, 68: 1321—1325.

Sillitoe, R.H., 1973b. The tops and bottoms of porphyry copper deposits. *Econ. Geol.*, 68: 799—815.

Simmons, B.D. and Geological Staff, Falconbridge Copper Ltd., 1973. Geology of the Millenbach massive sulphide deposit, Noranda, Quebec. *Can. Inst. Min. Metall. Bull.*, 66(739): 67—78.

Simpson, P.R., Plant, J. and Cope, M.J., 1977. Uranium abundance and distribution in some granites from northern Scotland and southwest England as indicators of uranium provinces. *Proc., Int. Symp. Inst. Min. Metall.*, London, pp. 126—139.

Sinclair, A.J., 1974. Selection of threshold values in geochemical data using probability graphs. *J. Geochem. Explor.*, 3: 129—149.

Sinclair, A.J., 1976. *Applications of Probability Graphs in Mineral Exploration.* Association of Exploration Geochemists, Rexdale, Ont., 95 pp.

Sinclair, I.G.L., 1977. Primary dispersion patterns associated with the Detour zinc-copper-silver deposit at Lac Brouillan, Province of Quebec, Canada. *J. Geochem. Explor.*, 8: 139—151.

Slawson, W.F. and Nackowski, M.P., 1959. Trace lead in potash feldspars associated with ore deposits. *Econ. Geol.*, 54: 1543—1555.

Smirnov, V.I., 1959. Experience in the metallogenic regional zonation of the U.S.S.R. *Izv. Acad. Sci. U.S.S.R., Geol. Ser.*, 4: 1—15.

Smith, T.E. and Turek, A., 1976. Tin-bearing potential of some Devonian granitic rocks. *Miner. Deposita*, 11: 234—245.

Sopuck, V.J., 1977. *A Lithogeochemical Approach in the Search for Areas of Felsic Volcanic Rocks Associated with Mineralization in the Canadian Shield.* Ph.D. Thesis, Queen's University, Kingston, Ont. (unpublished).

Sopuck, V.J., Lavin, O.P. and Nichol, I., 1980. Lithogeochemistry as a guide to identifying favourable areas for the discovery of volcanogenic massive sulphide deposits. *Can. Inst. Min. Metall. Bull.*, 73(823): 152—166.

Spence, C.D., 1967. The Noranda area. *Can. Inst. Min. Metall. Centennial Field Excursion Guideb.*, pp. 36—39.

Spence, C.D. and De Rosen-Spence, A.F., 1975. The place of sulfide mineralization in the volcanic sequence at Noranda, Quebec. *Econ. Geol.*, 70: 90—101.

Spitz, G. and Darling, R., 1978. Major and minor element lithogeochemical anomalies surrounding the Louvem copper deposit, Val d'Or, Quebec. *Can. J. Earth Sci.*, 15: 1161—1169.

Spurr, J.E., 1923. *The Ore Magmas. A Series of Essays on Ore Deposition.* McGraw-Hill, New York, N.Y., 2 volumes.

Stanton, R.L., 1972. *Ore Petrology.* McGraw-Hill, New York, N.Y., 713 pp.

Steed, G.M. and Tyler, P., 1979. Lithogeochemical halos about Gortdrum copper-mercury orebody, County Tipperary, Ireland. In: *Prospecting in Areas of Glaciated Terrain, 1979.* Institution of Mining and Metallurgy, London, pp. 30—39.

Steed, G.M., Annels, A.E., Shrestha, P.L. and Tater, P.S., 1976. Geochemical and biogeochemical prospecting in the area of Ogofau gold mines, Dyfed, Wales. *Inst. Min. Metall., Trans., Sect. B*, 85: 109—117.

Stollery, G., Borcsik, M. and Holland, H.D., 1971. Chlorine in intrusives: a possible prospecting tool. *Econ. Geol.*, 66: 361—367.

Sutulov, A., 1974. *Copper Porphyries.* University of Utah, Salt Lake City, Utah, 200 pp.

Szabo, N.L., Govett, G.J.S. and Lajti, E.Z., 1975. Dispersion trends of elements and indicator pebbles in glacial till around Mt. Pleasant, N.B., Canada. *Can. J. Earth Sci.*, 12: 1534—1556.

Tatsumi, T. and Clark, L.A., 1972. Chemical composition of acid volcanic rocks genetically related to formation of the Kuroko deposits. *J. Geol. Soc. Jpn.*, 78: 191—201.

Tauson, L.V., 1965. Factors in the distribution of the trace elements during crystallization of magmas. *Phys. Chem. Earth*, 6: 216—249.

Tauson, L.V., 1967. Geochemistry of rare elements in igneous rocks and metallogenic specialization of magmas. In: A.P. Vinogradov (Editor), *Chemistry of the Earth's Crust, Vol. 2*. Israel Program for Scientific Translation, Jerusalem, pp. 248—259.

Tauson, L.V., 1974. The geochemical types of granitoids and their potential ore capacity. In: M. Stemprok (Editor), *Metallization Associated with Acid Magmatism, Vol. 1*. Geological Survey, Prague.

Tauson, L.V. and Kozlov, V.D., 1973. Distribution functions and ratios of trace-element concentrations as estimators of the ore-bearing potential of granites. In: M.J. Jones (Editor), *Geochemical Exploration 1972*. Institution of Mining and Metallurgy, London, pp. 37—44.

Tauson, L.V. and Kravchenko, L.A., 1956. Characteristics of lead and zinc distribution in minerals of the Caledonian granitoids of the Susamyr batholith in the central Tian-Shan. *Geochemistry*, 1: 78—88.

Taylor, R.G., 1979. *Geology of Tin Deposits*. Elsevier, Amsterdam, 543 pp.

Taylor, S.R., 1964. Abundance of chemical elements in the continental crust: A new table. *Geochim. Cosmochim. Acta*, 28: 1273—1285.

Taylor, S.R., 1965. The application of trace element data to problems in petrology. *Phys. Chem. Earth*, 6: 133—213.

Taylor, S.R., 1968. Geochemistry of andesites. In L.H. Ahrens (Editor), *Origin and Distribution of the Elements. Int. Ser. Monog. Earth Sci.*, 30: 559—583.

Tennant, C.B. and White, M.L., 1959. Study of the distribution of some geochemical data. *Econ. Geol.*, 54: 1281—1290.

Theodore, T.G. and Blake, D.W., 1975. Geology and geochemistry of the Copper Canyon porphyry copper deposit and surrounding area, Lander County, Nevada, *U.S. Geol. Surv. Prof. Paper*, 798-B: 86 pp.

Theodore, T.G. and Nash, J.T., 1973. Geochemical and fluid zonation at Copper Canyon, Lander County, Nevada. *Econ. Geol.*, 68: 565—570.

Theodore, T.G., Silberman, M.L. and Blake, D.W., 1973. Geochemistry and K-Ar ages of plutonic rocks in the Battle Mountain mining district, Lander County, Nevada. *U.S. Geol. Surv. Prof. Paper*, 798-A: 24 pp.

Thurlow, J.G., Swanson, E.A. and Strong, D.F., 1975. Geology and lithogeochemistry of the Buchans polymetallic sulfide deposits, Newfoundland. *Econ. Geol.*, 70: 130—144.

Tilling, R.I., Gottfried, D. and Rowe, J.J., 1973. Gold abundance in igneous rocks: bearing on gold mineralization. *Econ. Geol.*, 68: 168—186.

Tischendorf, G., 1973. The metallogenetic basis of tin exploration in Erzgebirge. *Inst. Min. Metall., Trans., Sect. B*, 82: 9—24.

Tolstoi, M.I. and Ostafiichuk, I.M., 1963. Some regularities of the distribution of the elements in rocks and their geochemical significance. *Geochemistry*, 10: 986—991.

Tono, N., 1974. Minor elements distribution around Kuroko deposits in northern Akita, Japan. In: S. Ishihara (Editor), *Geology of Kuroko Deposits. Soc. Min. Geol. Jpn., Min. Geol. Spec. Issue*, 6: 399—420.

Turek, A., Tetley, N.W. and Jackson, T., 1976. A study of metal dispersion around the Fox orebody in Manitoba. *Can. Inst. Min. Metall. Bull.*, 69(770): 104—110.

Turekian, K.K. and Wedepohl, K.H., 1961. Distribution of the elements in some major units of the earth's crust. *Geol. Soc. Am. Bull.*, 72: 175—195.

Turneaure, F.S., 1955. Metallogenetic provinces and epochs. *Econ. Geol., 50th Anniv. Vol.*, 1: 38—98.

Vinogradov, A.P., 1962. Average content of chemical elements in the major types of igneous rocks of the earth's crust. *Geochemistry*, 7: 641—664.

Vistelius, A.B., 1960. The skew frequency distribution and the fundamental law of geochemical processes. *J. Geol.*, 68: 1—22.

Vogt, J.H.L., 1931. On the average composition of the earth's crust with particular reference to the contents of phosphoric and titanic acid. *Skr. Nor. Vidensk.-Akad. Oslo, 1,*
7: 1—48.

Wahl, J.L., 1978. *Rock Geochemical Exploration at the Heath Steele and Key Anacon
Deposits, New Brunswick.* Ph.D. Thesis, University of New Brunswick, Fredericton,
N.B. (unpublished).

Wahl, J.L. and Govett, G.J.S., 1978. Exploration rock geochemical techniques for massive
sulphides of the Bathurst type (New Brunswick, Canada). Paper presented at 7th
International Exploration Geochemistry Symposium, Denver, Colo., 1978.

Warren, H.V. and Delavault, R.E., 1960. Aqua regia extractable copper and zinc in plutonic rocks in relation to ore deposits. *Inst. Min. Metall., Trans.* 69: 495—504.

Washington, H.S., 1925. The chemical composition of the earth's crust. *Am. J. Sci.,*
209: 353—378.

Watterson, J.R., Gott, G.B., Neuerburg, G.J., Lakin, H.W. and Cathrall, J.B., 1977. Tellurium, a guide to mineral deposits. *J. Geochem. Explor.,* 8: 31—48.

Wedepohl, K.H., 1969. Composition and abundance of common igneous rocks. In: K.H.
Wedepohl (Editor), *Handbook of Geochemistry, Vol. I.* Springer-Verlag, Berlin, pp.
227—247.

Wedepohl, K.H., 1971. *Geochemistry.* Holt, Rinehart and Winston, New York, N.Y.,
231 pp.

Wedepohl, K.H., 1969—1979. *Handbook of Geochemistry, Vol. II.* Springer-Verlag, Berlin.

Wells, A.F., 1962. *Structural Inorganic Chemistry.* Clarendon Press, Oxford, 3rd ed.,
1055 pp.

Wells, J.B., Stoiser, L.R. and Elliott, J.E., 1969. Geology and geochemistry of the Cortez
gold deposit, Nevada. *Econ. Geol.,* 64: 526—537.

Whitehead, R.E.S., 1973a. Environment of stratiform sulphide deposition; Variation in
Mn:Fe ratio in host rocks at Heat Steele Mine, New Brunswick, Canada. *Miner. Deposita,* 8: 148—160.

Whitehead, R.E.S., 1973b. *Application of Rock Geochemistry to Problems of Mineral
Exploration and Ore Genesis at Heath Steele Mines, New Brunswick.* Ph.D. Thesis,
University of New Brunswick, Fredericton, N.B. (unpublished).

Whitehead, R.E.S. and Govett, G.J.S., 1974. Exploration rock geochemistry — detection
of trace element halos at Heath Steele Mines (N.B., Canada) by discriminant analysis.
J. Geochem. Explor., 3: 371—396.

Whittaker, E.J.W. and Muntus, R., 1970. Ionic radii for use in geochemistry. *Geochim.
Cosmochim. Acta,* 34: 945—956.

Williams, R.J.P., 1959. Deposition of trace elements in basic magma. *Nature,* 184: 44.

Wilson, I.R. and Jackson, N.J., 1977. Contributed remarks to: Aspects of trace metal and
ore deposition in Cornwall by R.P. Edwards. *Inst. Min. Metall., Trans., Sect. B,* 86:
61—62.

Wolfe, W.J., 1974. Geochemical and biogeochemical exploration research near Early Precambrian porphyry-type molybdenum-copper mineralization, northwestern Ontario,
Canada. *J. Geochem. Explor.,* 3: 25—41.

Wolfe, W.J., 1975. Zinc abundance in Early Precambrian volcanic rocks: its relationship
to exploitable levels of zinc in sulphide deposits of volcanic-exhalative origin. In:
I.L. Elliott and W.K. Fletcher (Editors), *Geochemical Exploration 1974.* Elsevier,
Amsterdam, pp. 261—278.

Zemann, J., 1969. Crystal chemistry. In: K.H. Wedepohl (Editor), *Handbook of Geochemistry, Vol. I.* Springer-Verlag, Berlin, pp. 12—36.

REFERENCES INDEX

SUBJECT INDEX

430

Anomalies (*continued*)
extent and contrast
—, copper porphyry, 191
—, massive sulphides, 260, 278
general, 9, 29, 30
ratio, 43
recognition, 29—47
—, background, threshold, 30, 31, 32
—, interpretation, Soviet procedures, 40—44
—, log-transformation, 36
—, patterns, 39, 40
—, population partition, 36—39
—, spatial distribution, 32—36
sub-ore, 40
supra-ore, 40
vein-type deposits
—, more than 6 km, 114, 115, 116
—, 3—6 km, 116—123
—, 1—3 km, 124—133
Antimony
in porphyry-type deposit host rock
—, Ely (U.S.A.), 210, 211, 212
—, geochemical response, summary, 334, 335
in vein and replacement deposit host rocks
—, Cobalt (Canada), 233
—, Coeur d'Alene (U.S.A.), 115
—, Cortez (U.S.A.), 129, 130
—, Costerfield (Australia), 240
—, general, 337
—, geochemical response, summary, 338
—, Mavro Vouno, Mykonos (Greece), 121, 234, 235
—, Taylor mining district (U.S.A), 119, 121
Antimony deposits in granite, Canadian Cordillera, 57
Appalachian, 17, 174
Argilic alteration, 181, 182, 188, 209, 325
Arizona (U.S.A.), 80, 83, 86, 87, 124, 199, 200, 204, 205, 212, 213, 214
Armstrong (Canada), 164, 168, 170
Arsenic
in massive sulphide host rocks
—, Akita district (Japan), 266
—, Brunswick No. 12 (Canada), 272
in porphyry-type deposit host rocks
—, Ely (U.S.A.), 210, 211, 212
—, geochemical response, summary, 234, 235

in vein and replacement deposit host rocks
—, Bradina (Canada), 118
—, Cobalt (Canada), 233
—, Coeur d'Alene (U.S.A), 115
—, Cortez (U.S.A.), 129, 130
—, Costerfield (Australia), 240
—, Detroit mining district (U.S.A.), 120
—, general, 337
—, geochemical response, summary, 338
Gortdrum (Ireland), 123
—, Ogafau (U.K.), 237, 238
—, Sam Goosly (Canada), 118, 119
Atlantis II Deep (Red Sea), 174, 175
Atmophile, 370
Austin Brook (Canada), 164
Australia, 94, 95, 96, 98, 99, 100, 101, 102, 136, 139, 144, 176, 222, 240, 262, 303, 313, 315, 316, 317, 318, 320, 321, 322, 323, 324, 342, 397, 399, 400, 401, 407
Azuero (Panama), 72

Background, 9, 30—36, 44
and log-transformation, 35, 36
Balabac Island (Philippines), 139
Balkanides (Yugoslavia), 83
Balkan Peninsula (Yugoslavia), 83
Banded Mine Sequence (Australia) 313, 314, 315
Barium
in biotite, 81
in granitic rocks
—, average, 100
—, Caledonian granites, Scotland, 65, 66
—, Erzgebirge (East Germany), 105
—, Galway pluton, Lettermullan (Ireland), 215, 216
—, geochemical response, summary, 333, 334, 335
—, granitoids, various, 53, 54
—, New Brunswick (Canada), 105
—, Nova Scotia (Canada), 105
—, southwest England, 103, 104
in massive sulphide host rocks
—, Buchans, 174
—, geochemical response, summary, 340, 341
—, Kuroko, general (Japan), 266
—, Mount Morgan (Australia), 314

Mercury
  in modern sediments, Red Sea, 174, 175
  in porphyry-type deposit host rocks
  —, Bethlehem JA (Canada), 189, 191
  —, Copper Canyon (U.S.A.), 207, 208
  —, Ely (U.S.A.), 210
  —, Lornex (Canada), 189, 190
  —, Valley Copper (Canada), 189, 191
  in vein and replacement deposit host rocks
  —, Cortez (U.S.A.), 129, 130
  —, Detroit mining district (U.S.A.), 120
  —, Gortdrum (Ireland), 123
  —, Mavro Vouno, Mykonos (Greece), 121, 234
  —, Nantymwyn (U.K.), 235
  —, Ogofau (U.K.), 235
  —, Taylor mining district (U.S.A.), 119, 120
Metallogenic province, 20
Metallogenetically specialized intrusions, 51, 52
Mexico, 74, 75, 401
Mihrap Dägi (Turkey), 310
Millenbach (Canada), 145, 149, 250, 251, 262, 340, 401
Millenbach andesite (Canada), 250
Minas de Oro (Honduras), 70, 71, 72, 401
Mineral Butte (U.S.A.), 199, 200, 201, 334, 401
Mineral occurrences (unspecified)
  Blackjack Lake (Canada), 161, 163
  De Vries Lake (Canada), 161
  Hardisty Lake (Canada), 161
  Mazenod Lake (Canada), 161, 163
Mineral Park (U.S.A.), 213, 401
Mineral Range (U.S.A.), 76, 82
Mineral-selective leaches, intrusive rocks, 69, 70, 87, 88, 89; see also Sulphide-selective leaches
Mineral separates, 69, 70
  elements in, 191, 192, 193
Mineral zoning (massive sulphides), 251, 264
Mississippi Valley deposits (U.S.A.), 20
Mobility of elements, 43
Molybdenum
  in intrusive rocks
  —, Caledonian granites, Scotland, 65, 66

  —, Coed-y-Brenin (U.K.), see Harlech Dome, this entry
  —, granitoids, 52, 53, 59
  —, Harlech Dome (U.K.), 62
  in massive sulphide host rocks
  —, Akita district (Japan), 266
  —, Prince Lyell (Australia), 317
  in porphyry-type deposit host rocks
  —, Bethlehem JA (Canada), 191
  —, Copper Canyon (U.S.A.), 207, 208
  —, Ely (U.S.A.), 211, 212
  —, geochemical response, summary, 334, 336
  —, Highmont (Canada), 189, 191
  —, Kalamazoo (U.S.A.), 205
  —, Lornex (Canada), 189, 190, 191
  —, Mineral Butte (U.S.A.), 200, 201
  —, Nojal Peak (U.S.A.), 212, 213
  —, Rialto (U.S.A.), see Nojal Peak, this entry
  —, Setting Net Lake (Canada), 198, 199
  —, Valley Copper (Canada), 188, 189, 191
  —, Wirepatch, Breckenridge (U.S.A.), 209
  in vein and replacement deposit host rocks, 120
Molybdenum deposits
  deposits and occurrences
  —, Detroit mining district (U.S.A.), 120
  —, Nojal Peak (U.S.A.), 212, 401
  —, Rialto (U.S.A.), see Nojal Peak, this entry
  —, Setting Net Lake (Canada), 198, 199, 402
  —, Wirepatch, Breckenridge (U.S.A.), 209, 210, 403
  general
  —, in Caledonian granites, Scotland, 65, 66
  —, North America, 20
  —, potential in palingenic granites, 53
  —, Siberia (U.S.S.R.), 52
Montana (U.S.A.), 124, 125, 126, 127, 128
Montezuma (U.S.A.), 115, 338, 401
Moongan Rhyolite (Australia), 323
Moonlight Valley (U.S.A.), 201, 202, 203, 400
Morenci (U.S.A.), 213, 401
Mount Allen (Canada), 56, 59, 401

453

—, Ajo (U.S.A.), 213
—, Battle Mountain (U.S.A.), 208, 219
—, Bethlehem JA (Canada), 193, 194, 195
—, Chilean deposits, general, 217, 223, 224
—, Copper Canyon (U.S.A.), 208, 209, 219
—, Copper Mountain—Ingerbelle (Canada), 197, 198, 218, 220
—, El Teniente (Chile), 218, 220, 221, 222
—, general, 213, 214
—, geochemical response, summary, 334, 336
—, Guichon Batholith (Canada), 187, 218
—, Kalamazoo (U.S.A.), 204
—, Los Bronces (Chile), 218, 222
—, Mineral Park (U.S.A.), 213
—, Morenci (U.S.A.), 213
—, Rio Blanco (Chile), 218, 220, 222
—, Santa Rita (U.S.A.), 213
—, Valley Copper (Canada), 193, 194, 218
—, Wirepatch, Breckenridge (U.S.A.), 209
radiometric measurement of, 213
in vein and replacement deposit host rocks
—, Nantymwyn (U.K.), 235
—, Ogofau (U.K.), 235, 237
Potassium, water-soluble, in massive sulphide host rocks
Brunswick No. 12 (Canada), 284, 285, 286
Heath Steele (Canada), 279, 285, 286
Potassic alteration, 188, 205, 336
Potato Hills (Canada), 59, 401
Precambrian Shields, major element composition, 18
Primary dispersion, definition, 1
Prince Lyell (Australia), 315, 316, 317, 318, 319, 327, 341
Propylitic alteration, 62, 181, 182, 195, 205, 209, 221, 326, 336
Providencia (Mexico), 74, 75, 401
Proximal, 291, 299, 301, 302, 303
Puerto Rico, 70, 72, 181, 399, 401

Quebec (Canada), 145, 250, 251, 258, 259, 260, 261, 342

Queensland (Australia), 100, 101, 219, 313, 315, 316, 323, 324
Quemont (Canada), 145, 149
Que, S. R. (Canada), 164

Rae Lake (Canada), 162, 163
Rainy Lake (Canada), 161, 162
Rammelsberg (Germany), 139
Ranua (Finland), 68
Red Sea
brines, 140
dispersion in sediments, 174, 175, 340
Reconnaissance exploration, see Exploration, reconnaissance
Regional scale exploration, see Exploration, regional scale
Resonance (bonding), 22
Rhydymwyn (U.K.), 235, 236
Rialto (U.S.A.), 213; see also Nojal Peak
Rio Blanco (Chile), 217, 218, 220, 222, 334, 401
Rio Guayabo (Panama), 72
Rio Pito (Panama), 72, 401
Rio Tinto (Spain), 139
Rio Vivi (Puerto Rico), 72, 401
Rocky Turn (Canada), 164, 170
Roseberry (Australia), 139
Rubidium
in andesite, 220, 223
in biotite, 83
in feldspar, 83, 85
in granitic rocks
—, Anchor mine (Australia), 98, 99, 219
—, average, 100, 218
—, barren and mineralized, 62, 221
—, Caledonian granites, Scotland, 65, 66
—, Erzgebirge (East Germany), 105
—, Galway pluton, Luttermullan (Ireland), 215, 216, 218
—, geochemical response, summary, 333, 335
—, granitoids, various, 53, 54
—, New Brunswick (Canada), 105, 219
—, Nova Scotia (Canada), 105, 219
—, Queensland (Australia), 101, 219
—, southeast England, 104, 219
and hydrothermal alteration, 214
in massive sulphide host rocks
—, Broken Hill (Australia), 263

454

Rubidium
  in massive sulphide host rocks (continued)
  —, geochemical response, summary, 340, 341, 342
  —, Noranda (Canada), 150, 151
  —, Prince Lyell (Australia), 317, 319
  —, Wainaleka (Fiji), 325, 326
  in muscovite, 83
  in porphyry-type deposit host rocks
  —, Battle Mountain (U.S.A.), 208, 219
  —, Bethlehem JA (Canada), 189, 191, 193, 194, 195
  —, Chilean deposits general, 217, 223, 224
  —, Copper Canyon (U.S.A.), 208, 209, 219
  —, Copper Mountain—Ingerbelle (Canada), 197, 199, 218, 220
  —, El Teniente (Chile), 218, 220, 221, 222
  —, general, 213
  —, geochemical response, summary, 334, 336
  —, Guichon Batholith (Canada), 186, 187, 218
  —, Kalamazoo (U.S.A.), 205
  —, Los Bronces (Chile), 218, 222
  —, Rio Blanco (Chile), 218, 220, 222
  —, Valley Copper (Canada), 188, 189, 191, 193, 194, 218
  —, Wirepatch, Breckenridge (U.S.A.), 209
  and potassium as measure of fractionation, 214
  and strontium as measure of fractionation, 214
  in vein and replacement deposit host rocks
  —, Nantymwyn (U.K.), 235, 236
  —, Ogofau (U.K.), 235—238
Rusty Peak (Canada), 59

Sam Goosly (Canada), 116, 117, 118, 119, 338, 401
Samples and sampling
  interval, 46, 134, 171, 225, 247, 268, 303, 333, 343
  minimum weight, 23—27, 91, 225, 268, 333
  —, and analytical variance, 26
  —, and content of mineral of interest, 27

number for defined probability of anomaly, 395
number to recognize mineralized intrusion, 90, 91
pattern, 46
—, of random spatial data, 31
—, of systematic spatial data, 39
type, depending on host mineral, 275
variability, 161
Sanctuary Lake (Canada), 145, 146, 147, 403
San Francisco, Basin and Range (U.S.A.), 76, 82, 335, 402
San Francisco (Honduras), 72, 402
San Lorenzo (Puerto Rico), 71, 72
San Manuel (U.S.A.), 204, 334
Santa Rita, Arizona (U.S.A.), 86, 89, 335, 338, 402
Santa Rita, New Mexico (U.S.A.), 75, 213
Sardinia (Italy), 113, 335, 402
Saskatchewan (Canada), 256, 257
Scheelite Dome (Canada), 59, 402
Scotland, 63, 64, 65, 85, 86; see also U.K.
Searchlight District (U.S.A.), 231, 233, 338, 402
Seawater, 140, 141, 142
Secondary dispersion, in rock, 121, 122, 323, 324
Sedgwick (U.S.A.), 243
Selenium in copper porphyry host rocks, 205
Setting Net Lake (Canada), 198, 199, 334, 402
Shap adamellite (U.K.), 84
Shield areas, composition, 18
Siberia, eastern (U.S.S.R.), 52
Siderophile, 21, 370
Sierrita (U.S.A.), 212, 214, 334, 402
Sierrita Mountains (U.S.A.), 84, 86
Silica
  in granites, 102
  in massive sulphide host rocks
  —, Agricola Lake (Canada), 147
  —, Boliden (Sweden), 253, 254
  —, Broken Hill (Australia), 263
  —, Caribou (Canada), 292
  —, Fox (Canada), 255, 256
  —, general, 142, 154, 155, 156, 178
  —, Jay Copper (Canada), 256
  —, Kakagi Lake (Canada), 159, 160
  —, Kuroko, general, 265
  —, Mattabi (Canada), 155

458

Terre-Neuve (Haiti), 70, 71, 72, 402
Textural variation, massive sulphides, 158—161, 261, 262
Thallium
  in massive sulphide host rocks, 266
  in porphyry-type deposit host rocks, 205
Thingmuli (Iceland), 381, 382, 383, 384
Tholeiitic rocks, see Massive sulphides
Thorium in Caledonian granites, Scotland, 65
Thorium deposits in Caledonian granites, Scotland, 66
Threshold, 30—36, 157
Timmins (Canada), 139, 147, 148
Tin
  in biotite
  —, general, 97, 99
  —, Tasmania (Australia), 97
  —, U.K., 83, 84
  enrichment in felsic rocks, 51
  in feldspar, 83, 84
  in granitic rocks
  —, Amur (U.S.S.R.), 95, 96
  —, Anchor mine (Australia), 98
  —, average, 100
  —, Caledonian granites, Scotland, 66
  —, eastern Australia, 96
  —, Erzgebirge (East Germany), 105
  —, geochemical response, summary, 335, 336
  —, in granitoids, various, 53
  —, New Brunswick (Canada), 105
  —, Nova Scotia (Canada), 105
  —, Queensland (Australia), 100, 101
  —, southwest England, 102, 104
  —, summary data, 109
  —, Tasman geosyncline (Australia), 95, 97
  —, U.S.S.R., 55
  in massive sulphide host rocks, 310
  in muscovite, 84
  recognition of tin granites, 93—99, 106—111
Tin deposits
  deposits and occurrences
  —, Anchor Mine (Australia), 97—100, 102, 107, 108, 397
  —, Annan River (Australia), 100, 397
  —, Erzgebirge (Germany), 103, 105, 399
  —, Herberton (Australia), 102, 400

  —, Mareeba (Australia), 100, 108, 401
  —, Mount Pleasant (Canada), 103, 401
  —, southwest England, 61, 398
  general
  —, Asia, 20
  —, in Caledonian granites, Scotland, 65, 66
  —, disseminated, 93
  —, in granites, 13
  —, lode, vein, replacement, 12, 93
  —, Malaysia, 20
  —, potential in palingenic granites, 53
  —, potential in plumasitic leucogranite, 53
  —, Siberia (U.S.S.R.), 52
  —, Tasman geosyncline (Australia), 94
Tin Holding Capacity, 99
Tintic (U.S.A.), 76, 82, 228, 229, 230, 402
Titanium
  in biotite, 77, 81
  in intrusive rocks, 62
  in massive sulphide host rocks, 254, 263
  in phlogopite, 75
Tombstone Mountain (Canada), 59
Tommie Lake (Canada), 161, 162, 402
Transbaikalia (U.S.S.R.), 55
Transitional granite (Ireland), 215, 216
Trend surface, 184, 188, 203
Tres Hermanas (U.S.A.), 124, 336, 402
Troodos volcanic series (Cyprus), 171, 305
True Fissure (U.S.A.), 125, 402
Tucson (U.S.A.), 204
Tungsten
  in granitoids, Canadian Cordillera, 59, 60
  in granitoids, various, 52, 53
  in vein and replacement deposit host rocks, 129, 130
Tungsten deposits
  deposits and occurrence
  —, Cantung (Canada), 56, 398
  —, Mount Allen (Canada), 56, 401
  —, Mount Carbine (Australia), 100, 401
  —, (Mo, Sn, Bi, base metals) Mount Pleasant (Canada), 103, 401
  —, southwest England, 102, 103, 398
  general
  —, potential in palingenic granites, 53